Data Engineering and Data Science

Scrivener Publishing
100 Cummings Center, Suite 541J
Beverly, MA 01915-6106

Advance in Data Engineering and Machine Learning

Series Editor: M. Niranjanamurthy, PhD, Juanying XIE, PhD, and Ramiz Aliguliyev, PhD

Scope: Data engineering is the aspect of data science that focuses on practical applications of data collection and analysis. For all the work that data scientists do to answer questions using large sets of information, there have to be mechanisms for collecting and validating that information. Data engineers are responsible for finding trends in data sets and developing algorithms to help make raw data more useful to the enterprise.

It is important to have business goals in line when working with data, especially for companies that handle large and complex datasets and databases. Data Engineering Contains DevOps, Data Science, and Machine Learning Engineering. DevOps (development and operations) is an enterprise software development phrase used to mean a type of agile relationship between development and IT operations. The goal of DevOps is to change and improve the relationship by advocating better communication and collaboration between these two business units. Data science is the study of data. It involves developing methods of recording, storing, and analyzing data to effectively extract useful information. The goal of data science is to gain insights and knowledge from any type of data — both structured and unstructured.

Machine learning engineers are sophisticated programmers who develop machines and systems that can learn and apply knowledge without specific direction. Machine learning engineering is the process of using software engineering principles, and analytical and data science knowledge, and combining both of those in order to take an ML model that's created and making it available for use by the product or the consumers. "Advances in Data Engineering and Machine Learning Engineering" will reach a wide audience including data scientists, engineers, industry, researchers and students working in the field of Data Engineering and Machine Learning Engineering.

Publishers at Scrivener
Martin Scrivener (martin@scrivenerpublishing.com)
Phillip Carmical (pcarmical@scrivenerpublishing.com)

Data Engineering and Data Science

Concepts and Applications

Edited by
Kukatlapalli Pradeep Kumar
Aynur Unal
Vinay Jha Pillai
Hari Murthy
and
M. Niranjanamurthy

Scrivener Publishing

WILEY

This edition first published 2023 by John Wiley & Sons, Inc., 111 River Street, Hoboken, NJ 07030, USA and Scrivener Publishing LLC, 100 Cummings Center, Suite 541J, Beverly, MA 01915, USA

For more information about Scrivener publications please visit www.scrivenerpublishing.com.

Wiley Global Headquarters
111 River Street, Hoboken, NJ 07030, USA

For details of our global editorial offices, customer services, and more information about Wiley products visit us at www.wiley.com.

Library of Congress Cataloging-in-Publication Data

ISBN 9781119841876

Front cover images supplied by Pixabay.com
Cover design by Russell Richardson

Set in size of 11pt and Minion Pro by Manila Typesetting Company, Makati, Philippines

Printed in the USA

10 9 8 7 6 5 4 3 2 1

Contents

Preface xv

1 Quality Assurance in Data Science: Need, Challenges and Focus 1
Jasmine K.S., Ajay D. K. and Aditya Raj
1.1 Introduction 1
1.1.1 Quality Assurance and Testing 1
1.1.2 Data Science and Quality Assurance 2
1.1.3 Background 2
1.2 Testing and Quality Assurance 3
1.2.1 Key Terminologies Associated With Testing 3
1.3 Product Quality and Test Efforts 4
1.3.1 Testing Metrics 4
1.3.2 How to Improve the Business Value to Products Using Test Automation 7
1.3.3 Data Analysis and Management in Test Automation 8
1.3.4 Data Models in Data Science 8
1.4 Data Masking in Data Model and Associated Risks 8
1.5 Prediction in Data Science 9
Case Study 9
1.6 Role of Metrics in Evaluation 20
1.7 Quantity of Data in Quality Assurance 20
1.8 Identifying the Right Data Sources 20
1.8.1 Need to Gather Up-to-Date Data 20
1.8.2 Synthesising Existing Advanced Technologies for Continuous Business Improvements 21
1.9 Conclusion 21
References 21

2 Design and Implementation of Social Media Mining – Knowledge Discovery Methods for Effective Digital Marketing Strategies **23**
Prashant Bhat and Pradnya Malaganve
2.1 Introduction 24
2.1.1 Objectives of the Study 26
2.2 Literature Review 26
2.3 Novel Framework for Social Media Data Mining and Knowledge Discovery 29
2.4 Classification for Comparison Analysis 34
2.5 Clustering Methodology to Provide Digital Marketing Strategies 38
2.5.1 Status (Text Form) 39
2.5.2 Images (Photos) 40
2.5.3 Video Post 40
2.5.4 Link Post 40
2.6 Experimental Results 43
2.7 Conclusion 45
References 46

3 A Study on Big Data Engineering Using Cloud Data Warehouse **49**
Manjunath T. N., Pushpa S. K., Ravindra S. Hegadi and Ananya Hathwar K. S.
3.1 Introduction 50
3.2 Comparison Study of Different Cloud Data Warehouses 51
3.2.1 Amazon Redshift 52
3.2.2 High-Level Architecture of Amazon Redshift 52
3.2.3 Features of Amazon Redshift Cloud Data Warehouse 54
3.2.4 Pricing of Amazon Redshift Cloud Data Warehouse 55
3.3 Snowflake Cloud Data Warehouse 55
3.3.1 High-Level Architecture of Snowflake Cloud Data Warehouse 56
3.3.2 Features of Snowflake Cloud Data Warehouse 57
3.3.3 Snowflake Cloud Data Warehouse Pricing 58
3.4 Google BigQuery Cloud Data Warehouse 58
3.4.1 High-Level Architecture of Google BigQuery Cloud Data Warehouse 59
3.4.2 Features of Google BigQuery Cloud Data Warehouse 60
3.4.3 Google BigQuery Cloud Data Warehouse Pricing 60
3.5 Microsoft Azure Synapse Cloud Data Warehouse 61

3.5.1 Microsoft Azure Synapse Cloud Data Warehouse Architecture 61
3.5.2 Features of Microsoft Azure Synapse Cloud Data Warehouse 63
3.5.3 Pricing of Microsoft Azure Synapse Cloud Data Warehouse 64
3.6 Informatica Intelligent Cloud Services (IICS) 64
3.6.1 Informatica Intelligent Cloud Services Architecture 65
3.6.2 Salient Features of Informatica Intelligent Cloud Services 66
3.6.3 Informatica Intelligent Cloud Services Pricing Model 67
3.7 Conclusion 67
Acknowledgements 67
References 67

4 Data Mining with Cluster Analysis Through Partitioning Approach of Huge Transaction Data 71
Sampath Kini K. and Karthik Pai B.H.
4.1 Introduction 72
4.2 Methodology Used in Proposed Cluster Analysis System 75
4.2.1 Design of Algorithms 78
4.3 Literature Survey on Existing Systems 80
4.3.1 Experimental Results 80
4.4 Conclusion 82
References 82

5 Application of Data Science in Macromodeling of Nonlinear Dynamical Systems 85
Nagaraj S., Seshachalam D. and Jayalatha G.
5.1 Introduction 86
5.2 Nonlinear Autonomous Dynamical System 89
5.3 Nonlinear System - MOR 90
5.3.1 Proper Orthogonal Decomposition 91
5.4 Data Science Life Cycle 92
5.4.1 Problem Identification 92
5.4.2 Identifying Available Data Sources and Data Collection 93
5.4.3 Data Processing 93
5.4.4 Data Exploration 93
5.4.5 Feature Extraction 93

5.4.6 Modeling 93
5.4.7 Model Performance Evaluation 94
5.5 Artificial Neural Network in Modeling 94
5.5.1 Machine Learning 94
5.5.2 Biological Neuron Model 95
5.5.3 Artificial Neural Networks 96
5.5.4 Network Topologies 96
5.5.4.1 NARX Neural Network 98
5.5.5 ANN Modeling Using Mathematical Models 98
5.6 Neuron Spiking Model Using FitzHugh-Nagumo (F-N) System 99
5.6.1 Linearization of F-N System 99
5.6.2 Reduced Order Model of Linear System 101
5.6.3 Finite Difference Discretization of F-N System 101
5.6.4 MOR of F-N System Using POD-Galerkin Method 102
5.7 Ring Oscillator Model 104
5.7.1 Model Order Reduction of Ring Oscillator Circuit 106
5.7.2 Ring Oscillator Circuit Approximation Using Linear System MOR 106
5.7.3 POD-ANN Macromodel of Ring Oscillator Circuit 107
5.8 Nonlinear VLSI Interconnect Model Using Telegraph Equation 108
5.8.1 Macromodeling of VLSI Interconnect 109
5.8.2 Discretisation of Interconnect Model 110
5.8.3 Linearization of VLSI Interconnect Model 111
5.8.4 Reduced Order Linear Model of VLSI Interconnect 112
5.9 Macromodel Using Machine Learning 112
5.9.1 Activation Function 114
5.9.2 Bayesian Regularization 114
5.9.3 Optimization 115
5.10 MOR of Dynamical Systems Using POD-ANN 115
5.10.1 Accuracy and Performance Index 116
5.11 Numerical Results 117
5.11.1 F-N System 117
5.11.2 Ring Oscillator Model 120
5.11.3 Reduced Order POD Approximation of Ring Oscillator 120
5.11.3.1 Study of POD-ANN Approximation of Ring Oscillator for Variation in Amplitude of Input Signal and for Different Input Signals 121

5.11.3.2 POD-ANN Approximation of Ring Oscillator for Variation in Frequency 123
5.11.4 POD-ANN Approximation of VLSI Interconnect 125
5.12 Conclusion 126
References 130

6 Comparative Analysis of Various Ensemble Approaches for Web Page Classification 137
J. Dutta, Yong Woon Kim and Dalia Dominic
6.1 Introduction 138
6.2 Literature Survey 139
6.3 Material and Methods 144
6.4 Ensemble Classifiers 146
6.4.1 Bagging 147
6.4.1.1 Bagging Meta Estimator 147
6.4.1.2 Random Forest 147
6.4.2 Boosting 147
6.4.2.1 AdaBoost 148
6.4.2.2 Gradient Tree Boosting 148
6.4.2.3 XGBoost 148
6.4.3 Stacking 148
6.5 Results 148
6.5.1 Bagging Meta Estimator 149
6.5.2 Random Forest 152
6.5.3 AdaBoost 156
6.5.4 Gradient Tree Boosting 160
6.5.5 XGBoost 162
6.5.6 Stacking 164
6.5.7 Comparison with Single Classifiers 168
6.6 Conclusion 169
Acknowledgement 170
References 170

7 Feature Engineering and Selection Approach Over Malicious Image 173
P.M. Kavitha and B. Muruganantham
7.1 Introduction 173
7.2 Feature Engineering Techniques 176
7.2.1 Methodologies in Feature Engineering 177
7.2.2 Strides in Feature Engineering 178
7.2.3 Feature Extraction 179
7.2.4 Feature Selection 180

7.2.5 Feature Engineering in Image Processing 181
7.2.6 Importance of Feature Engineering in Image Processing 181
7.3 Malicious Feature Engineering 182
7.4 Image Processing Technique 183
7.4.1 Steps Involved in Image Processing Technique 183
7.4.2 Image Processing Task 183
7.4.2.1 Image Enhancement 183
7.4.2.2 Image Restoration 184
7.4.2.3 Coloring Image Processing 184
7.4.2.4 Wavelets Processing and Multiple Solutions 184
7.4.2.5 Image Compression 184
7.4.2.6 Character Recognition 184
7.4.2.7 Characteristics of Image Processing 185
7.5 Image Processing Techniques for Analysis on Malicious Images 185
7.6 Conclusion 191
References 191
Blog 193

8 Cubic-Regression and Likelihood Based Boosting GAM to Model Drug Sensitivity for Glioblastoma 195
Satyawant Kumar, Vinai George Biju, Ho-Kyoung Lee and Blessy Baby Mathew
8.1 Introduction 196
8.1.1 Glioblastoma 197
8.2 Literature Survey 198
8.3 Materials and Methods 201
8.3.1 Methodology 202
8.3.1.1 Generalized Additive Models (GAMs) 202
8.3.1.2 Model-Based Boosting – Boosted GAM 205
8.3.2 Datasets Description 207
8.4 Evaluations, Results and Discussions 209
8.4.1 Akaike Information Criterion (AIC) 210
8.4.2 Adjusted R-Squared 210
8.4.3 Discussion 210
Conclusion 220
References 220

9 Unobtrusive Engagement Detection through Semantic Pose Estimation and Lightweight ResNet for an Online Class Environment **225**
Michael Moses Thiruthuvanathan, Balachandran Krishnan and Madhavi Rangaswamy
9.1 Introduction 226
9.2 Related Work 230
9.2.1 Analysis for a Classroom Environment 230
9.2.2 Pose Estimation 231
9.2.3 Face Alignment and Landmark Estimation 232
9.2.4 Deep Networks for Emotional Analysis 233
9.3 Proposed Methodology 234
9.3.1 Data Description 234
9.3.2 Facial Detection and Recognition 234
9.3.2.1 Face Detection 235
9.3.2.2 Facial Landmark Detection 235
9.3.3 Emotion Quantification 236
9.3.4 Pose Estimation 238
9.3.4.1 Facial Pose Estimation 238
9.4 Experimentation 241
9.5 Results and Discussions 245
Conclusion 250
References 250

10 Building Rule Base for Decision Making – A Fuzzy-Rough Approach **255**
Sabu M. K., Neeraj Krishna M. S. and Reshmi R.
10.1 Introduction 256
10.2 Literature Review 258
10.3 Discretization of the Dataset Using Fuzzy Set Theory 260
10.4 Description of the Dataset 260
10.5 Process Involved in Proposed Work 261
10.6 Experiment 262
10.7 Evaluation Result 267
10.8 Discussion 273
Conclusion 275
References 276

11 An Effective Machine Learning Approach to Model Healthcare Data **279**
Shaila H. Koppad, S. Anupama Kumar and Mohan Kumar
11.1 Introduction 280
11.2 Types of Data in Healthcare 281
11.3 Big Data in Healthcare 283
11.4 Different V's of Big Data 284
11.5 About COPD 285
11.6 Methodology Implemented 290
Conclusion 297
References 298

12 Recommendation Engine for Retail Domain Using Machine Learning Techniques **303**
Chandrashekhara K. T., Gireesh Babu C. N. and Thungamani M.
12.1 Introduction 304
12.2 Proposed System 304
12.2.1 Classification of Suppliers 306
12.2.2 Recommendation for Buyer 307
12.2.3 Forecasting Using ARIMA Model 309
12.3 Results 312
12.3.1 ARIMA Forecasting 312
12.4 Conclusion 313
References 314

13 Mining Heterogeneous Lung Cancer from Computer Tomography (CT) Scan with the Confusion Matrix **317**
Denny Dominic and Krishnan Balachandran
13.1 Introduction 317
13.2 Literature Review 319
13.3 Methodology 320
13.3.1 Description of the Data 320
13.3.2 Image Preprocessing 321
13.3.3 Image Segmentation 322
13.3.4 Image Processing 322
13.3.5 Zero Component Analysis (ZCA) Whitening 322
13.3.6 Local Binary Pattern (LBP Feature) 323
13.3.7 LESH Vector 324
13.3.8 Local Energy Map and Orientation Map 324
13.3.9 Training with Deep Learning Methods 325

13.4 Result 326
13.4.1 Lorenz Curve 327
13.4.2 Confusion Matrix 328
13.4.3 Gini Coefficient 330
13.5 Conclusion and Future Scope 332
References 332

14 ML Algorithms and Their Approach on COVID-19 Data Analysis 335
Kambaluru Ashok, Penumalli Anvesh Reddy and Kukatlapalli Pradeep Kumar
14.1 Introduction 336
14.2 DataSet 336
14.2.1 Labeled Datasets 337
14.2.2 Unlabelled Datasets 337
14.2.3 COVID-19 Data 337
14.3 Types of Machine Learning Algorithms 338
14.3.1 Supervised Learning 338
14.3.2 Unsupervised Learning 345
14.3.3 Semi-Supervised Learning 347
14.3.4 Reinforcement Learning 347
14.4 Conclusion 348
References 348

15 Analysis and Design for the Early Stage Detection of Lung Diseases Using Machine Learning Algorithms 351
Sindhu Madhuri, Mahesh T. R., Vivek V., Shashikala H. K. and C. Saravanan
15.1 Introduction 352
15.2 Machine Learning Algorithms 358
15.2.1 Linear Regression 358
15.2.2 Logistic Regression 359
15.2.3 Decision Tree 360
15.2.4 Random Forest 362
15.2.5 Naïve Bayes 362
15.2.6 Support Vector Machine (SVM) 363
15.3 Evaluation Metrics and Comparative Results for Early Detection of Lung Diseases 364
15.3.1 Accuracy (A) 366
15.3.2 Precision (P) and Recall (R) 366
15.3.3 Mean Squared Error (MSE) 366

15.3.4 Matthews Correlation Coefficient (MCC) 367
15.4 Conclusion 369
References 369

16 Estimation of Cancer Risk through Artificial Neural Network 373
K. Aditya Shastry, Sanjay H. A., Balaji N. and Karthik Pai B. H.
16.1 Introduction 373
16.2 Case Studies Related to Cancer Risk Estimation Using ANN 375
16.2.1 ANN Technique for Early LC Detection 378
16.2.2 ANNs in Image Processing for Early Diagnosis of BC 382
16.3 Datasets Used in Cancer Risk Estimation 388
16.3.1 Datasets Related to Breast Cancer 388
16.3.2 Dataset Related to Lung Cancer 389
16.3.3 BC Coimbra Data Set and BC Wisconsin (Diagnostic) Data 389
16.3.4 Comparison of ANN Techniques with Other Methods for Cancer Risk Estimation 390
16.4 Discussion 397
16.5 Future Scope 400
16.6 Conclusion 400
References 401

17 Applications and Advancements in Data Science and Analytics 409
T. Mamatha, A. Balaram, B. Rama Subba Reddy, C. Shoba Bindu and M. Niranjanamurthy
17.1 Data Science and Analytics in Software Testing 410
17.2 Applications of Data Science and Analytics 411
17.3 Selenium Testing Tool in Data Science 419
17.3.1 Basic Techniques for Testing Voice-Based Applications 419
17.3.2 Image Web Scraping Using Selenium 421
17.4 Challenges and Advancements in Data Science 425
17.5 Data Science and Analytics Tools 430
17.6 Conclusion 438
References 439

About the Editors 441

Index 443

Preface

Data engineering is the aspect of data science that focuses on practical applications of data collection and analysis. The data science field is incredibly broad, encompassing everything from cleaning data to deploying predictive models. It complements emerging areas such as machine learning, artificial intelligence and cyber security analytics. Decision making is a vital component is an industry economy that is motivated by algorithms in data science allied fields.

Data-driven discovery is revolutionizing the modeling, prediction, and control of complex systems. This book brings together machine learning, engineering mathematics, and mathematical physics to integrate modeling and control of dynamical systems with modern methods in data science. It highlights many of the recent advances in scientific computing that enable data-driven methods to be applied to a diverse range of complex systems, such as turbulence, the brain, climate, epidemiology, finance, robotics, and autonomy. The book introduces the foundational concepts of data engineering – data models, value chain models and team dynamics. Aimed at advanced undergraduate and beginning graduate students in the engineering and physical sciences, the book presents a range of topics and methods from introductory to state of the art.

This book followed a sequential approach of development of an artifact which initiates from feasibility studies, communication and evolving towards construction and deployment. Similar entities are captured from the chapters received from various authors and a beautiful lineage of data items has been crafted.

1

Quality Assurance in Data Science: Need, Challenges and Focus

Jasmine K.S.*, Ajay D. K. and Aditya Raj

Dept. of MCA, RV College of Engineering®, Bengaluru, Karnataka, India

Abstract
It is widely accepted that quality is assured for a product through the process of testing. With the rapid development in the area of data science, research is going on with proper management of data and with its right usage, test engineers can learn about their users. One can predict the associated risks and with a focus on data masking based on the data model. Prescriptive and predictive analysis can be more accurate if the techniques are developed and the accuracy is measured using metrics. Preparing data with required quality and identifying the possible resources are challenging tasks faced by a data scientist. The effective and systematic use of advanced technologies like high-speed hardware and network computing, cloud computing, cross platform tools, etc., continues to be an elusive goal for many organizations. In this context, the chapter investigates the feasibility of novel and practical solutions in this aspect.

Keywords: Quality assurance, testing, data science, data analysis, decision making

1.1 Introduction

1.1.1 Quality Assurance and Testing

In the traditional software development approach, Quality Assurance was done at the later stages of the development process and feedback was collected for improvement. In almost all organizations there exists a Quality

**Corresponding author*: jasmineks@rvce.edu.in

Kukatlapalli Pradeep Kumar, Aynur Unal, Vinay Jha Pillai, Hari Murthy, and M. Niranjanamurthy (eds.) Data Engineering and Data Science: Concepts and Applications, (1–22) © 2023 Scrivener Publishing LLC

Assurance team, responsible for identifying the product defects and resolving them before release of the product.

But in the Agile development approach, the Quality Assurance (QA) team works collaboratively with the development team with a shared responsibility in improving the quality of the product. Irrespective of the project and domain QA plays an important role in delivering high-quality products. It helps developers establish goals and define quality standards. The process also helps in identifying and resolving defects in the products before releasing them to market.

1.1.2 Data Science and Quality Assurance

Data is available in structured and unstructured format. There is a need of methods and algorithms to extract knowledge and inference so that available data can be converted into wisdom to apply in any application domain. In the context of data science, QA is a process of analyzing and modeling the available data to ensure high quality and meeting quality standards. Models have to be created with a variety of data sets and robustness tests have to be conducted for trained models similar to any use case for real-world scenarios [6].

1.1.3 Background

Data science basically puts efforts to verify the general consistency among relevant data and applies scientific methods to ensure the data quality. Domain knowledge is necessary to ensure not only the quality of data but also to avoid errors and inconsistencies in the data collected in the Quality assurance process [1]. After collecting the data, interpretation of data is also a huge challenge. So it is essential that quality assurance should be a continuous process throughout the product development phase, starting from data collection till the delivery of the product. Expert opinion will add value to it. In order to validate the results obtained after data analysis, domain experts play an important role. Finding and resolving quality-related issues once the product is delivered will result in unnecessary wastage of time, money and effort [2, 9]. In order to perform data analytics in an efficient manner, a variety of data sets is essential [3]. Frequency, time and many iterations are part of data analytics. So there is a need for systems which focus on data quality [4, 10]. Software quality and hardware quality also play an equal role [5].

1.2 Testing and Quality Assurance

Testing is a process which ensures the intended behavior of any product within the given time frame and also helps to avoid additional efforts, time and cost overruns. System testing ensures not only product quality but also process quality.

1.2.1 Key Terminologies Associated With Testing

1.2.1(a) Defect count: A defect count is one of the measurements of product quality. It shows the number of undetected errors. The defect rate can be computed by defective products observed divided by the number of units tested. For example, if 5 out of 10 tested units are defective, the defect rate is 5 divided by 10.

1.2.1(b) Test case execution: During the test execution process the developed code is executed with generated test cases and actual output obtained is compared with the expected output. After the test execution process, bugs and test execution status are maintained for further measures. Following are steps involved in test execution:

Step 1: Gathering testing requirements
Step 2: Test plan
Step 3: Test design
Step 4: Test execution
Step 5: Defect reporting and tracking
Step 6: Defect Mapping

1.2.1(c) Test execution classification: The type of testing can be classified based on the purpose for which it has to be executed: for example, performance testing, defects testing, regression testing, etc. During the test execution process, the main focus is to verify how much the actual results varied from the expected results.

1.2.1(d) Pre-requisites for test execution phase

1. Test Plan, test cases development and the test data should be ready
2. Test environment setup and its validation
3. Clear understanding of exit criteria

1.2.2(e) Post-requirements for test execution phase

1. Validation of actual results with expected results
2. Defects should be closed or deferred
3. Test execution and defects summary report
4. Prioritize the test plans based on the identified risks
5. Continue the process

1.2.2(f) Test velocity

Testing velocity shows the number of tests one is running per day/weekly/monthly, etc. It also shows the difference between the planned and actual time.

1.3 Product Quality and Test Efforts

1.3.1 Testing Metrics

Testing metrics are required to measure and estimate process and product quality and to improve the efficiency of the overall testing process. Two major categories are defect metrics and productivity metrics.

Example of productivity metrics: Mileage of any vehicle compared to its ideal mileage recommended by the manufacturer. Figure 1.1 shows the test velocity and loop count also we can see the thread properties.

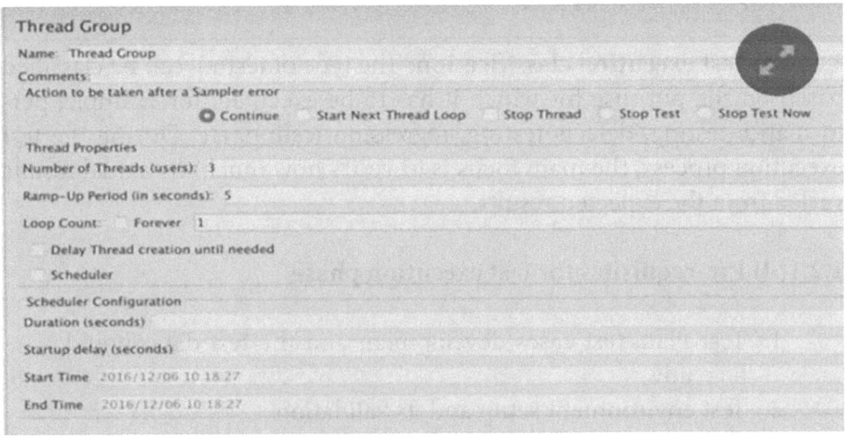

Figure 1.1 Test velocity, loop count.

Example of defect metrics: Number of defects found, accepted, rejected and deferred in the payment process using a credit card.
A few automated testing tools: Selenium, JMeter, Appium, Junit

The chapter demonstrates a few testing aspects using JMeter. It helps to test and analyze overall performance under different load types.

Description of Figure 1.1:

- **Name:** Provide a custom name or, if you prefer, simply leave it named the default "Thread Group".
- **Thread count:** The number of threads indicates the number of users under test.
- **Ramp-up Period:** How quickly (in seconds) you want to add users. For example, if set to zero, all users will begin immediately. If set to 10, one user will be added at a time until the full amount is reached at the end of the 10 seconds. Let's say 5 seconds.
- **Loop Count:** It indicates the number of times the test should repeat.

Figure 1.2 demonstrates thread delay of 300 milliseconds.

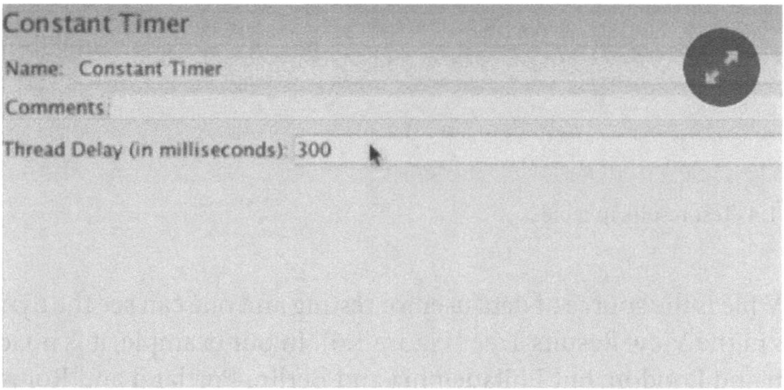

Figure 1.2 Thread delay.

In Figure 1.3, successful test results show in green icons. Figure 1.3 also shows load time, connect time, errors, the request data, the response data, etc. Figure 1.4 demonstrates the Test results in table and and Figure 1.5 shows the Test report summary.

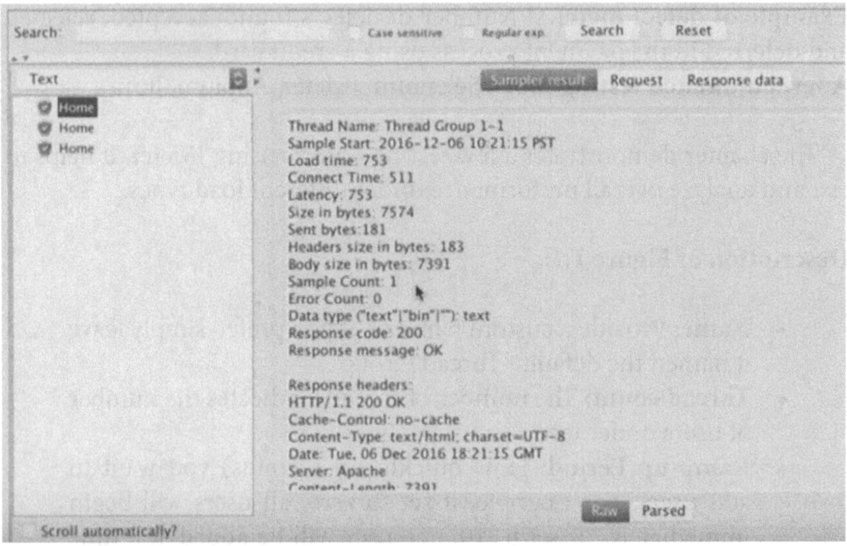

Figure 1.3 Test results.

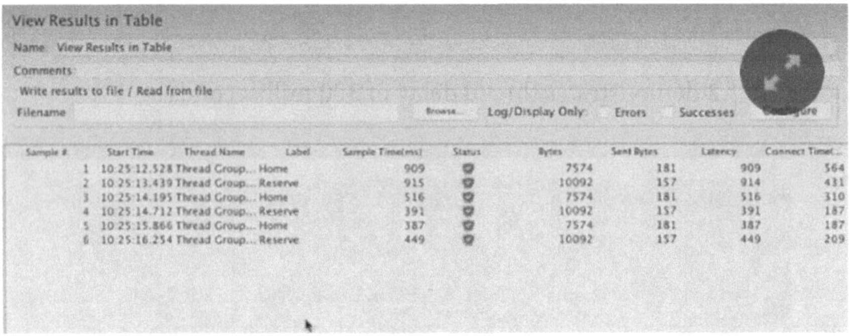

Figure 1.4 Test results in table.

CSV file is the source of data user for testing and one can see the dynamic results in the View Results Tree (Figure 1.6). In our example, it is no longer Boston and London, but Philadelphia and Berlin, Portland and Rome, etc.

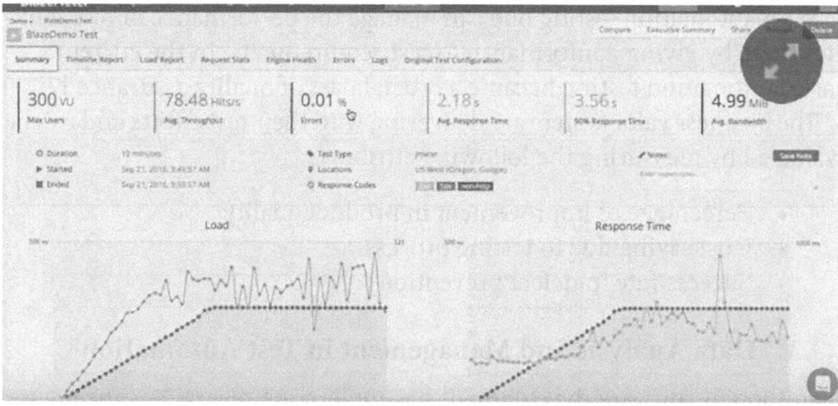

Figure 1.5 Test report summary.

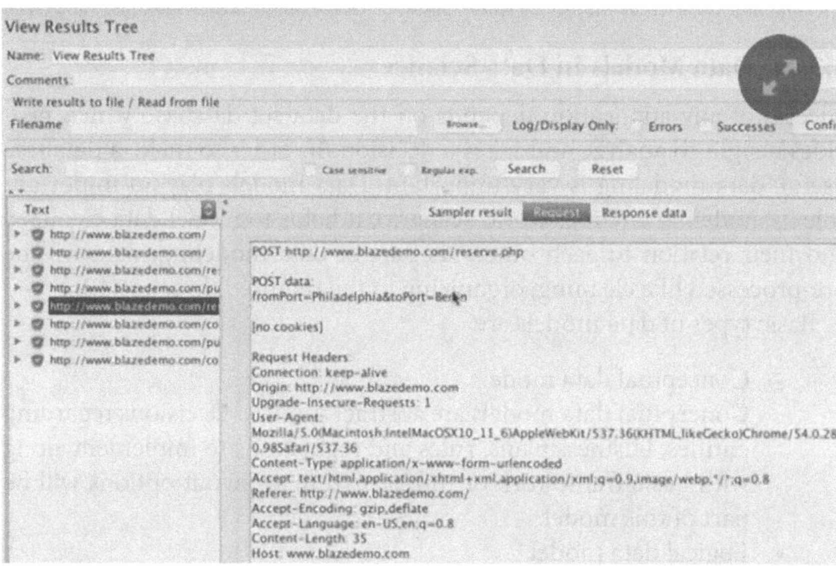

Figure 1.6 View results tree.

1.3.2 How to Improve the Business Value to Products Using Test Automation

With the advancements in the area of automation testing, in order to ensure software quality, not only the developed software has to meet the functional requirements specifications, it also has to meet the non-functional requirements specifications such as operational efficiency, reliability, security, maintainability, availability, code efficiency, etc., through which one can increase the business value.

With automation testing one can manage the performance of test activities there by giving conformance to test requirements. In the current scenario, automation testing became a crucial part of quality assurance [7].

The business value testers are delivering with their test efforts and can be evaluated by measuring the following attributes:

- Percentage of improvement in product quality
- Cost saving due to testing process
- Success rate in defect prevention

1.3.3 Data Analysis and Management in Test Automation

Even though software development has undergone several advancements in the last few years, the importance of evaluating test execution results is unchanged, especially in the case of automated tests.

1.3.4 Data Models in Data Science

Success of any application depends on the data set/database which provides insight to analyze and takes a decision. In order to make a database useful, data modeling is very important. Data models play an important role in modeling any real-world scenario. It helps to model data elements and their relation to each other. As part of data modeling, data will be pre-processed like cleaning, organizing in the required direction [8].

Basic types of data models are:

- Conceptual data mode
  Conceptual data models are abstract models. Decisions regarding entities, business plans, rules and regulations to implement along with basic framework or infrastructure or layout options will be part of this model.
- Logical data model
  In this type of data model, importance is given for relational factors such as properties, or data attributes. This model is mainly useful in data warehousing plans.
- Physical data model
  This model focuses on data points and their relationships to make a final actionable blueprint.

1.4 Data Masking in Data Model and Associated Risks

When handling sensitive data such as financial records, one has to take care of the risk associated with data breaches to a minimum by addressing possible sources of risk. Each organization should have comprehensive strategy for reducing data breach risks and mitigating their effect in order to protect their brand and reputation. In the conventional approach, data protection is done with encryption [9].

With the approach of data masking, one can replace the characters of sensitive data with similar but meaningless characters. It is also associated with de-masking process, during which the application maps the changed numbers back to the original ones. It is normally implemented as a cloud-based application. This type of masking can be applied to PAN number, Aadhar number, etc. Applications will be designed to work with specific data formats which can handle the masked data.

1.5 Prediction in Data Science

Data analytics can be broadly classified majorly into three types such as Descriptive, Predictive and Prescriptive Analytics [12].

Descriptive Analytics - This model helps in hypothesis generation, variable transformation and any root cause analysis of specific behavioral patterns.
Predictive Analytics - This type of model is used to predict the future behavior of the dependent variable. Predictive analytics answers the question of what is likely to happen.
Prescriptive Analytics - This type of analytics is required to avoid the catastrophic consequences of a pandemic like Covid-19.

Case Study

Aim: To predict the prices of StainlessSteelPrice
Time Period: 3 months and 6 months

In this case study, the aim was to predict the price of stainless steel for the next three months using machine learning algorithms and compare different models on the basis of accuracy.

Solution Approach
The first step started with exploratory data analysis which checks the insights of historical data and is considered as descriptive analytics. It includes checking of consistent data, inconsistent data (outliers), missing value, measure of central tendency (mean, median and mode), measure of

dispersion (Variance, standard deviation and range), graphical representation like Box plot which tells the value of outlier, identifies whether the data is symmetrical, determines how tightly data is grouped and also verifies the data is skewed. After feature selection from different attributes of the data set with the help of correlation matrix and univariate selection is applied followed by feature engineering which includes handling missing values and handling outliers. Modeling with the help of statistical method, ARIMA, is also performed here [11]. Figure 1.7 shows the Data set for analysis and Figure 1.8 represents the Graph shows analyzed data.

Models Used for Visualization

- Boxplot
  Boxplot graphically depicts groups of numerical data through their quartiles.
- Regression Model
  The linear regression model is very useful in forecasting, time series modeling and finding the cause and effect relationship between variables.
- ARIMA Model
  One can forecast a time series using the series past values in ARIMA model.

```
import numpy as np
import pandas as pd
import matplotlib as mpl
import matplotlib.pyplot as plt
%matplotlib inline
df= pd.read_csv('/tmp/aq/Stainless-Steel-Prices-Forecasty-Assignment.
csv',index_col=0,parse_dates=True,sep =',')
df.head(5)
```

Date	StainlessSteelPrice	Steel_Inventory_M_USD	Steel_Orders_M_USD	IronOre_Global_USD	CoalAustralia_Global_USD	Energy_Index	Nickel_Global_USD
2013-07-01	0.827352	20575	10568	127.191304	82.222360	227.022942	13750.31522
2013-08-01	0.760221	20609	10821	137.055000	82.221429	232.234194	14308.26190
2013-09-01	0.784715	20761	10774	134.185714	83.295918	234.335594	13801.39286
2013-10-01	0.769293	20735	11291	132.572727	85.443634	229.393932	14117.65217
2013-11-01	0.750242	20854	11559	136.323810	88.363265	225.437049	13684.01190

Zinc_Global_USD	Freight_Transport_Index	WTI_Crude_Global_USD	Baltic_Dry_Index	BCI_China	BCI_Europe	BCI_US	CLI_China	CLI_Europe	CLI_US
1837.619565	-0.4	104.545217	1062	98.80018	99.64265	100.1044	101.1423	99.43615	100.6230
1896.392857	0.5	106.546364	1132	99.04874	99.91766	100.3962	101.1938	99.64082	100.6545
1846.880952	0.1	106.313809	2003	99.27196	100.18490	100.5363	101.2066	99.84383	100.6646
1884.836957	-0.1	100.500435	1504	99.40488	100.36330	100.5771	101.1835	100.02300	100.6666
1866.416667	2.1	93.812381	1821	99.36221	100.49830	100.5459	101.1283	100.17150	100.6744

Figure 1.7 Data set for analysis.

```
df.index.freq = 'MS'
df.describe()
```

WTI_Crude_Global_USD	Baltic_Dry_Index	BCI_China	BCI_Europe	BCI_US	CLI_China	CLI_Europe	CLI_US	PMI_US_Manufacturing	Copper_Global_USD
84.000000	84.000000	84.000000	84.000000	84.000000	84.000000	84.000000	84.000000	84.000000	84.000000
61.017507	1067.785714	98.710869	100.569427	100.175521	99.563679	100.023602	99.814380	54.026190	6037.075714
21.407539	443.697935	0.848983	1.123901	0.876723	2.027340	1.567463	1.367807	4.097578	783.645475
16.807273	317.000000	94.501530	95.996640	98.102130	85.383290	90.343960	92.280250	41.500000	4471.790000
47.738636	699.500000	98.271312	100.427200	99.535878	99.225065	99.696710	99.278723	51.275000	5645.152500
54.912545	1058.500000	98.675055	100.562400	100.312350	99.854750	100.294100	100.030800	54.250000	6024.555000
68.464674	1354.500000	99.364373	101.040100	100.803675	100.518350	100.500175	100.679550	57.400000	6801.285000
106.546364	2378.000000	99.923670	102.323700	101.573800	101.206600	101.709300	100.968000	60.800000	7291.470000

	StainlessSteelPrice	Steel_Inventory_M_USD	Steel_Orders_M_USD	IronOre_Global_USD	CoalAustralia_Global_USD	Energy_Index	Nickel_Global_USD
count	84.000000	84.000000	84.000000	84.000000	84.000000	84.000000	84.000000
mean	0.810715	20070.559524	9831.595238	79.962940	81.844151	143.408951	12712.275477
std	0.143444	1764.033210	1240.333905	23.874776	19.443360	47.136073	2694.729168
min	0.499859	16688.000000	7650.000000	40.882609	53.428929	52.771306	8298.500000
25%	0.743778	18980.000000	8904.000000	61.484190	67.111352	111.607197	10578.975000
50%	0.812157	20190.000000	9947.000000	74.280682	78.677679	131.088857	12721.351135
75%	0.893690	21458.500000	11062.500000	92.409524	96.985695	163.923895	14105.350543
max	1.075014	23464.000000	11825.000000	137.055000	125.065877	237.208789	19401.075000

```
color_list = [ "blue", "orange", "green", "red", "purple", "brown", "pink", "gray" "olive",
"cyan",]
def Visualize(data):
features= list(df.select_dtypes(include=[np.number]).columns.values)
feature_size=len(features)
fig, axes = plt.subplots(
nrows=int(np.ceil(feature_size/2)), ncols=2, figsize=(14, feature_size*2),
dpi=80, facecolor="w", edgecolor="k"
)
for i in range(feature_size):
key = features[i]
c = color_list[i % (len(color_list))]
t_data = data[key]
t_data.head()
ax = t_data.plot(
ax=axes[i // 2, i % 2],
color=c,
title="{}".format(key),
rot=25,
)
ax.legend([key])
plt.tight_layout()
Visualize(df)
```

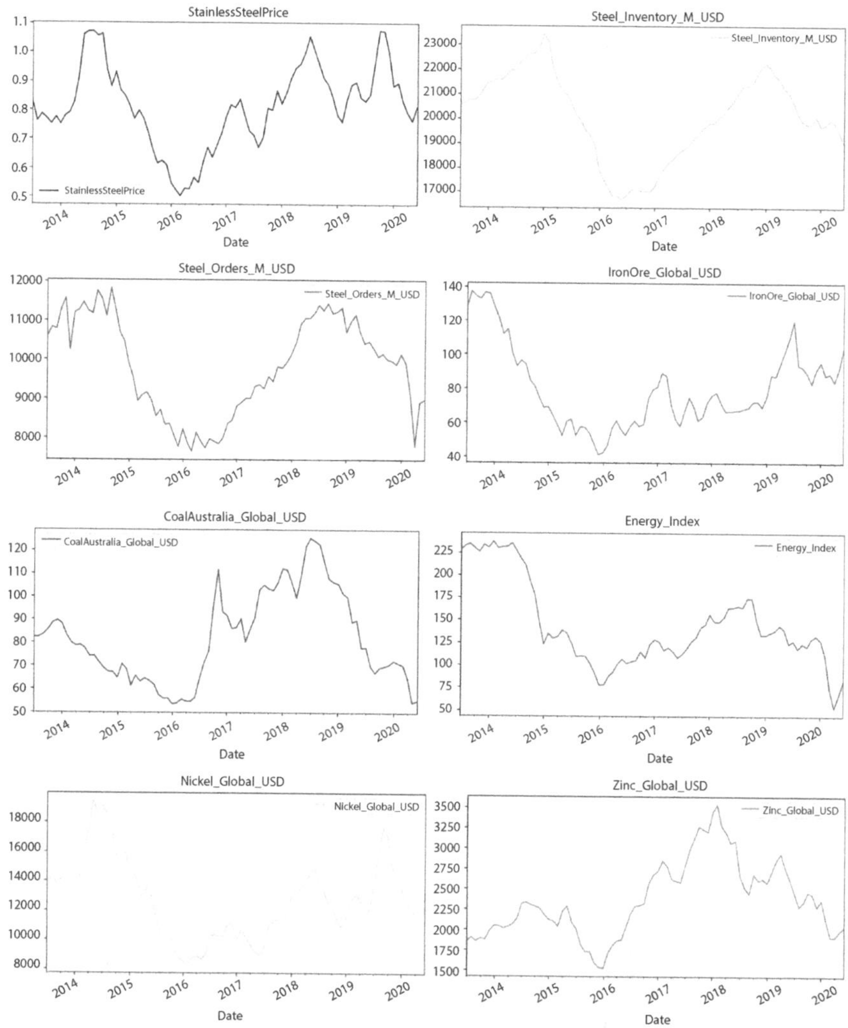

Figure 1.8 Graph shows analyzed data. *(Continued)*

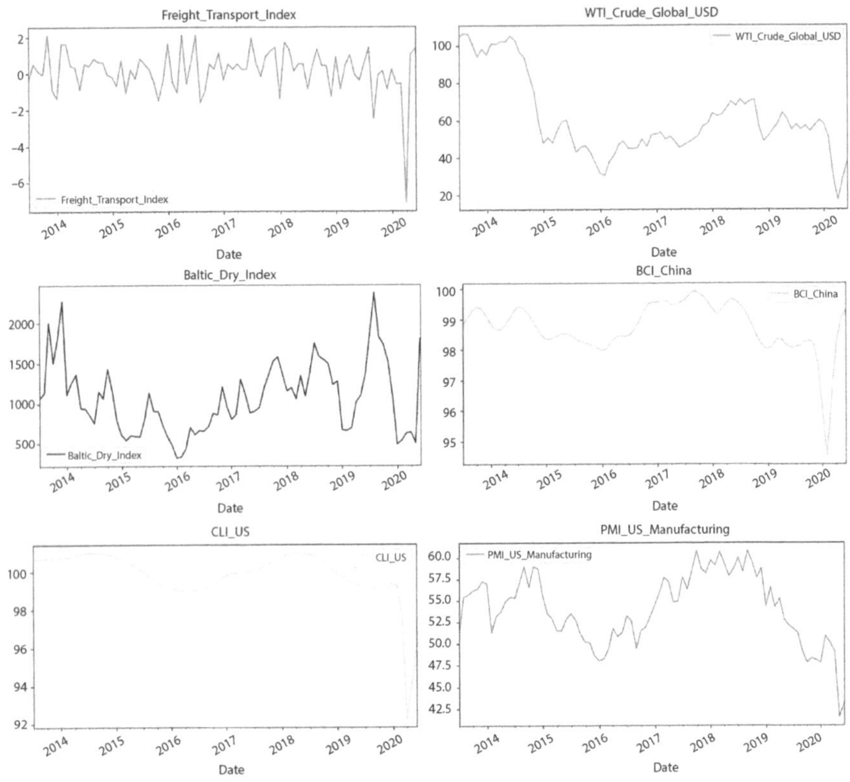

Figure 1.8 (Continued) Graph shows analyzed data.

```
#BOX Plot
# Tell us the values of outliers
# Identify if data is symmetrical
# Determine how tightly data is grouped
# see if data is skewed
plt.figure(figsize=(12,8))
plt.boxplot(df)
plt.xlabel('Date')
plt.ylabel('df')
Text(0, 0.5, 'df')
```

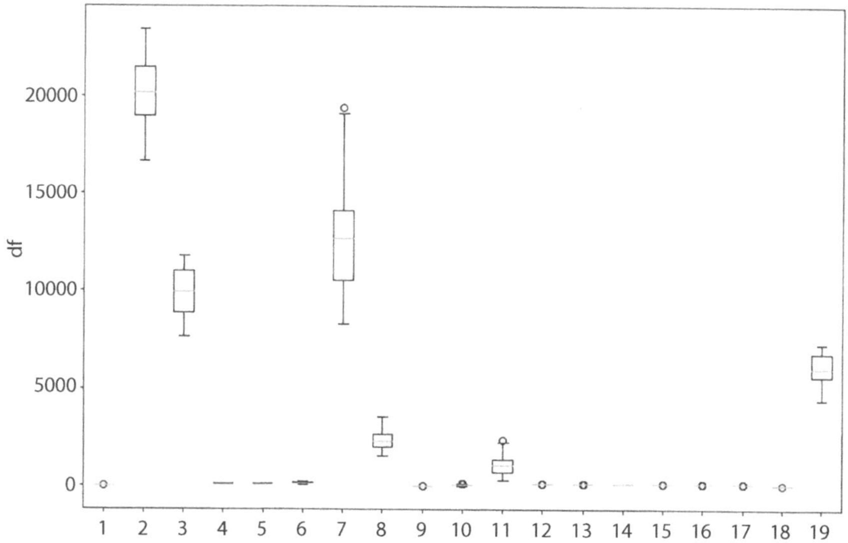

Figure 1.9 Boxplot data.

The above Figure 1.9 represents the Boxplot data here we can easily analyze the values.

```
#ggcorrplot(cor())
plt.matshow(df.corr(method='spearman'),vmax=1,vmin=-1,c-
map='PRGn')
plt.title('without resampling', size=10)
plt.colorbar()
plt.show()
```

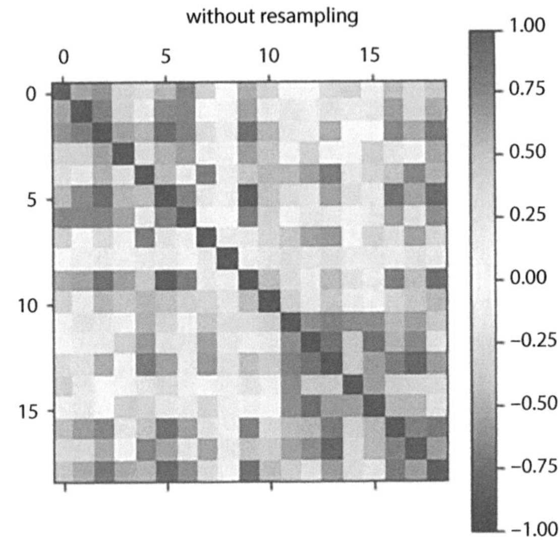

Figure 1.10 Correlation data plot.

```
import statsmodels.api as sm
from statsmodels.tsa.stattools import acovf,acf,pacf,pacf_yw,pacf_ols
from pandas.plotting import lag_plot
lag_plot(df['StainlessSteelPrice'])
```

```
<AxesSubplot:xlabel='y(t)', ylabel='y(t + 1)'>
```

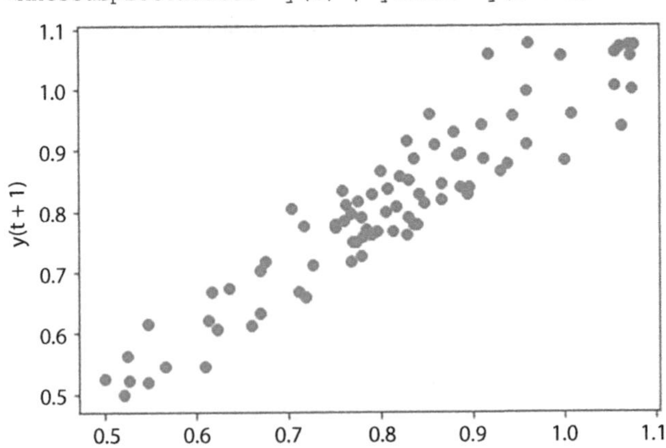

Figure 1.11 Lag plot.

```
from statmodels.graphics.traplots import plot_acf, plot_pacf
plot_acf(df['StainlessSteelPrice'], lags=40);
```

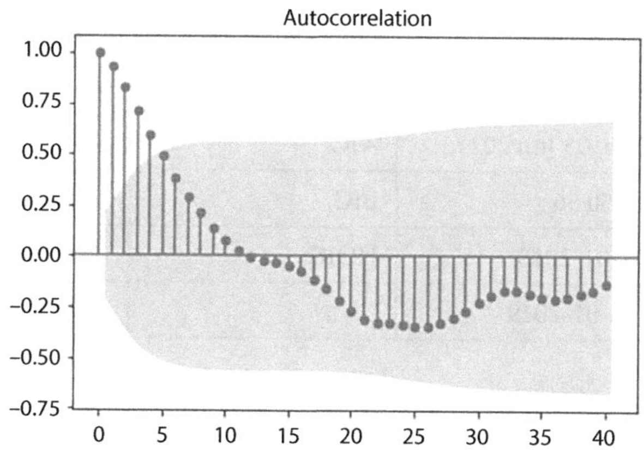

Figure 1.12 Autocorrelation.

```
plot_pacf(df['StainlessSteelPrice'],lags=35);
```

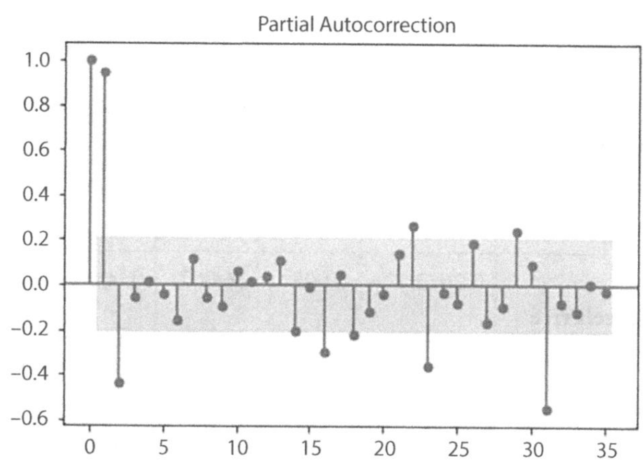

Figure 1.13 Partial autocorrelation.

The above Figure 1.10 represents Correlation data plot, Figure 1.11 shows the Lag plot of the given data, Figure 1.12 represents the Autocorrelation and Figure 1.13 represents the Partial autocorrelation, we can easily analyze the data.

Table 1.1 ARMA model results.

Dep. Variable:	StainlessSteelPrice	No. Observations	78
Model:	ARMA(5, 4)	Log Likelihood	131.623
Method:	css-mle	S.D. of innovations	0.043
Date:	Thu, 03 Jun 2021	AIC	-241.246
Time:	11:50:56	BIC	-215.322
Sample:	07-01-2013	HQIC	-230.868
	-12-01-2019		

	coef	**std err**	**z**	**P>|z|**	**[0.025**	**0.975]**
Const	0.8018	0.022	36.169	0.000	0.758	0.845
ar.L1. StainlessSteelPrice	0.5693	nan	nan	nan	nan	nan
ar.L2. StainlessSteelPrice	1.0398	0.039	26.499	0.000	0.963	1.117
ar.L3. StainlessSteelPrice	0.1129	nan	nan	nan	nan	nan
ar.L4. StainlessSteelPrice	-1.0213	nan	nan	nan	nan	nan
ar.L5. StainlessSteelPrice	0.2672	0.005	56.007	0.000	0.258	0.277
ma.L1. StainlessSteelPrice	0.7260	0.112	6.478	0.000	0.506	0.946
ma.L2. StainlessSteelPrice	-0.5262	0.083	-6.322	0.000	-0.689	-0.363
ma.L3. StainlessSteelPrice	-1.0750	0.072	-15.018	0.000	-1.215	-0.935
ma.L4. StainlessSteelPrice	-0.1248	0.110	-1.137	0.256	-0.340	0.090

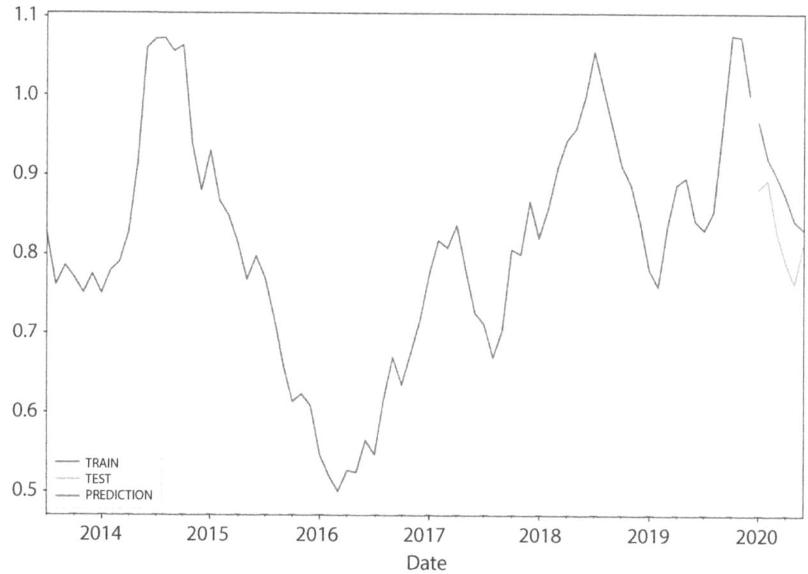

Figure 1.14 ARIMA model - train-test-prediction.

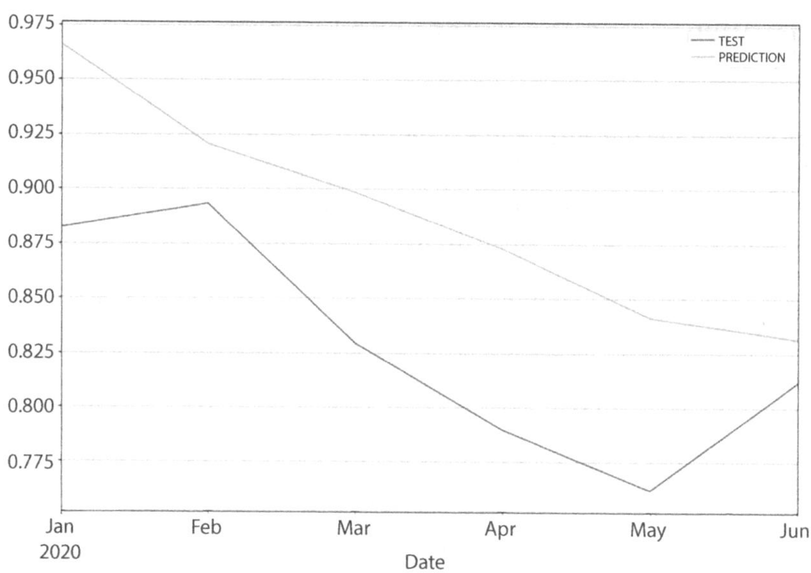

Figure 1.15 ARIMA Model - Test & prediction.

The above Figure 1.14 represents the ARIMA model , we can easily analyze the train, test and prediction. Figure 1.15 shows the ARIMA Model, here we can analyze Test and prediction.

1.6 Role of Metrics in Evaluation

In order to measure, quality of model, evaluation metrics are used. It will be suitable for either a statistical or machine learning model. The metrics are required to measure environmental and sustainable impacts, and the scope of their application should be wide in order to cover variety of data. As a result, metrics are to be developed and used in a continuous fashion.

1.7 Quantity of Data in Quality Assurance

It is required to keep a threshold value to the quality score and benchmark them based on their relevance. This is very crucial for business success. High-quality data improves confidence in decision making. Data quantity plays an important role in deciding data quality because as the data size increases, decision making will be more challenging and will lead to the right one with the large set of available data.

1.8 Identifying the Right Data Sources

Everyone knows that quantity and quality of data is important for right decision making but at the same time, the data content also plays a crucial role. So the question, "Are we collecting the right data?" is very relevant. Each data set provides a unique point of view to business operations. There is a chance of duplicate data, data with gaps, and biased data, so one has to ensure all these points while gathering data from multiple resources. By aggregating data from different data sources, one can make a holistic view towards business operations. If you come across conflicting data, it indicates that data is gathered from multiple sources and it can be looked into from multiple perspectives.

1.8.1 Need to Gather Up-to-Date Data

Every piece of data has a time stamp showing when it was created. Data starts aging from time to time. It is important to understand when the data was collected and how current the data is. In order to ensure its relevance, it is essential to trace the origin of data, and also it is recommended to collect data from the source system where it is created. Digital business processes require real-time data to be effective. This data can be analyzed to provide real-time visibility towards business forecasting and operations.

1.8.2 Synthesising Existing Advanced Technologies for Continuous Business Improvements

The definition of the term "software quality" evolved through many years. DevOps and cloud computing technologies play important roles in this direction. But adopting DevOps requires the knowledge of Automation Testing to increase the effectiveness and coverage of software testing. Availability of high-speed internet made the trend of storing and maintaining digital data easy. With the usage of social networking, the data generating apps generate a large amount of data that can be estimated as Terabytes daily. There is a need of some techniques and standards to extract some meaningful results from this large data set. High-speed computing hardware, distributed and parallel computing can also play crucial roles in assuring quality data towards continuous business improvements.

1.9 Conclusion

The chapter focuses on the importance of data masking to predict the associated risks. The chapter also deals with data analytics methods like prescriptive and predictive approaches and demonstrates how the quality of the data influences the desired outcome. In writing this chapter, we are convinced that the insights provided will be of crucial value to any organization dependent on quality assurance in data science as the basis of their business success. This information can be adopted in their business strategy to leverage their potential to its maximum.

References

1. S. Farooqui and W. Mahmood, A survey of Pakistan's SQA Practices: a Comparative Study, in *29th International Business Information Management Association Conference, Vienna, Austria, 2017.*
2. Rakesh Kumar, Birth Subhash, Maria Fatima, Waqas Mahmood, Quality Assurance for Analytics, *International Journal of Advanced Computer Science and Applications*, Vol. 9, No. 8, 2018 [Online], pp. 160-166.
3. F. Lambert, Tesla Autopilot confuses markings toward barrier in recreation of fatal Model X crash at exact same location, 3 April 2018. [Online]. Available:

https://electrek.co/2018/04/03/tesla-autopilot-crash-barrier-markings-fatal-model-x-accident/. [Accessed 4 April 2018].
4. SeattleDataGuy, Data Quality Is Not as Sexy as Data Science, 15 September 2017. [Online]. Available: https://medium.com/@SeattleDataGuy/good-data-quality-is-key-for-great-data-science-and-analytics-ccfa18d0fff8. [Accessed 4 April 2018].
5. F. J. Buckley and R. Poston, Software Quality Assurance, *IEEE Transactions on Software Engineering*, pp. 36-41, 1984.
6. Michel Dumontier and Tobias Kuhn, Data Science – Methods, infrastructure, and applications, *Data Science* 1 (2017) 1–5 1 DOI 10.3233/DS-170013 IOS Pres
7. Y. Chen, A.J. Elenee and G. Weber, IBM Watson: How cognitive computing can be applied to big data challenges in life sciences research, *Clin Ther* 38 (2016), 688–701. doi:10.1016/j.clinthera.2015.12.001.
8. D. Demner-Fushman and N. Elhadad, Aspiring to unintended consequences of natural language processing: A review of recent developments in clinical and consumer-generated text processing, *Yearb Med Inform* 10 (2016), 224–233. doi:10. 15265/IY-2016-017.
9. https://acuvate.com/blog/challenges-faced-by-data-scientists/ [Date:12/11/2021]
10. Robert Hoehndorf and Núria Queralt-Rosinach, Data Science and symbolic AI: Synergies, challenges and opportunities, *Data Science* 1 (2017) 27–38 27 DOI 10.3233/DS-170004 IOS Press
11. Kumar Manoj, Anand Madhu, An application of time series ARIMA forecasting model for predicting sugarcane production in India, *Studies in Business and Economics Studies in Business and Economics*, Vol. 9, Issue 1, April 2014, pp. 81- 94.
12. Vaibhav Kumar, Predictive Analytics: A Review of Trends and Techniques, International *Journal of Computer Applications*, 182(1), pp.31-37, July 2018. DOI:10.5120/ijca201891743

2

Design and Implementation of Social Media Mining – Knowledge Discovery Methods for Effective Digital Marketing Strategies

Prashant Bhat and Pradnya Malaganve*

Department of Computer Science, School of Computational Sciences and IT, Garden City University, Bengaluru, India

Abstract

Nowadays, Data Science and Data Engineering have become an essential part of any business, given the enormous amount of data storage that are raised. Most industries are implementing Data Science techniques to get higher revenue and in order to increase customer satisfaction in their respective businesses, which also influences Digital Marketing and online business. Considering these things, we have implemented a strategic methodology and conducted number of experiments on the same. Data Mining tools, algorithms and methods are closely related to Data Science and Data Engineering. Hence, we have used machine learning algorithms in our experiments. Facebook social media dataset is used in the work which influences all the experiments conducted. The data contains the advertisement details of user interactions on every post uploaded regarding cosmetic products. In the present work, we have proposed a novel framework for knowledge discovery from Facebook advertisement data using Data Mining techniques and methodologies. One of the experiments shows the comparison analysis of different Classification algorithms and in the experiment, we have used different Decision Tree classifiers and Naive Bayes Classifiers to check Classification accuracy and other parameters, and as a result found that Decision Tree classifiers give greater accuracy while classifying Facebook data. Another experiment follows Clustering Data Mining technique and we have used Expectation maximization Clustering, based on which we have tried to suggest good Marketing Strategy for Facebook

**Corresponding author*: pradnyamalaganve@gmail.com

Kukatlapalli Pradeep Kumar, Aynur Unal, Vinay Jha Pillai, Hari Murthy, and M. Niranjanamurthy (eds.) Data Engineering and Data Science: Concepts and Applications, (23–48) © 2023 Scrivener Publishing LLC

post publishers. These experiments have been conducted on the Facebook data to produce good marketing strategies for Digital Marketing, and the findings show which kind of content publishing on Facebook can get more views and customer interactions on it, whether it is a video, photo, link or a status.

Keywords: Classification, cluster, marketing strategy, facebook, data mining, digital marketing

2.1 Introduction

As the entire world is suffering from the Covid-19 pandemic at present, people fear interacting with each other face to face. In order to save lives, government has taken essential steps, such as lockdown, social distancing, sanitizing, increasing immunity in everyone, vaccination and several rules like no one should go out of their homes till the Covid-19 chain breaks and until we feel we can breathe peacefully again. In such a situation, social media has played a very helpful and enormous role to spread awareness about everything and most importantly in connecting people live [9]. And because of the digital world, one cannot imagine his/her life without social media and social channels. Although people cannot meet each other directly, they are using social media like Facebook, Twitter, Instagram, WhatsApp, etc., very widely and often. By this the companies are making unimaginable profits in the present days. Apart from these, other applications such as WebEx, Google Meet, Microsoft Teams, TeamViewer and other developments are being used by all educational institutions, schools, colleges and universities for academic purposes. Even for purchasing food and other goods, people have to use the marketing applications such as Amazon, Flipkart, Zomato, BigBasket, etc. After observing all these things, it is concluded that the world is ruled by digital media. And someday, the human being will get addicted to digital media entirely [4].

Hence, since people throughout the entire world are making use of digital media for their everyday needs, the amount of data stored on the cloud amounts to several terabytes every day, and this contains huge noise and useless data. In order to clean it and to extract knowledge out of it, data science techniques and Data Mining algorithms are very useful.

In our work, we conducted several experiments on a cosmetic company's Facebook data set. The Facebook page of the company was generated for marketing advertisement of Cosmetic products and to spread information like special offers available on each product, to elaborate the product, i.e., the description about cosmetic products and a few more details about the products. The content uploaded to describe the products detail is in multimedia form. A few of our experiments show which kind of content uploading on Facebook may lead to more customer interaction. To achieve this, we have used Cluster techniques. Here interaction does not mean just giving a reply or comments but it includes each and every click on the post and page, likes, comments, shares and all the activities made on the post or page of the company's Facebook data. As most of the researches are done using Data Mining Classification techniques, in our experiment also we have used Classification techniques to provide good Classification accuracy. We derived two different algorithms for Facebook data set and concluded which algorithm best suits to classify Facebook data set with good Classification accuracy rate [24].

The international broadcasting of social media was activated by the exponential evolution of Internet users, leading to an entirely new atmosphere for consumers to convey ideas and comment or advice about products and facilities. According to Statista, the number of social network users will surge from 2.44 billion to 3.5 billion from 2018 to 2021, an upsurge of around 150% in three years. Given its rapid expansion [12], social media has become the most significant media network for marketers to reach their customers. Companies comprehended quickly regarding the potential of using Internet-based social media to impact clients, incorporating social media business interaction in their approaches for leveraging their industries. Knowing and understanding the influence of advertisement is a vital issue to be included in a worldwide social media approach [10]. There have been a number of experiments focused on discovering the associations between digital publications on social networks and the influence of such publications measured by client's interactions. Also, some experiments dedicated to research on executing predictive systems which can efficiently be used to predict the progression of a post before publishing it.

A system which can predict the effect of distinctly published posts can deliver a treasured benefit when a company determines to interact via any social media, adapting the elevation of products and services. Promotion managers and those who make the advertisement could make informed decisions on the accessibility of the published posts, hence positioning approaches toward improving the impact of posts, promoting from the estimates made. Also described, those social media publications are extremely associated to brand building. Data Mining offers a stimulating method for mining predictive knowledge from a mixture of data. In our work, we studied the applications of social media and Data Mining, specifically for assessing market trends from user involvements, and most of the work has been concentrated on the ideas and interactions of users via social network, with an emphasis on collecting data from various network groups or from personal posts. Also, our work is focused on predicting the effect of publishing distinct posts on a company's Facebook page [11]. The effect is calculated through various existing methods related to consumer views and interactions. The resultant outcomes help the website administrator or the content uploading manager decide which kind of post should be published so that the maximum user interactions can be achieved. For validating the adopted method in the present work, we have used a cosmetics company with a well-known brand, containing 500 posts uploaded by the company on its Facebook brand page [6]. Hence, the data set of posts has been used as an input to the Data Mining technique [12].

2.1.1 Objectives of the Study

- Implementing a novel framework using Data Mining techniques and algorithms in order to discover the knowledge
- Evaluating comparison analysis of well-known Data Mining Classifiers to choose the best suitable classifier for Facebook data
- Implementing Cluster model to predict the impact of uploaded content or post using their characteristics
- Describing the relationship between the discovered knowledge, presenting the influence of input characteristics and their impact on the published posts

Section 2.3 describes the resources used, the algorithms and techniques preferred for the experiments. The section is focused on explaining the explicit background on the methodological features of the Data Mining technique which also includes prediction demonstrating and knowledge discovery.

2.2 Literature Review

Authors Mohammed and Ahmed Burhan's [14] work shows the comparison of three distinct data mining algorithms for classifications, namely "Decision Tree, Naïve Bayes and Support Vector Machine", applied on social media dataset by NYC, to generate the best result of the classification algorithm which can classify the variables according to the stages. The result of the authors [14] work conveys that Support Vector Machine performs best compared to other two techniques used in the work.

Twitter is a famous social media which carries a huge number of statements regarding happiness, dejection, communal information, and so on. In such cases when people feel anger, they express it in the tweet. Several tweets may hold Indonesian swear characters. It can be a serious cause because people of Indonesia may not use swear words. Twitter has given data about tweets by account, new topics, and advance keyword. Authors W. B. Zulfikar, M. Irfan, C. N. Alam and M. Indra [15] have analysed several tweets related to political data, political event, and well-known Indonesian persons because the presumed tweet carries Indonesian swear words. The resultant tweets processed via text mining and analysed using "Naive Bayes Classifier, Nearest Neighbor, and Decision Tree" to distinguish Indonesian swear words. Authors [15] experiment discovered the highest accurate classification method, which means the method can distinguish the actual meaning of "Indonesian swear word" present in the tweets.

Largely scalable computing atmosphere gives an opportunity of bringing out different data-intensive natural language processing and Data Mining jobs. Text classification is one with some matters inspected by many data scientists. The authors Tomas Pranckevičius and Virginijus Marcinkevičius [16] have examined "Naïve Bayes, Random Forest, Decision Tree, Support Vector Machines, and Logistic Regression" algorithms built using Apache Spark. Authors [16] have analysed short text on product review data of Amazon and focused on comparing and analysing the classifiers by calculating the classification accuracy considering the size of training data, and the number of n-grams.

Every day millions of videos are uploaded on YouTube by people around the world, hence enormous data is gathered in the cloud. Authors Riyan Amand and Edi Surya Negara [17] have found Data Mining Classification as a solution for digging up information from this data. Many time, users on YouTube may not find the exact category of video by the title they give. Hence, authors [17] have made an experiment to classify YouTube data associated with search text. The author's [17] work focuses on comparing three methods for classifying YouTube data into "Kesenian and Sains" category. The algorithms used are "Random Forest, SVM and Naive Bayes". The outcome indicated that the accuracy rate of the algorithm "random forest" was the best by 0.004%, and the label encoder was not applied to the target class, hence it did not affect Classification accuracy. Authors [17] found that, Naïve Bayes method gives maximum accuracy with 88% and 90% of average precision.

Sentiment analysis approach centres the data on the study and conception of the feelings from the text forms. It inevitably characterizes the expression of feelings as negative, positive or natural about anything.

Different sources like social sites, medical, news and video review can be used for data analysis. Authors Rahman and M. S. Hossen [18] have gathered movie review data in their experiment and applied Data Mining classifiers such as: "Bernoulli Naïve Bayes (BNB), Decision Tree (DE), Support Vector Machine (SVM), Maximum Entropy (ME) and Multinomial Naïve Bayes (MNB)" in the research. Authors [18] have analysed all the derivatives and concluded that MNB gives better accuracy, precision and F-score results. At the same time SVM gives high result for "recall" compared to others.

Author Chiu, Y.P. [19] proposed a content-based framework which is designed to operate on social media data. The framework integrated the characteristics of brand pages and data on user's reactions on brand pages. Official brand page data from 2,076 pages was taken for the experiment and a total of 500,000 interaction data had been processed. "Decision tree" was used to classify brand pages and find out characteristics that help in distinguishing the brand pages. The author's [19] approach has a simple training process with no restriction to any classifier algorithm and gives improved results on data sets extracted from different Facebook brand pages. The results showed the diversity of social site user criteria in calculating whether a task is interesting to users. These characteristics show that brand pages differentiate themselves from others' views and willingness to get engage with users. The outcomes are useful for learners and researchers to improve a recommendation system for social media.

Authors Vineeth M.S., RamKarthik K., Phaneendra Reddy M.S., Surya N., and Deepthi L.R. [20] have made an attempt to detect the community in social networks. The authors [20] have performed comparative analysis by considering six different algorithms to detect the communities in social network by considering different variables like run-time, size of the cluster, normalized mutual data, generated score and average score.

Social Media Data represents Big Data and is categorized by composite and unstructured forms, which is difficult to analyse. Finding a way to facilitate the analysis of such data and deriving marketing strategies is a challenging task. One of Data Mining method Clustering gives a wide range of techniques that categorize noisy data and finds knowledge from huge data. Authors Benslama T. and Jallouli R's [21] goal is to clarify a few doubts in this regard, such as What are the methods for accumulating Social Media Data? What are the business decisions produced by Social Data Clustering? Hence, the authors [21] have presented a review article on Social Media data Analysis in regard of marketing and business using Clustering techniques.

2.3 Novel Framework for Social Media Data Mining and Knowledge Discovery

Web: The amount of data storage on the web is increasing every day, hence the value of every industrial data. Extracting the data from any website is vital to achieve the goals in online marketing. Past data of a trade is very important in predicting the changes and further improvements.
Resource discovery: In our research, data has been extracted from a renowned cosmetic company's Facebook advertisement page. An API (Application Programming Interface) was generated prior using Python code.
Data Repository: Extracted data was gathered in Excel form, i.e., using .CSV (Comma Separated Value) file format.
Data pre-processing: Encoding the data, data integration, data reduction, transforming data, cleaning the data, selecting the data, handling the data; these are the terms associated with data pre-processing. When the data has been extracted from the website, it may contain most of the noise, hence it will be unrefined format. To bring it in particular structure and in order to clear the noise, data pre-processing techniques exist such as, data selection, data handling, cleaning etc. These help in generating refined flawless data [9]. The data set used in our experiments contained six missing values initially. Hence, the missing values are filled with a global value i.e., 0 (zero). We have use Python coding for pre-processing the data.
Data Mining Process: The process of searching interesting patterns and discovering knowledge from unstructured, fuzzy or large amount of data is considered as Data Mining. A Data Mining method naturally contains segments such as data understanding, data groundwork, i.e., preparation, demonstrating, and valuation. The data set designated in Table 2.1 comprises nineteen features, in that eighteen attributes are extracted from Facebook page and the attribute "Total interactions" is generated manually that may be used to calculate a post's performance. Hence, any of the above attributes can be considered as an output to predict. In the same way, business trades are making effective use of Data Mining process to transform raw data into good information to increase their quality of business and in order to get good marketing strategies. Thus, business people are understanding their followers and consumers. In the present work, we have used Data Mining techniques and algorithms to provide better online marketing strategies which influence publishing posts on social media. The methods that come under Data Mining process are "Classification, Clustering, Outlier Detection, Regression and Association Rule Mining".

All these methods carry a number of algorithms which give effective mining results. The methods are associated with "Unsupervised, Supervised and Semi supervised learning methods". Table 2.1 shows the differences between "Supervised method, Unsupervised method and Semi-supervised method".

Apart from three methods shown in Table 2.1, another method is, Re-information learning, where no training dataset is used. This method can be used most likely in gaming and Artificial Intelligence [10].

Evaluate Mining Results: Evaluating the outcomes found using Data Mining methods and algorithms is a fundamental step in Data Science and machine learning. The results differ from one model to another. Understanding the performance measures and efficiency of any model is not an easy task; it may depend on the information scenario or the input data. Usually, Data Mining methods are used for predicting, hence it is very important to select a correct and reliable evaluation measure.

Table 2.1 Differences between supervised, unsupervised and semi-supervised methods.

Supervised method	Unsupervised method	Semi-supervised method
The method deals with Regression and Classification techniques	Deals with Clustering and Association rule mining techniques	The method is also termed as deep learning and it deals with all techniques that are: "Classification, Clustering, Association rule mining and Regression".
Data is labelled that makes the task easier for any algorithm to predict and understand	Data is unlabelled that gives inheriting structures and detects different patterns using algorithms	Some data is labelled and most of the time data is unlabelled. It can be the mixture of supervised and unsupervised methods.
Examples can be: Decision Tree, Support Vector Machine, Linear Regression	Examples can be: K-Means, Hierarchical, PCA	Self-training, Semi Supervised SVM, Mixture models

Knowledge Discovery: The ultimate goal of the Data Mining process is to develop new ideas or explicit knowledge from the database and this is a technique used to find novel, valid and potentially useful information from raw data. It conveys smart and accurate decisions for business, marketing, education and many other fields.

In Figure 2.1, the proposed framework contains a Data Mining experimental approach at its core, resulting in a data-driven process. Hence, it

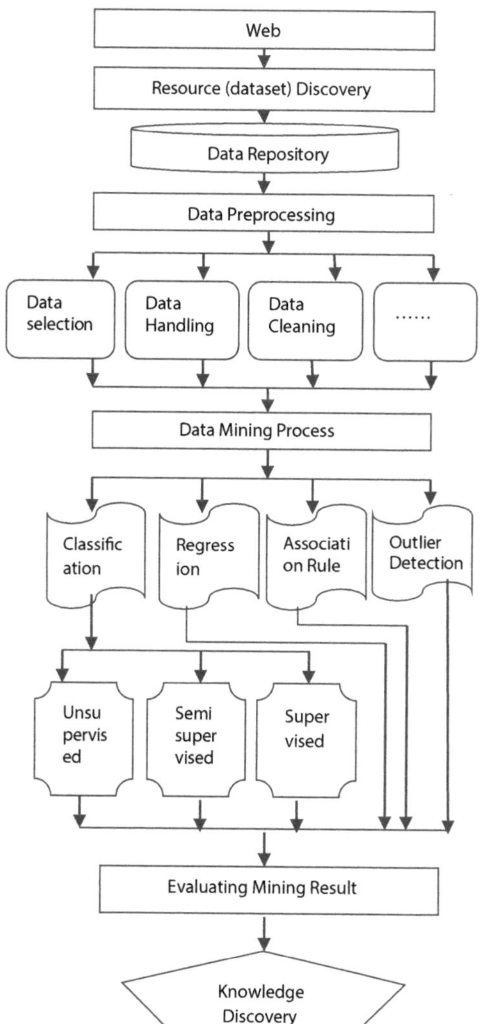

Figure 2.1 Novel framework proposed for knowledge discovery.

collected an illustrative data set of posts published on Facebook page. The posts uploaded in the Facebook page are of a well-known cosmetic brand. As a result, the dataset has a total of 500 posts uploaded. What should be considered here is: Facebook was the most used social media and had an average of 1.30 billion active users per month in 2014. Figure 2.2 is a slice of input data used in the current work. The extracted dataset includes different kinds of features:

- Identification—characteristics which allow recognizing individual post
- Content—the numeric and nominal data about the post
- Categorization—characteristics which describe the post
- Performance—calculation of measuring the impact of published post

Table 2.2 denotes every characteristic gathered in the data set. Most of the data was considered directly from the company's Facebook page except two instances i.e., "total interactions" and "category". The column "Total interactions" was generated manually by adding total number of likes, comments and shares of the posts. The categorization was an appeal by the company's senior marketing managers as it is associated with the kinds of operations performed by the cosmetic company. "Category" carries three distinct numeric variables, i.e., 1, 2 and 3.

Wherein 1 denotes regarding special offers on the brand and about the contest, 2 denotes direct advertisement, i.e., explicit brand content, and 3 denotes non-explicit content about the brand.

Page total	Type	Category	Post Mont	Post Weel	Post Hour	Paid	Lifetime F	Lifetime F	Lifetime E	Lifetime F
139441	Photo	2	12	4	3	0	2752	5091	178	109
139441	Status	2	12	3	10	0	10460	19057	1457	1361
139441	Photo	3	12	3	3	0	2413	4373	177	113
139441	Photo	2	12	2	10	1	50128	87991	2211	790
139441	Photo	2	12	2	3	0	7244	13594	671	410
139441	Status	2	12	1	9	0	10472	20849	1191	1073
139441	Photo	3	12	1	3	1	11692	19479	481	265
139441	Photo	3	12	7	9	1	13720	24137	537	232
139441	Status	2	12	7	3	0	11844	22538	1530	1407
139441	Photo	3	12	6	10	0	4694	8668	280	183
139441	Status	2	12	5	10	0	21744	42334	4258	4100
139441	Photo	2	12	5	10	0	3112	5590	208	127
139441	Photo	2	12	5	10	0	2847	5133	193	115
139441	Photo	2	12	5	3	0	2549	4896	249	134

Figure 2.2 Facebook dataset used in the experiment.

Table 2.2 Dataset description.

Name of attribute	Data type	Description
Page total likes	Numeric	Number of users who liked the company's page
Type	String	Type of content (Link, Photo, Video, Status)
Category	Numeric	Manual content characterization
Post month	Numeric	Post published month
Post week	Numeric	Post published week
Post hour	Numeric	Post published hour
Paid	Numeric	Company paid to the Facebook for advertising
Life time post total reach	Numeric	Number of unique users who saw the page post
Life time post total impression	Numeric	Number of times a post from a page is displayed
Life time engaged users	Numeric	Number of unique users clicked anywhere in the post
Life time post consumers	Numeric	Number of users clicked anywhere in the post
Life time post consumptions	Numeric	Numbers of clicks anywhere in the post
Lifetime Post Impressions by people who have liked your Page	Numeric	Number of impressions by users who have liked the page
Lifetime Post reach by people who like your Page	Numeric	Number of unique users saw a page post because they have liked it

(Continued)

Table 2.2 Dataset description. (*Continued*)

Name of attribute	Data type	Description
Lifetime People who have liked your Page and engaged with your post	Numeric	Number of unique users liked and clicked anywhere in the post
Comment	Numeric	Number of comments of the post
Like	Numeric	Number of likes of the post
Share	Numeric	Number of shares of the post
Total interactions	Numeric	Sum of comments, likes and shares of the post

2.4 Classification for Comparison Analysis

The astonishing advancement of data on social media websites such as Facebook, Twitter, YouTube, Yahoo, etc., establishing data into classes is of principal importance for enhancing user experience and website consumption. In this information stage of development, multimedia information is swiftly shared by the users via social media websites. Different sorts of multimedia data are shared on social media sites and viewed by billions of people around the globe. The Classification/partitioning of social media data in terms of number of comments, [23] shares, likes, user interactions, etc., and analysis of Facebook social media data as an unstructured multifaceted data is a challenging task. The work presents an effective Classification model to classify each category of Facebook data used in the experiment based on the metadata attributes as splitting measures. We have made an attempt to extract metadata from Facebook web pages and based on the extracted metadata, the Facebook data are classified/partitioned into several categories using Data Mining Classification techniques such as Decision Tree Classifiers and Bayes Classifiers. The Classification outcomes are compared with each other and analysed based on the accuracy rate, TP, FP, TN, FN, Gini index and some other measures. Classification/partitioning of Facebook data are essential jobs with different applications in multimedia data search and information retrieval procedure, although gathering metadata required for Classification method is prohibitively expensive. We have proposed a framework shown in Figure 2.3, which elaborates the Metadata-based Classification technique for Facebook data [13].

Social Media has vast data shared by billions of users around the globe, and there are a lot of difficulties involved in organizing and arranging it systematically. The same challenge is faced by marketers while extracting social media data and understanding it. To tackle this problem and to manage vast data, different Data Mining approaches have been developed such as "Classification, association rule, Clustering, etc." The current experiment shows how Data Mining Classification methods are used to label the Facebook data into classes accordingly. Several Classifiers exist in Data Mining, namely: Bayes, Rules, Function, Decision Tree Classifiers and Lazy Classifiers, etc. [7]. All the Classifiers have a number of Classification algorithms. In the present work, we have worked with Bayes and Decision Tree Classifiers to evaluate the comparison between both Classifiers, to examine the knowledge delivered by the Classifiers and to help the online marketer by providing the best suitable Data Mining algorithm which can classify Facebook advertisement data with high accuracy. The novelty in the experiment is to demonstrate which Data Mining algorithm works best in classifying/categorizing Facebook data in an appropriate manner with greater Classification accuracy. In order to achieve this, we have referred to several works conducted earlier and found that no work existed that can provide the best Data Mining algorithm to classify the Facebook data with highest accuracy. To fill this gap, we have used eight different Data

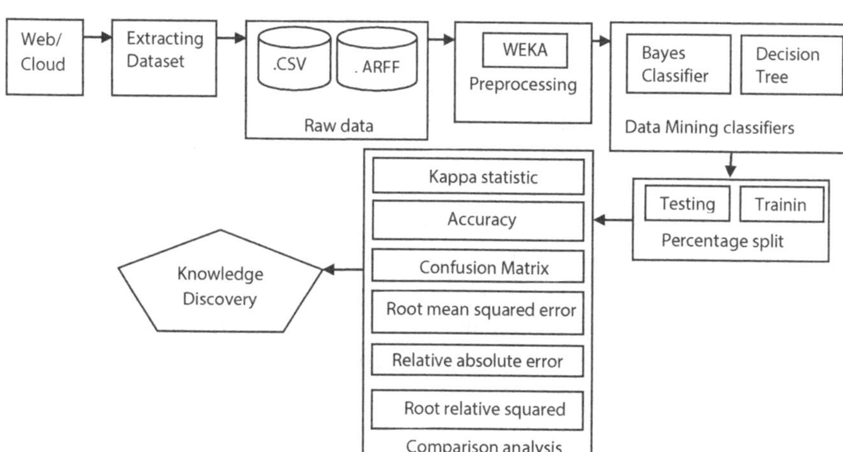

Figure 2.3 Proposed framework for metadata-based classification techniques for facebook dataset.

Mining algorithm methods to classify Facebook data. Two well-known classifiers are used: Bayes and Decision Tree Classifiers. Under Bayes, four Classification algorithms are considered in the experiment [8]: "BayesNet, NaiveBayes, NaiveBayesMultinominalText and NaiveBayesUpdatable" to check the Classification Accuracy of Meta Data 'Type'. As described in the Table 2.1, 'Type' is a Meta Data of Cosmetic Company's Facebook page dataset that holds four different kinds of data such as "status, link, photo and video". Under Decision Tree classifiers: "J48, LMT [11], RandomForest and RandomTree" are considered for calculating the Classification Accuracy of Meta Data 'Type' from the dataset described in Table 2.2. The goal of the experiment is to provide best Classification algorithm which can categorize all four kinds of data or attributes in respective classes with highest accuracy. The Facebook dataset used in the present experiment is divided as two splits, training and testing, using "percentage split method". One part of the dataset is called as training and another set is called as testing set. Data Mining Classification work bench WEKA is used for calculating the accuracy of correctly classified variables; apart from this "kappa statistics, root mean squared error, relative absolute error, root relative squared error and generated Confusion Matrix" is also derived to describe different classes such as "link, photo, status and video" from Meta Data 'Type'. Table 2.3 is generated for indicating Classification errors of 8 algorithms used in the experiment. Table 2.3 represents the number of correctly classified variables and number of incorrectly classified variables out of 500 total variables existing in the dataset.

As the data used in the present work is a company's cosmetic product advertisement information, our work is valuable to the data scientists to analyze Facebook advertisement data, and also the work is essential for business analysts to provide marketing-related advice by analyzing the data, the results may help them by suggesting which Data Mining algorithm and methodology is best to classify the content posted on the social media. Ultimately, knowledge can be achieved from huge noisy data.

Table 2.3 gives the result of Correctly classified variables, incorrectly classified variables, kappa statistics, mean absolute error, root mean squared error, relative absolute error and root relative squared error of all eight Data Mining algorithms used in the work. After analysing all the results from respective algorithms, we can observe that the algorithms come under decision tree classifier are giving a maximum accuracy rate of 91.76%, 90.58%, 93.52% and 87.05% for decision tree Classifiers J48, LMT, Randomforest and Random tree algorithms, respectively. This can be compared to Bayes classifiers algorithms BayesNet, NaiveBayes, NaiveBayesMultinominalText and NaiveBayesUpdatable algorithms, which have given less Classification accuracy of 75.88%, 53.52%,

Table 2.3 Classification results of bayes and decision tree classifiers.

Name of the classifier	Classification algorithm	Correctly classified variables	Incorrectly classified variables	Kappa statistics	Mean absolute error	Root mean squared error	Relative absolute error	Root relative squared error
Bayes Classifier	BayesNet	129 75.88%	41 24.11%	0.3235	0.1319	0.3308	96.2373%	124.6664%
	NaiveBayes	91 53.52%	79 46.47%	0.2318	0.2322	0.4649	169.4273%	175.2081%
	NaiveBayesMultinominalText	143 84.11%	27 15.88%	0	0.137	0.2653	100%	100%
	NaiveBayesUpdatable	91 53.52%	79 46.47%	0.2318	0.2322	0.4649	169.4273%	175.2081%
Decision Tree Classifier	J48	156 91.76%	14 8.23%	0.6204	0.0604	0.1978	44.0817%	74.5644%
	LMT	154 90.58%	16 9.41%	0.6294	0.0467	0.2023	34.0796%	76.2536%
	Random Forest	159 93.52%	11 6.47%	0.7191	0.0659	0.1672	48.0583%	63.0035%
	Random Tree	148 87.05%	22 12.94%	0.4625	0.0647	0.2544	47.2206%	95.8746%

84.11% and 53.52%, respectively. Hence the conclusion regarding Facebook Data Mining according to our experiments is that Decision tree classifier algorithms are best to categorize the Facebook advertisement data with greater accuracy rate; thus, it helps data scientists while classifying and making predictions on Facebook data and also helps business analysts to analyse the data as well as to suggest best marketing strategies to marketers.

The process followed while achieving the above results is shown in Figure 2.2. It proposes a novel framework for Metadata-based Classification Techniques for Facebook Dataset. Figure 2.2 describes that the cloud, also known as the web, is occupied with huge data. Hence in the first step, the data is extracted from the cloud/web. However, the data is noisy and raw and needs work in arranging it, hence it needs to be refined before using further. Once the raw data is refined, it is stored as .ARFF or .CSV files. The unrefined data is preprocessed using WEKA or Rapid Miner preprocessor. Two Data Mining Classifiers are used and each contains four distinct Data Mining algorithms to calculate the Classification Accuracy and other measures of Meta Data 'Type'. To Classify, to check the accuracy and other results of the data, the entire dataset cannot be considered as training or testing set that leads in giving 100% accuracy for all the datasets when applied. Hence a popular partition method is used, i.e., percentage split method to get the efficient Classification results. It helps splitting the dataset into two parts; 66% of the entire dataset is considered as training set data and the remaining 34% of dataset is testing data. Finally generated "Classification Accuracy, kappa statistics, root mean squared error, relative absolute error and root relative squared error", by calculating different estimates like "TP value (True Positive), FP value (False Positive), Precision, Recall, F-Measure, MCC, ROC area, PRC area and Class". Based on these outcomes of all eight algorithms, all the measures are compared, understood and analyzed in depth to discover the knowledge by providing best Classifier for Facebook data which labels the data in suitable classes and leads to convey the knowledge from social media data.

2.5 Clustering Methodology to Provide Digital Marketing Strategies

While analyzing and dealing with marketing data, it's not realistic to deal with every customer independently. Though collecting every customer's data is an essential step of Data Mining, it is not possible to interact and analyze millions of individual customer's data at the same time. Decision making in a strategic situation cannot be done practically. In Data Mining, several approaches exist to fix data segmentation challenges. Cluster technique is one such approach; it categorizes unlabeled data based on the attribute behavior and other characteristics. The technique is the most accessible and explainable approach to use machine learning on business and marketing data. Many times, it is thought that segmenting data or partitioning the data can be done manually but the fact is, it is impossible when the data is huge, unrefined and noisy; it is not possible to read and understand it. Thus, the ultimate way to segment the data into a particular group is possible only using machine learning and Data Mining techniques.

As mentioned in the earlier sections of the chapter, Facebook is an open source for communication, for conveying the ideas, to signify individual skills and to achieve admiration around the globe. According to a 2020 survey, the number of Facebook users is around 2.80 billion. Hence business marketers realize that Facebook social media is one of the best practicalities for Digital Marketing to reach a huge audience and to persuade them adopt or understand one's thought. As people are tremendously interested in new trends and always look for modernization in one or other way, it is indispensable to bring novel marketing ideas and strategies for all.

The dataset is all about a cosmetic company's products; the company has generated a Facebook page and posted various kinds of data such as "photo, status, video and link". To understand and analyse the data, we have chosen Data Mining methodology, that is, "Clustering" and used expectation maximization technique [22] to Cluster the data in three kinds of Clusters. Based on the Clusters, we have come up with the outcome as to which type of data posts can achieve a good number of user engagement and user interaction towards it. Also, we determined which type of post content can provide best Digital Marketing Strategy. The experiment benefits are in knowing the cosmetic product users and their prospects [5].

Types of post contents on Facebook: Once the business and marketing page is created on a social platform such as Facebook, an important object to consider initially is, the kind of post to be uploaded on the business site in order to attract users' attention.

2.5.1 Status (Text Form)

Friendster was the site that introduced the basic form of a post on social media that is "Text". Now the status form is uploaded using text format most of the time and generally it does not include videos or images, which shows the standardization of a status on social media. Different tactics to make the user engage with published content can include asking them to share their view and suggestions regarding the products and getting ideas from them to enhance product quality via short queries. And knowing their likes and dislikes about the products and services is the best way to keep good relations with users and followers.

2.5.2 Images (Photos)

The image posts are also employed to spark customer engagement. According to web analysis, posted images get higher engagement than other kinds of posts. Using an image is a better mode to describe the products while having upright visual appearance or could be cherished by users. The image post in the social media site does not have to be simply a photo; it can be different type of image, graphic or illusion, pictorial representation [12].

2.5.3 Video Post

If the video has tremendous visual appeal, then even a short video information can also be fascinating. However, since the videos have an auto play option on Facebook, even the smallest videos can also fascinate someone's eyes on looking at it. Videos play a very important role in advertising and showcasing about cosmetic products; they helps in viewing different shades and all other characteristics of the product. It can be the way of usage, the quantity to be used or anything.

2.5.4 Link Post

Link provides the way to interact with different pages existed in the web. It can be inside a Facebook site or outside the Facebook page. Earlier work and web analysis says that a link is not as popular as an image and video post but may achieve better user interaction than text content. In advertising cosmetic products also, it may work well while visiting different user's profiles of those who have already used the same products and to know their experience regarding the cosmetic brands and products. It's an even better way to visit different professionals' sites. And a good practice is to always write the description or give a caption about the link whenever posted for better understanding.

The experiment is conducted using Cluster methodologies to give the best marketing strategies based on the posts uploaded to Facebook and the popularity assessment of the kind of content used in the posts. The predicted result varies from the existed web analysis results and conclusion. To achieve these outcomes, we have taken four distinct attributes from the dataset those are influenced by "Type" attribute are, "Total Interactions (sum of likes, comments and shares on each post), Life Time Post Total Reach, Life Time Engaged Users and Life Time People who have Liked a Page and Engaged with your Post".

Algorithm 1 Derived Algorithm for Expectation Maximization

Input Dataset x_i, where (1< =i< =n), k (number of Clusters)

Output Set of instances $\Theta_j C_j$ where (1< =j< =k)

1: Randomly choose starting instances $\Theta_{j,}$ W_j=0.5

2: While (1)

3: Expectation Step // Assignment step

For each instance x_i 1< =i< =n calculate probability of Jth distribution given that x_i, Θ

$$\text{Prob}\,(x_i\,|\,\Theta_j) = \frac{1}{\sqrt{2\pi\sigma j}} e^{-}(xi - \mu j)/2\sigma j2 \tag{2.1}$$

$$\text{Prob}\,(J^{th}\ \text{distri}\,|\,x_i, \Theta_j) = W_j\ \text{prob}\,(x_i, \Theta_j) \Big/ \sum_{j=1}^{k} Wj * \text{prob}\,(x_i, \Theta_j) \tag{2.2}$$

4: Maximization step

Using probability of Jth distribution calculate x_i, Θ

Find the new estimation of the attribute which maximize the assumed likelihood

$$\mu_j = \sum_{i=1}^{n} (x_i\ \text{prob}\,(J^{th}\ \text{distri}\,|\,x_i, \Theta) \Big/ \sum_{i=1}^{n} \text{prob}\,(\text{Jth distri}\,|\,x_i, \Theta)) \tag{2.3}$$

5: If instances are not changed go to step 6

Else repeat from step 2

6: Stop

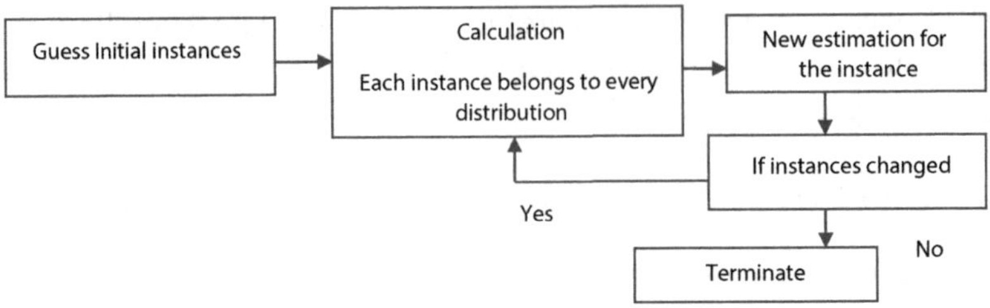

Figure 2.4 Flow diagram for expectation maximization technique to derive the clusters.

Figure 2.5 derived algorithm for expectation maximization technique.

Figure 2.4 designates the flow diagram of Expectation Maximization method that is used in the present work to understand the Clusters derived from the algorithm represented in Figure 2.5. The introduced algorithm is executed using Python programming in Jupiter Notebook. Expectation maximization technique is an efficient iterative way to calculate the maximum possibility estimate when the data is incomplete or the variables are missing. In maximum possibility estimation, it is vital to evaluate the model attributes for which the observed instances are alike in nature. Each iteration or loop of expectation maximization method is derived using two procedures. E step and M step [3].

The E step: "E step is the expectation step"; in this step, the missing instances are estimated provided that the experimental dataset and present estimate of the model instances.
The M step: "M step is the maximization step", where "the possibility module is maximized under the prediction that the missing instances are known.

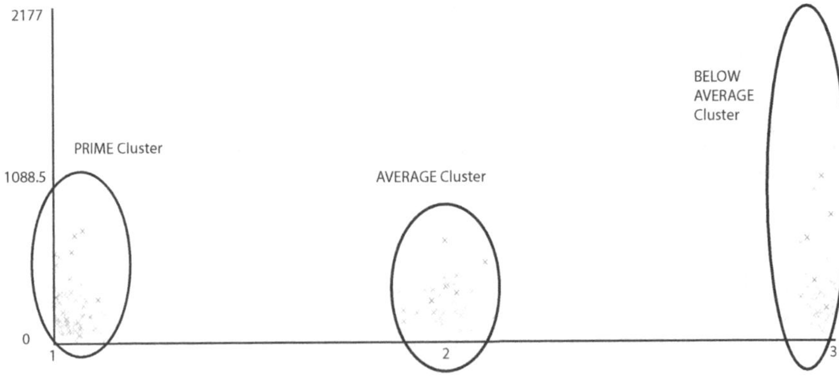

Figure 2.5 Generated clusters are represented in graphical format.

The estimate of the missing instances from the E step are used instead of actual missing instances".

2.6 Experimental Results

Figure 2.5 shows all three Clusters generated in graphical format with respective X and Y coordinates. The first column of Table 2.4 contains 'Type' attribute of the Facebook dataset with its variables (Content type). The rest of the columns represent Cluster 0, Cluster 1 and Cluster 2 which are defined as "PRIME, AVERAGE and BELOW AVERAGE", respectively. By analysing this formation and the values generated for each Content, we have observed and discovered which type of Content updating on Facebook can generate more user interaction and user engagement, which is again influenced by four distinct attributes of the Facebook dataset: "Total Interactions on each post, Life time post total reach, Life time engaged users, Life time people who have liked a page and engaged with your post". Finally, the minimum and maximum interaction of users on every category is shown in Figure 2.6, Figure 2.7, Figure 2.8, and Figure 2.9.

Most of the web analysis has shown that image and video posts do get a greater number of user interactions but, in our experiment, we have shown that uploading a status post which is in text format can also get a greater number of user interactions and user engagement. As shown in Table 2.4. PRIME is a Cluster which holds only such variables whose total interaction ratio falls between 121 to 2,177 and the respective graph is presented in Figure 2.6. Total interactions are the sum of number of likes, comments and shares on the posts Clustered under PRIME Cluster. In AVERAGE Cluster, the total interactions ratio falls between 7 to 771, which is less than PRIME categorization. Under BELOW AVERAGE Cluster, the total

Table 2.4 Clusters formed using expectation maximization method.

Type (attribute)	Prime (19)	Average (42)	Below average (109)
Status	07	08	00
Photo	11	23	108
Video	01	01	00
Link	00	10	00

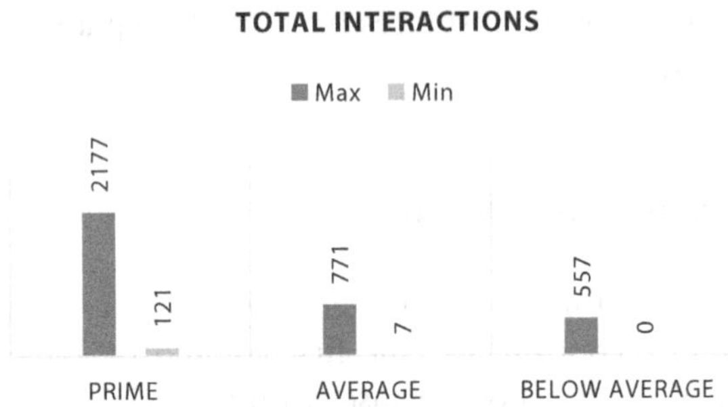

Figure 2.6 Variations in number of total interactions between three clusters.

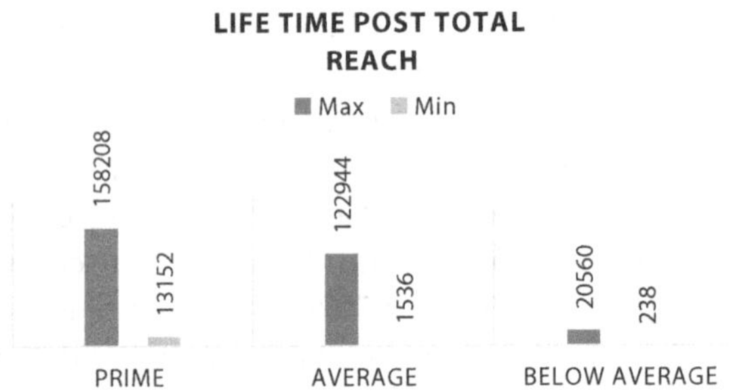

Figure 2.7 Variations in number life time post total reach between three clusters.

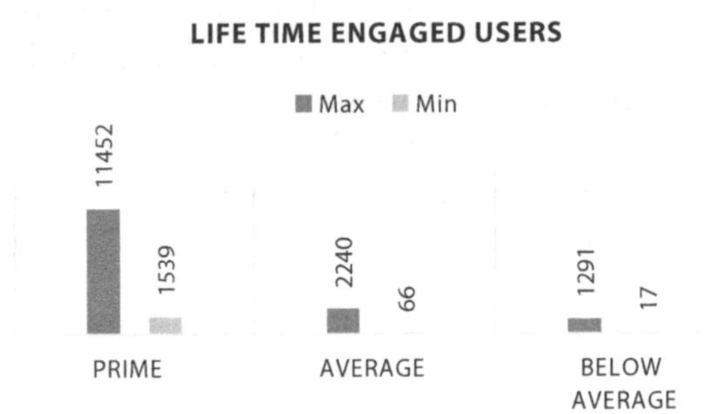

Figure 2.8 Variations in number of life time engaged users between three clusters.

interactions ratio falls between 0 to 557 which is less than the other two categorizations. Also, Figure 2.7, Figure 2.8, and Figure 2.9 contain the highest user interactions on different instances such as "Life time post total reach, Life time engaged users and Life time people who have liked the page and engaged with your post" under PRIME Cluster only. Wherein

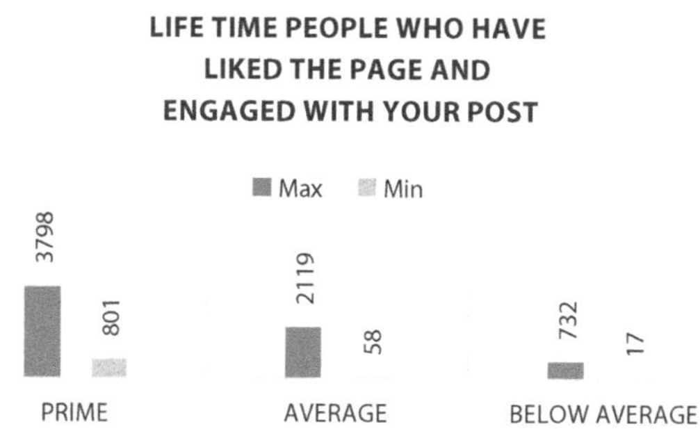

Figure 2.9 Variations in number of life time people who have liked the page and engaged with your post between three clusters.

the AVERAGE Cluster contains all less interacted Contents, and BELOW AVERAGE Cluster contains lowest interacted Contents. Hence, the result that, as shown in Table 2.4, PRIME Cluster carries most of the posts which are status (7) though the number of photos is 11, that is because of the total number of photos present in the entire dataset is 426 out of 500, which is far more compared to other type of Content; status posts are 45, link posts are 22 and video posts are 7. Still the testing set has considered and Clustered a greater number of status posts under PRIME Cluster and maximum number of link posts under AVERAGE Cluster. With these findings, we would like to give the Marketing Strategy [2] while uploading the posts as: "uploading delightful and attractive words as a status post and even uploading the link post can gain a greater number of customers to be engaged with the post, product and brand [1]. Hence, it results in accomplishing a greater number of customers towards the product and company with what type of Content one will post in the company's Facebook page.

2.7 Conclusion

In this chapter we have concentrated on three novel approaches. The first provides a new framework for extracting social media data till knowledge discovery. The second is proving a best suitable Classification algorithm to classify the Facebook social media data with highest Classification accuracy. The third approach is providing best Marketing Strategy by proving that even posting a status and link posts also achieve a greater number of user interactions on a product advertisement post on Facebook.

References

1. Moro, S., *et al.* "Predicting social media performance metrics and evaluation of the impact on brand building: A Data Mining approach", *Journal of Business Research*, 2016. http://dx.doi.org/10.1016/j.jbusres.2016.02.010
2. Gil Appel, Lauren Grewal, Rhonda Hadi and Andrew T. Stephen. "The future of social media in marketing", *Journal of the Academy of Marketing Science* 48:79–95, online Publication: 12 Oct Y:2019.
3. Jaon Garriga, John R. B. Palmer, Aitana Oltra and Frederic Bartumeus. "Expectation-Maximization Binary Clustering for Behavioural Annotation", March 22 2016. https://doi.org/10.1371/journal.pone.0151984
4. K. Lin, S. Wu, L. Chen and P. Yang. "Finding the Key Users in Facebook Fan Pages via a Clustering Approach", *IEEE International Conference on Information Reuse and Integration. San Francisco, CA, USA*, 2015, pp. 556-561, doi: 10.1109/IRI.2015.89.
5. Jan-Willem van Dam and Michel van de Velden. "Online profiling and Clustering of Facebook users", *Decision Support Systems* 70 (2015) 60–72, 0167-9236/ 2014 Elsevier.
6. M.K. Ganeshan and Dr. C. Vethirajan. "Emerging Trends In Digital Marketing In India", ISSN: 2394-3114. Vol. 40 Issue 12, Feb 2020.
7. Harsh H. Patel, Purvi Prajapati, "Study and Analysis of Decision Tree Based Classification Algorithms", *IJCSE* Vol. 6, Issue 10, Oct. 2018 E-ISSN: 2347-2693
8. B. Tang, H. He, P. M. Baggenstoss and S. Kay, "A Bayesian Classification Approach Using Class-Specific Features for Text Categorization", *IEEE Transactions on Knowledge and Data Engineering*, vol. 28, no. 6, pp. 1602-1606, 1 June 2016, doi: 10.1109/TKDE.2016.2522427.
9. Prashant Bhat, Pradnya Malaganve and Prajna Hegade, "A New Framework for Social Media Content Mining and Knowledge Discovery", *IJCA* (0975 – 8887) Vol. 182, no. 36, January 2019
10. Siddu P. Algur and Prashant Bhat, "Web Video Mining: Metadata Predictive Analysis using Classification Techniques", I.J. *Information Technology and Computer Science*, 2016, 2, 68-76. Published Online February 2016 in MECS.
11. Bhat P., Malaganve P. (2021) Effect of J48 and LMT Algorithms to Classify Movies in the Web — A Comparative Approach. In: Saini H.S., Sayal R., Govardhan A., Buyya R. (eds.) *Innovations in Computer Science and Engineering. Lecture Notes in Networks and Systems*, vol 171. Springer, Singapore. https://doi.org/10.1007/978-981-33-4543-0_58
12. Prashant Bhat, Pradnya Malaganve, "Review on Social Media Content Mining and KDD" © 2018 IJRAR July 2018, Volume 5, Issue 3 www.ijrar.org (E-ISSN 2348-1269, P-ISSN 2349-5138)
13. Prashant Bhat, Pradnya Malaganve, "Metadata based Classification Techniques for Knowledge Discovery from Facebook Multimedia Database",

International Journal of Intelligent Systems and Applications (IJISA), Vol. 13, no. 4, pp. 38-48, 2021. DOI: 10.5815/ijisa.2021.04.04

14. Mohammed, Ahmed Burhan, "Decision Tree, Naïve Bayes and Support Vector Machine Applying on Social Media Usage in NYC/Comparative Analysis" *Tikrit Journal of Pure Science*, [S.l.], v. 22, n. 9, p. 94-99, Oct. 2018. ISSN 2415-1726.
15. W. B. Zulfikar, M. Irfan, C. N. Alam and M. Indra, "The comparation of text mining with Naive Bayes classifier, nearest neighbor, and decision tree to detect Indonesian swear words on Twitter", 2017 *5th International Conference on Cyber and IT Service Management (CITSM)*, 2017, pp. 1-5, doi: 10.1109/CITSM.2017.8089231.
16. Tomas Pranckevičius, Virginijus Marcinkevičius, "Comparison of Naïve Bayes, Random Forest, Decision Tree, Support Vector Machines, and Logistic Regression Classifiers for Text Reviews Classification", *Baltic J. Modern Computing*, Vol. 5 (2017), No. 2, 221-232 http://dx.doi.org/10.22364/bjmc.2017.5.2.05
17. Riyan Amand and Edi Surya Negara, "Analysis and Implementation Machine Learning for YouTube Data Classification by Comparing the Performance of Classification Algorithms", *JOIN (Jurnal Online Informatika)* Vol. 5, no. 1, June 2020: 61-72 DOI:10.15575/join.v5i1.505 p-ISSN:2528-1682 e-ISSN: 2527-9165
18. Rahman and M. S. Hossen, "Sentiment Analysis on Movie Review Data Using Machine Learning Approach", *International Conference on Bangla Speech and Language Processing (ICBSLP)*, 2019, pp. 1-4, doi: 10.1109/ICBSLP47725.2019.201470.
19. Chiu, Y.-P., "Social Recommendations for Facebook Brand Pages", *J. Theor. Appl. Electron. Commer. Res.* 2021, *16*, 71-84. https://doi.org/10.4067/S0718-18762021000100106
20. Vineeth M.S., RamKarthik K., Phaneendra Reddy M.S., Surya N., Deepthi L.R., "Comparative Analysis of Graph Clustering Algorithms for Detecting Communities in Social Networks", In: Hu YC., Tiwari S., Trivedi M., Mishra K. (eds.) *Ambient Communications and Computer Systems. Year: 2020Advances in Intelligent Systems and Computing*, vol. 1097. Springer.
21. Benslama T., Jallouli R., "Clustering of Social Media Data and Marketing Decisions", In: Bach Tobji M.A., Jallouli R., Samet A., Touzani M., Strat V.A., Pocatilu P. (eds.) *Digital Economy. Emerging Technologies and Business Innovation*. ICDEc 2020. *Lecture Notes in Business Information Processing*, vol 395. Springer, Cham.
22. Singh, I., Jindal, "R. Expectation maximization clustering and sequential pattern mining based approach for detecting intrusive transactions in databases", *Multimed Tools Appl* 80, 27649–27681 (2021). https://doi.org/10.1007/s11042-021-10786-3
23. Khan M.K., Rahman N., "Study of Users Attitude and Classification of Comments and Likes from Facebook Using RapidMine", In: Jeena Jacob I.,

Kolandapalayam Shanmugam S., Piramuthu S., Falkowski-Gilski P. (eds.) *Data Intelligence and Cognitive Informatics. Algorithms for Intelligent Systems.* Springer, Singapore. https://doi.org/10.1007/978-981-15-8530-2_53

24. S. Veni, Garcia-Lopez Y.J., Reseenamol N. A., "Sentiment Analysis of Facebook Comments Using Various Machine Learning Techniques", Fernando Antonio Lamas Delgado, Rosy Eugenia Reyes Pinilla, Mercedes Gaitán-Angulo, *Linguistica Antverpiensia*, 2021 Issue-1 www.hivt.be ISSN: 0304-2294.

3

A Study on Big Data Engineering Using Cloud Data Warehouse

Manjunath T. N.[1]*, Pushpa S. K.[1], Ravindra S. Hegadi[2] and Ananya Hathwar K. S.[1]

[1]Department of ISE, BMS Institute of Technology and Management, Bengaluru, Karnataka, India
[2]Department of CS, Central University of Karnataka, Kalaburagi, Karnataka, India

Abstract

In the current smart, Internet of Things world, millions of devices are connected around us for effective communication. A huge volume of data is getting generated, varying from gigabytes to brontobytes with a variety of data formats, storing all the acquired data in an on-premise Data warehouse has limitations due to storage capacity and homogeneous data formats compatibility. The internet revolution has brought us an entirely new communication and cloud model. We now have fast wireless connectivity for data transfer, messaging and web surfing, and we also perform extract, transform and Load (ETL) activities to the cloud data warehouse based on the client's requirements. The internet revolution has been the core inspiration for many new business ideas and models. The rise in online retailing business is strictly dependent on internet connectivity. Even consumers find online shopping convenient and time friendly. Internet communication is a must for urban life. Mobile internet helps us keep things handy and organized. We now can transfer or share a heavy size file from a portable handy device in a couple of minutes. A lot of businesses use internet for e-commerce, media site of a company, social media, Customer relationship management, Employee productivity management, etc., and due to this evolution, business faces a lot of issues with the storing of a large amount of data and analysing it over time. This is where Data Warehouse was born. Data warehouse is a central repository which is built to in-house the data acquired from various source systems in the enterprise business ecosystem. The intention is to support the decision makers with appropriate reporting facility to arrive at the right decisions to the right time. An on-premise data warehouse is built to handle online analytical processing with moderate

**Corresponding author*: manju.tn@gmail.com

Kukatlapalli Pradeep Kumar, Aynur Unal, Vinay Jha Pillai, Hari Murthy, and M. Niranjanamurthy (eds.) Data Engineering and Data Science: Concepts and Applications, (49–70) © 2023 Scrivener Publishing LLC

cost for moderate business enterprises. After a while the businesses faced issues in installing the components to store lot of data on-premise and it was not cost-effective. To solve this problem, cloud data warehouse was evolved. We explored the best possible solutions to handle the limitations of on-premise data warehouse with comparative study.

Keywords: Big data, data warehouse, cloud, data engineering, ETL

3.1 Introduction

The cloud environment has become the baseline which allows clients to access services at any time. In a cloud computing system, resources are distributed throughout the globe to provide clients with speedier service. Clients/customers have a habit of using various devices to access information, including laptops, cell phones, PDAs, and tablets. A data warehouse is a repository for all of the data that an organization collects from numerous sources and organizes into one place, from which it can be accessed and analyzed to run the business. This information could come from a variety of sources, including various data streams, the Internet of Things, relational databases, and data management systems. On-premises data warehouses suffer from limited storage capacity, technical challenges, and high operational costs. Cloud-based data warehouses take advantage of on-premise server computing's key benefits like broad user access, seemingly endless storage, and enhanced computational capacity with the ability to scale while only paying for what is consumed [1, 2]. When business demands to host the on-premise data warehouse onto the cloud platform, then there is an integration of heterogeneous data from various sources either from on-premise or cloud-deployed applications. Then there arises a data integration solution to transform the data into usable knowledge. Legacy extract, transform, and load techniques cannot switch to moving a huge volume of big data to a cloud data warehouse because of low read write operations. Customers can integrate their data seamlessly and without further code with state-of-the-art web-based iPaaS and AI-powered technologies. Several multiple cloud-based data warehouse choices are available, with their own architecture and offerings/features for integration, analysis of data from heterogeneous sources. In the internet world business ecosystem, every organization has its data acquisitions systems at multiple places in multiple formats; this demands the processing of a huge set of data to load the data into cloud data warehouses, which helps businesses

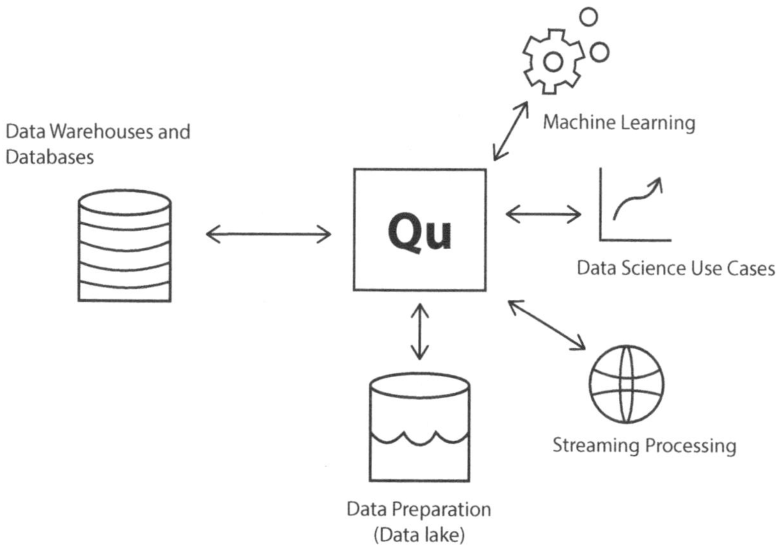

Fig. 3.1 Structure of cloud data warehouse.

in decision making. A data integration solution extracts, transforms and loads into cloud data warehouse for different domains such as telecommunication, healthcare, finance, securities and insurance [3]. Capturing the changes made to the data source required for the warehouse, restructuring keys maintains the metadata and refreshes the warehouse with updated data periodically [7, 9].

The structure of a cloud data warehouse is shown in Fig. 3.1 with all the components necessary for accessing, computing and storage and predominantly to address the mass ingestion activities. The cloud data warehouses which are available in the marketplace include Amazon Redshift, GCP - Google Big Query, Snowflake virtual warehouse, and MS - Azure SQL DWH, IBM DB2 and Informatica intelligent cloud services [4].

3.2 Comparison Study of Different Cloud Data Warehouses

We will compare the six popular cloud data warehouses, Amazon Redshift, Snowflake, Microsoft Azure SQL, Google Big Query, IBM DB2 Warehouse and Informatica – IICS (Informatica Intelligent cloud services) with respect to their architecture, features, capabilities, offerings and pricing.

3.2.1 Amazon Redshift

Amazon Redshift uses the power of massive parallel processing structure/architecture with the nodes used for data storage in terms of slices in the column format. Each node will have storage (RAM) and computing power. It has two nodes, leader and compute nodes; the leaders accept the query from the user and assign to the compute node for computation purpose.

3.2.2 High-Level Architecture of Amazon Redshift

The details of Amazon Redshift architecture have two components, client application and Data warehouse cluster (DWH-C), in the DWH-C we have leader node and compute node, these are connected via JDBC (Java database connectivity) and ODBC (oracle database connection) as depicted in Fig. 3.2 with all the details [13].

- **Amazon Redshift - Client Applications**
  It connects to load the data using ETL (Extract, Transform, Loading) Process, as well as reporting, data mining, and analytics applications for business intelligence (BI). Amazon Redshift is built using PostgreSQL and most SQL client apps will be operational with small modifications.

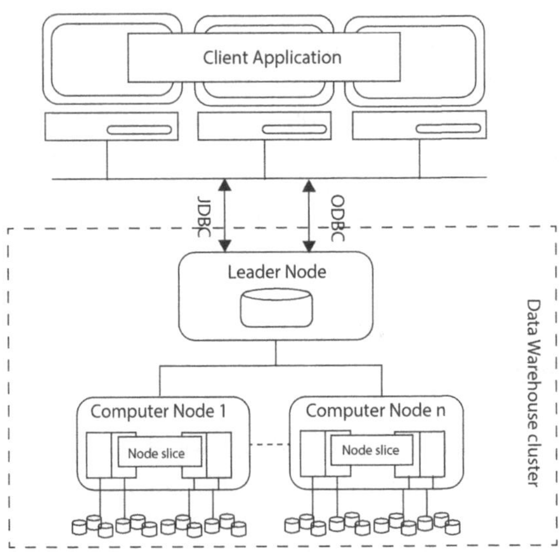

Fig. 3.2 Amazon Redshift architecture.

- **Connections**
  Amazon Redshift uses connection such as standard JDBC and ODBC drivers used for PostgreSQL to interface with client applications.
- **Clusters**
  A cluster is the most important component of an Amazon Redshift cloud data warehouse's infrastructure. Cluster is formed using more than one compute node and if a cluster has two or more compute nodes, an extra leader node manages external communication and coordinates with the compute nodes. Leader node directly contacts the client application and the computation nodes which are transparent.
- **Leader node**
  The leader node is responsible for handling all the duplex communication with associated clients and compute nodes. It will analyze and generate execution plans for queries which we execute to retrieve the data. The leader node will take the responsibility to compile the code and to distribute to various compute nodes in a cluster based on execution plan. The leader node is used exclusively for all other searches. SQL functions are implemented on the leader node in Amazon Redshift. If a query that employs any of these functions refers to tables on the compute nodes, it will yield an error otherwise.
- **Compute nodes**
  The execution plan is understood by the leader node and distributed to the individual compute nodes for compilation and returns transitional response to the leader node for aggregation for proper interpretation. Depending on the node type, every compute node has its own devoted CPU, RAM, and associated disc for storage. Scaling up the computational and storage resources of the cluster based on the workload by adding more nodes and upgrading the node. Amazon Redshift offerings can be customized to meet the business requirements.
- **Node Slices**
  Compute nodes are used to create slices where every sliced portion will have small portion of the memory in the node and disc space for storage, and it is responsible for processing a portion of the workload that has been allotted to it. The

leader node is responsible for transferring data to slices and allocating resources for queries and other database operations. The slices start working as per the schedule and the number of slices assigned per node will be determined by cluster size and nodes in it and choose one column as the distribution key when building a table. The table is loaded with data; after that the rows are distributed to the node slices using the table's distribution key. The Amazon Redshift cloud Data warehouse can leverage parallel processing to load data and perform SQL operations more quickly when a good distribution key is chosen.

- **Internal Network**
  To have intercommunication, Amazon Redshift uses "high-speed network" between the leader node and computing nodes, Amazon Redshift uses "high-bandwidth connections", close proximity, and unique "communication protocols". Client applications never have direct access to the computing nodes since they run on a different, separated network.
- **Databases**
  Every cluster will have one or more database and on the compute nodes, user data is saved. The leader node interacts with SQL client then leader node takes care of query execution with the help of compute nodes. Amazon Redshift is enhanced/optimized to handle high-performance analysis and reporting for decision support. It has all the capabilities as a traditional RDBMS, including online transaction processing (OLTP) activities like inserting and removing data. PostgreSQL is the database used by Amazon Redshift. There are a few key distinctions between Amazon Redshift and PostgreSQL that should be considered when designing and developing data warehouse systems. Amazon Redshift and PostgreSQL for details on how Amazon Redshift SQL varies from PostgreSQL [5, 6].

3.2.3 Features of Amazon Redshift Cloud Data Warehouse

The salient features of Amazon Redshift cloud Data warehouse are highlighted below:

a. Capability to integrate structured, semi-structured, unstructured data formats.
b. Data Retrieval capability from the data lake (S3) and operational databases using SQL.
c. Instant scaling of storage and compute resources with the possibility of separate billing.
d. Federated queries support.
e. High-performance queries due to optimized data storage (columnar storage, data compressions, etc.).
f. Materialized views support and result caching.
g. Database automation (automated provisioning, backups, table design, etc.).
h. Machine Learning (ML) capabilities for predicting the workload and optimizing the data warehouse performance [17].
i. Native connectors to integrate with the AWS ecosystem (including S3, AWS Glue, Amazon EMR, Amazon Kinesis Data Firehose, Amazon QuickSight, etc.).
j. Enhanced security using fine-grained access control which covers end to end.
k. Security compliant - SOC1, SOC2, SOC3, and PCI DSS Level 1 requirements, HIPAA as per global standards.

3.2.4 Pricing of Amazon Redshift Cloud Data Warehouse

The pricing of the Amazon Redshift cloud Data warehouse for end users to access the services on SaaS model delivery mode are shown below.

- On-demand pricing ranges from $0.25 to $13.04 per hour.
- Compared to on-demand pricing, reserved instance price to save up to 75% (a 3-year term).
- $0.024/GB/month for data storage (RA3 node types).

3.3 Snowflake Cloud Data Warehouse

Snowflake cloud Data warehouse uses architecture similar to Amazon Redshift with cluster and node approach. The major variance with Snowflake is that the compute and storage features are demystified; this is the unique feature. The benefit of using storage location changing depends on user requirement of computing resources. If the user does not want computing resources, the respective data can be moved to another storage which is cost effective. Because of this feature of separation with computing and storage resources, the Snowflake architecture is scalable based on the user requirements which enhances the elasticity property of the cloud data warehouse and works beyond the capacity of on-premise data warehouse [11, 12]. It also has the feature for mass ingestion where one can accommodate files, databases and streaming as well based on the user input type.

3.3.1 High-Level Architecture of Snowflake Cloud Data Warehouse

The Snowflake architecture has three different components, database storage, query processing and cloud services as shown in Fig. 3.3.

The details of every component are described below:

- **Database Storage**
  When ETL job runs, the data is moved into Snowflake, it is reorganized into a columnar format which is internally optimized and compressed. Snowflake saves the optimized data to the cloud. Snowflake is in charge of all aspects of data storage in the business, with various parameters such as data structures, size of the file, data compression, metadata

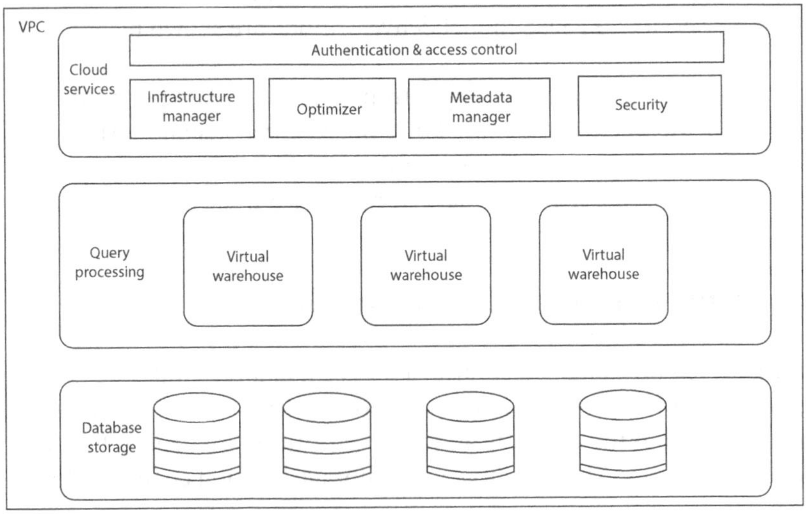

Fig. 3.3 Snowflake architecture.

statistics, and other aspects of data storage. Customers cannot see or access the data objects stored by Snowflake; they can only access them using Query Language Scripts which are executed through Snowflake [8, 18].

- **Query Processing**
  This is the processing layer which is used to execute queries where Snowflake employs "virtual warehouses (v-warehouse)" to process queries. Each v-warehouse is a multi-parallel processing compute cluster with numerous compute nodes acquired from a cloud service provider by Snowflake, mostly Amazon, Google and Microsoft Azure. Every v-warehouse is its own compute cluster, with no other virtual warehouses sharing compute resources.
- **Cloud Services**
  This layer is called as cloud service layer which is a collection of services which help Snowflake and coordinates its actions. The connect services allow Snowflake's different components to create an action for user requests, starting from query to response dispatch. Snowflake provisioned compute instances from the cloud provider for the cloud services layer. The services managed by cloud service layer which includes authentication, Infrastructure management, Metadata management, Query parsing and optimization and Access control.

3.3.2 Features of Snowflake Cloud Data Warehouse

The salient features of Snowflake cloud Data warehouse are listed below:

a. Data integration of structured, semi-structured and unstructured.
b. Querying - data retrieval using SQL.
c. Storage and compute resources can be scaled instantly, with distinct pricing options available.
d. Automation of databases (automated provisioning, backups, etc.)
e. Query caching, planning, parsing, and optimization are all automated.
f. Support for materialized views.
g. For reduced development costs, a broad ecosystem of supported third-party partners and technologies for data

integration, machine learning and data science, data security and governance, BI and data visualization are available [14].

h. Separate storage and computation resource scaling.
i. Data encryption from beginning to end.
j. Masking and tokenization of dynamic data.
k. Access control with finer granularity.
l. SOC2 Type 2 compliance, ISO 27001 compliance, PCI compliance, HIPAA compliance, FedRAMP compliance, and so on.

3.3.3 Snowflake Cloud Data Warehouse Pricing

The Snowflake cloud Data warehouse pricing for SaaS Model delivery is shown below.

- Price to utilize the Snowflake cloud services with no long-term commitment and billing is done based on seconds.
- Snowflake storage procurement with minimum of 60 seconds facility.
- Pricing can be obtained by contacting Snowflake directly.

3.4 Google BigQuery Cloud Data Warehouse

Google BigQuery is a component of Google platform designed to handle huge data in the cloud with serverless architecture where various machines are used by the cloud service to manage resources. In this case the client machines are not intricate in the resource management activities. BigQuery's "serverless architecture" provisions the out-of-date data population as well as streaming of data and enables mass ingestion for real-time data with the help of MPP-massively parallel processes with the capacity to handle millions of records together. This method stores data in a Colossus which is part of file management system, which organizes it into clusters made up of distinct nodes. In a columnar format, the files are dispersed in 64-megabyte increments. Requests are made from tree structure architecture between the several machines where the data is kept, allowing for faster response times [10, 16].

3.4.1 High-Level Architecture of Google BigQuery Cloud Data Warehouse

The high-level architecture of Google BigQuery cloud Data warehouse is depicted in Fig. 3.4. The Google BigQuery Cloud Data warehouse has three layers, namely Client Layer, Dremel (Compute Layer) and Colossus Layer (Storage Layer). Through Jupiter network we can access all the components and Borg is in charge of coordinating so that all of the engine's sections can operate together effortlessly.

- **Dremel (Compute Layer)**
  Dremel generates execution trees from SQL queries. The tree's leaves are known as slots, and they are responsible for reading data from storage and doing any necessary calculations. The tree's branches are "mixers," which execute the aggregation. Dremel dynamically distributes slots to queries, ensuring that numerous users' queries are treated equally.
- **Colossus (Storage Layer)**
  In Colossus, all of the data is compressed and stored in a columnar storage structure. Colossus is in charge of replication, recovery (in the event that drives fail), and

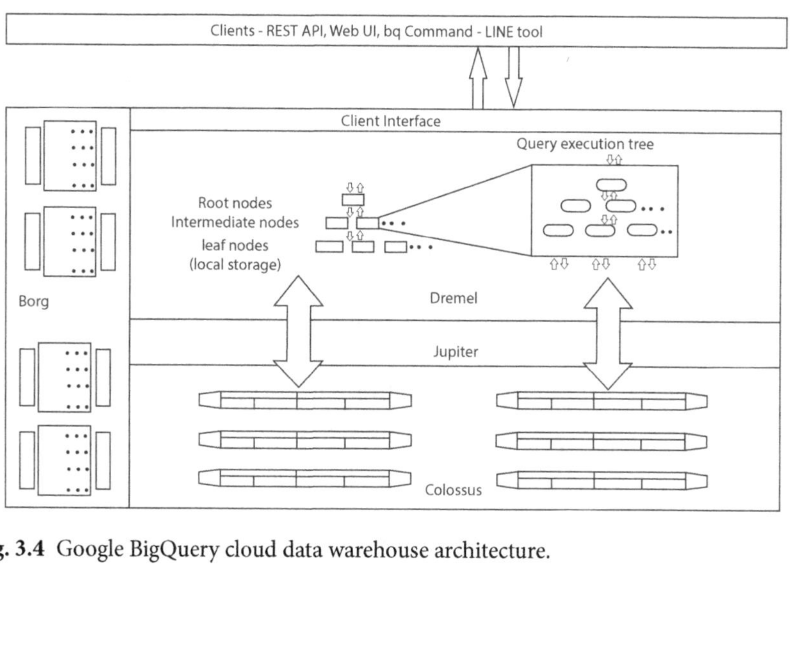

Fig. 3.4 Google BigQuery cloud data warehouse architecture.

distributed administration (so there is no single point of failure). Colossus allows customers to store several petabytes of data in a single location.

- **Jupiter Network**
  The petabit Jupiter network allows compute and storage to communicate with one another. The Shuffle is a term used to describe the Network.
- **Borg (Orchestration engine)**
  Borg is in charge of coordinating so that all of the engine's sections can operate together effortlessly. Borg controls the mixers and slots, as well as allocating hardware resources.

3.4.2 Features of Google BigQuery Cloud Data Warehouse

The Google BigQuery cloud Data warehouse has the salient features which will address the modern data loadings to accommodate the data generation from the internet world.

- Data Ingestion of all forms such as structured semi-structured, unstructured data.
- Using BigQuery Omni, SQL-based data searching across different clouds (AWS and Azure).
- Scaling, patching, and upgrades, as well as backups and recovery, are all automated.
- ML and AI integrations built-in (BigQuery ML, AI Platform, TensorFlow).
- Cloud BigTable, Google Cloud Storage, Cloud AI Notebooks, and Google Sheets integration.
- Support for materialized views.
- Separate storage and computation resource scaling.
- Hadoop/Spark workloads can read/write data directly from BigQuery with Storage API thanks to integration with the Apache big data ecosystem.
- Data encryption from beginning to end.
- Access control with finer granularity.
- HIPAA-eligible and global security standards compliant.

3.4.3 Google BigQuery Cloud Data Warehouse Pricing

The Google BigQuery cloud Data warehouse SaaS delivery model has pricing for all business enterprises.

- Storage: "$0.02/GB/month ($0.01/GB/month for long-term storage" - table partition that has not been changed in 90 days).
- Streaming inserts are $0.01 for 200 megabytes.
- Query performance: $5/TB, with the first TB/month free (pay-as-you-go).
- A dedicated reservation of 500 processing units starts at $10K per month (flat-rate pricing).

3.5 Microsoft Azure Synapse Cloud Data Warehouse

Microsoft Azure is based on massively parallel processing. Data is stored in relational databases that, because of this architecture, run fast SQL queries of enormous complexity. Additional tools in the Azure cloud ecosystem allow users to create automated pipelines for transmitting, processing and storing data at petabyte scale [15].

3.5.1 Microsoft Azure Synapse Cloud Data Warehouse Architecture

The Microsoft Azure synapse architecture is depicted in Fig. 3.5. The Microsoft Azure synapse cloud Data warehouse has three main

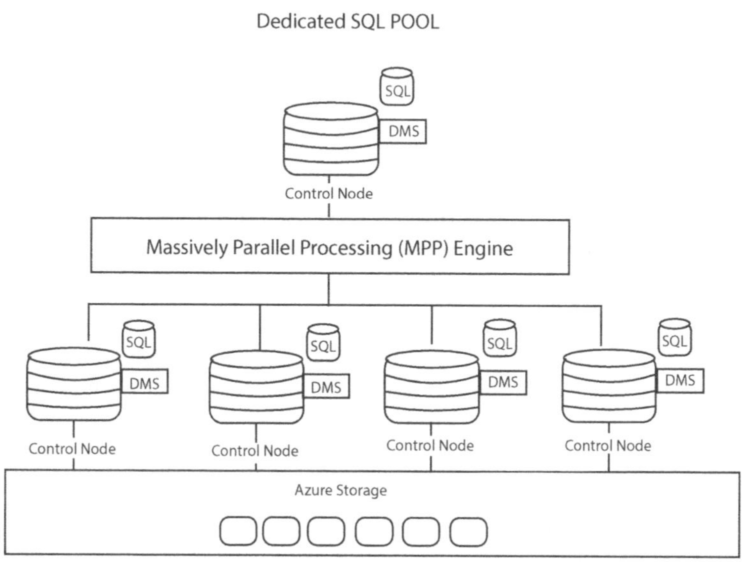

Fig. 3.5 Microsoft Azure synapse architecture.

components: MPP engine, control node and Azure storage. The details of the components are described below:

- **Azure Storage**
  Azure storage is used to keep the use data safe. Azure uses a separate business model for storage consumption because your data is kept and maintained by Azure Storage. One can retrieve the data from the data lake files with a serverless SQL pool, but one cannot retrieve and ingest data from them with a dedicated SQL pool. The data is imported into a devoted SQL pool and it is sharded into various distributions, which improves the system performance. When you define the table, you can choose which small portion of the entire large or sharding pattern to employ to distribute the data sharding is reinforced by round robin, Hash and Replicate methods.
- **Control Node**
  It is the brain of the google BigQuery construction and uses component of UI- web portal that communicates with all the other apps and connections. The distributed query engine of Synapse SQL will operate using control node to optimize and coordinate all similar queries. T-SQL query is used to the dedicate SQL pool, the Control node which converts into parallel queries for every distribution. The DQP engine runs on the Control node in a serverless SQL pool to enhance the efficiency and manage distributed execution of user queries by breaking them into smaller queries that are executed on Compute nodes.
- **Compute Nodes**
  Computing power is provided by the compute nodes where distributions are mapped to compute nodes for processing in a dedicated SQL pool. It reallocates the distributions to all the available Compute nodes when you get more resources through payment. The dedicated SQL pool's service level determines compute node requirements, which might range from 1 to 60. Every compute node has a unique node ID that may be seen in the system. Look for the node id column in system views that start with sys.pdw nodes to find the Compute node ID. See Synapse SQL system views for a list of these system views. Each Compute node in the serverless SQL pool is given a job and a set of files to run the jobs. Job/

task is a distributed for query execution unit for execution purpose and to ensure that adequate Compute nodes are used to process user queries, automatic scaling is in place.

- **Data Transfer Service**
  Data Transfer Service (DTS) is a data transport solution that manages data drive across compute nodes in a dedicated SQL pool. To ensure that parallel searches produce reliable results, some queries involve data transfer. During data migration is happening the necessary, DTS ensures that the correct data is sent to the correct location.
- **Distributions Across the Locations**
  DTS is a data transport technology that controls data transportation across compute nodes in a dedicated SQL pool. Some queries require data transfer to ensure that simultaneous searches produce valid results. DMS guarantees that the correct data is transferred to the correct destination when data migration is required. One of the data distributions is used by each of the 60 smaller searches. One or more of the 60 distributions are managed by each Compute node. Computation node with one distribution that exists in a dedicated SQL resource with maximal compute resources. All of the dispersals are handled in one compute node in a dedicated SQL pool with minimal computing resources.

3.5.2 Features of Microsoft Azure Synapse Cloud Data Warehouse

The salient features of Microsoft Azure synapse cloud Data warehouse which can accommodate the modern data loads. The features are listed below:

a. Querying organized, semi-structured, and unstructured data types with SQL.
b. Support for multiple languages such as T-SQL, Python, Scala, Spark SQL, .Net.
c. Separate storage and computing resources which can be scaled further.
d. Automation of databases (automated provisioning, backups, table design, etc.)

e. Caching of result sets, workload isolation, and clustered columnstore indexes are all available.
f. Automatic data backups and restoration points.
g. Native connectors to perform integration activities with the ecosystem given, i.e., Apache Spark, Power BI, Azure Machine Learning, Azure Stream Analytics, Azure Cognitive Services, Azure Cosmos DB, and other technologies.
h. Integration with third-party business intelligence products such as Tableau, SAS, and Qlik, among others.
i. End-to-end security using encryption, dynamic data masking, granular access control.
j. Compliant with ISO 27001/27002, PCI DSS, SOC 1 Type II, SOC II, SOC III, HIPAA-eligible.

3.5.3 Pricing of Microsoft Azure Synapse Cloud Data Warehouse

The Microsoft Azure synapse cloud Data warehouse has the SaaS delivery model for all the enterprises with the pricing listed below:

- Pricing for compute on-demand: $1.20/hour (DW100c) - $360/hour (DW30000c).
- When compared to on-demand computing, reserved instance can save up to 65 percent
- Data storage costs $122.88 per TB each month.

3.6 Informatica Intelligent Cloud Services (IICS)

Informatica Corporation has revamped the PowerCenter data integration activities as and when the cloud computing got matured to handle cloud hosted application data. Informatica Intelligent cloud services – IICS is built with the intelligence to handle the auto mapping of the columns and built various add-on connectors. Informatica cloud has various services such as application integration, data integration and respective consoles and to monitor all the activities and workload at one place. IICS has a secure agent which has five different components such as common integration components, process server, data integration server, mass ingestion and OI data collector. The new IICS which is designed to handle data integration activities from the cloud applications such as marketo, salesforce, NetSuite, workday and numerous on-premise databases source systems and integrated and loaded the data to the targets such as Amazon S3 bucket, Google BigQuery or Snowflake cloud warehouse and on-premise Data warehouse as well. The innovations in the cloud domain can help various business ecosystems to drive their business effectively and make resources free and can be used productively. Cloud Infrastructure empowers the business processes which is used to orchestrate and synchronize the business from various parts of the globe. The delivery model and business model are customizable based on the end-user requirements. In the current situation iPaaS solution would help many of the organization, Informatica Corporation used the AI/ML engine called CLIRE with micro-service architecture which enables the performance of data integration activities very effectively and in a robust way.

3.6.1 Informatica Intelligent Cloud Services Architecture

The Informatica intelligent cloud service architecture is depicted in Fig. 3.6. It addresses the on-premise data loading problems, and the advent of cloud technology and the entire business ecosystem is on the cloud because of its popularity and because it is easy to maintain and cheaper and is a highly secured environment to process all the ETL (Extract, Transform and Loading) Jobs. Informatica intelligent cloud services has

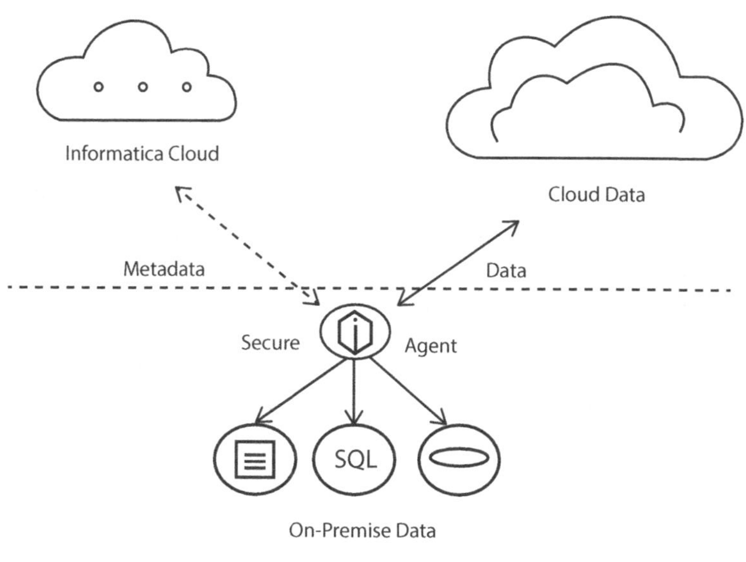

Fig. 3.6 Informatica intelligent cloud architecture.

various offerings to handle the data migration either from on-premise to on premise or cloud to cloud and to orchestrate various business operations. Secure Agent has five different components such as process server, common integration components, mass ingestion, data integration and OI data collector which are responsible to handle various operations such as data integration and mass ingestion of various formats. The function of various components of IICS is described below.

- **Informatica Cloud**
  The Informatica cloud can be accessed through the web browsers and through SAML authentication and this will serve as a data integration platform.
- **Informatica Cloud hosting facility**
  Hosting facility is where an organization's jobs and information are stored. It is kept in the PowerCenter database.
- **Informatica Cloud Application**
  The informatica cloud has various assets and application to perform data synchronization, data replication; taskflows are examples of tasks that applications can accomplish.
- **Informatica Cloud Secure Agent**
  Informatica cloud secure agent is a light weighted process used to connect to the Informatica cloud securely using SAML authentication. Secure agent has various components such as process server, common integration components, OI data collector, mass data ingestion and mass ingestion.

3.6.2 Salient Features of Informatica Intelligent Cloud Services

- An easy-to-use, drag-and-drop user experience with 100+ prebuilt functions and templates for mappings and transformations is provided by Informatica Intelligent Warehouse on Cloud.
- You can quickly scale up with simple, consumption-based pricing.
- Use CLAIRE AI-powered automation, suggestions, and auto-tuning and auto-scaling to speed up the creation and deployment of data pipelines.
- With serverless Spark processing, you can scale up to 10x performance.
- Workloads can be easily moved between cloud platforms.

3.6.3 Informatica Intelligent Cloud Services Pricing Model

The Informatica intelligent cloud service is an extension of PowerCenter and the service model is based on the SaaS delivery model with the below pricing model for enterprises or business forms.

- The monthly cost of Informatica Intelligent cloud starts at €2,400.ss
- There isn't a free version available.
- Informatica can provide you with other pricing options if you contact them directly.

3.7 Conclusion

Because it allows web applications to flexibly alter their infrastructure capacity on demand, the cloud is an appealing infrastructure solution for web applications. Cloud-based data warehousing has many advantages. Thanks to its elasticity, scalability, deployment time, reliability, and lower costs (due, for example, to elasticity), we believe that data warehousing systems in the cloud have a lot of potential. To make use of the cloud's advantages, a data warehousing system must be highly parallel and distributed, as well as meet a number of other requirements. The major cloud data warehouses, which are commonly used by most companies, are evaluated, and the benefits and disadvantages of each are discussed, as well as the pricing of each.

Acknowledgements

We would like to thank Mr. Satish B M, Project Lead, ITC Infotech, and Mr. Prabhuram, Data Analyst, Numerify, for giving their valuable suggestions in compiling this book chapter on the importance of cloud data warehouse in the internet world.

References

1. A. Khalil and M. Belaissaoui, "New approach for implementing big DataMart using NoSQL key-value stores," *2020 5th International Conference*

on Cloud Computing and Artificial Intelligence: Technologies and Applications (CloudTech), 2020, pp. 1-6, doi: 10.1109/CloudTech49835.2020.9365897.

2. R. Prasad Tripathy, M. Ranjan Mishra and S. R. Dash, "Next Generation Warehouse through disruptive IoT Blockchain," *2020 International Conference on Computer Science, Engineering and Applications (ICCSEA),* 2020, pp. 1-6, https://doi: 10.1109/ICCSEA49143.2020.9132906.
3. Mohamed A. Soliman *et al.*, A Framework for Emulating Database Operations in Cloud Data Warehouses, *Proceedings of the 2020 ACM SIGMOD International Conference on Management of Data,* June 2020, pp. 1447-1461, https://doi.org/10.1145/3318464.3386128
4. Mathieu B. Demarne *et al.*, Reliability Analytics for Cloud Based Distributed Databases, *Proceedings of the 2020 ACM SIGMOD International Conference on Management of Data,* June 2020, pp. 1479-1492, https://doi.org/10.1145/3318464.3386130
5. E. Guermazi, M. Ben Ayed and H. Ben-Abdallah, "Adaptive security for Cloud data warehouse as a service," *2015 IEEE/ACIS 14th International Conference on Computer and Information Science (ICIS),* 2015, pp. 647-650, doi: 10.1109/ICIS.2015.7166672.
6. I. Ting, C. Lin and C. Wang, "Constructing a Cloud Computing Based Social Networks Data Warehousing and Analyzing System," *2011 International Conference on Advances in Social Networks Analysis and Mining,* 2011, pp. 735-740, doi: 10.1109/ASONAM.2011.25.
7. "Windows Azure Platform: An Era for Cloud Computing", Madhurima *et al., (IJCSIT) International Journal of Computer Science and Information Technologies,* Vol. 2 (2), 2011, 621-623.
8. Moscoso-Zea, Oswaldo & Paredes-Gualtor, Joel & Luján-Mora, Sergio. (2018). A Holistic View of Data Warehousing in Education. *IEEE Access.* PP. 1-1. 10.1109/ACCESS.2018.2876753.
9. Ahmed A. Soror, Umar Farooq Minhas, Ashraf Aboulnaga, Kenneth Salem, Peter Kokosielis, and Sunil Kamath. Automatic virtual machine configuration for database workloads. In *Proc. ACM SIGMOD Int. Conf. on Management of Data,* 2008.
10. Ahmed A. Soror, Ashraf Aboulnaga, and Kenneth Salem. Database virtualization: A new frontier for database tuning and physical design. In *Proc. Workshop on Self-Managing Database Systems (SMDB),* 2007.
11. Gireesh Babu C.N., Thungamani M., Chandrashekhara K.T., Manjunath T.N. (2019) Real Time Hand Gesture Recognition for Differently-Abled Using Deep Learning. In: Santosh K., Hegadi R. (eds.) Recent Trends in Image Processing and Pattern Recognition. RTIP2R 2018. *Communications in Computer and Information Science,* vol 1035. Springer, Singapore. https://doi.org/10.1007/978-981-13-9181-1_29
12. J. Barr. AWS Import/Export Snowball – Transfer 1 petabyte per week using Amazon-owned storage appliances. aws.amazon.com/blogs/aws/

aws-importexport-snowballtransfer-1-petabyte-per-week-using-amazon-owned-storageappliances/, 2015.

13. Manjunath *et al.*, Design and Analysis of DWH and BI in Education Domain, *International Journal of Computer Science Issues (IJCSI)* 8 (2), 545.
14. RA Archana *et al.*, A novel data security framework using E-MOD for big data, *2015 IEEE International WIE Conference on Electrical and Computer Engineering (WIECON-ECE)*, pp. 546-551.
15. RK GK, BJ Rabi, TN Manjunath, "A study on dynamic data masking with its trends and implications", *International Journal of Computer Applications* 38 (6), 19-24.
16. GK Ravikumar, BJ Rabi, RS Hegadi, TN Manjunath, RA Archana, "Experimental study of various data masking techniques with random replacement using data volume", *International Journal of Computer Science and Information Security* 9 (8), 154.
17. Chandrashekhara K.T., Thungamani M., Gireesh Babu C.N., Manjunath T.N. (2019) Smartphone Price Prediction in Retail Industry Using Machine Learning Techniques. In: Sridhar V., Padma M., Rao K. (eds.) *Emerging Research in Electronics, Computer Science and Technology. Lecture Notes in Electrical Engineering*, vol 545. Springer, Singapore. https://doi.org/10.1007/978-981-13-5802-9_34
18. K. R. Swetha, Niranjanamurthy, M, Amulya, M. P., and Manu, Y. M., "Prediction of Pneumonia Using Big Data, Deep Learning and Machine Learning Techniques," *2021 6th International Conference on Communication and Electronics Systems (ICCES)*, 2021, pp. 1697-1700, doi: 10.1109/ICCES51350.2021.9489188.

4

Data Mining with Cluster Analysis Through Partitioning Approach of Huge Transaction Data

Sampath Kini K.[1]* and Karthik Pai B.H.[2]

[1]*Department of CSE, NMAM Institute of Technology Nitte, Karkala, Karnataka, India*
[2]*Department of ISE, NMAM Institute of Technology Nitte, Karkala, Karnataka, India*

Abstract
Data mining, data analysis and its engineering from transactional databases have been applied across various domains for years now. Similarly, clustering mechanism has been used to group elements having similar properties together. The traditional method makes processing of data objects on the whole set of data objects exist in the transaction store at once. This is going to cause performance concerns such as memory crash and huge elapsed time for the computing task. The clustering approach of data mining will be applied by knowledge workers in many real-life case studies. As data amount grows or scales up, a few methods were introduced to meet the needs of scaling up further. But still, some of the methods have met the needs of scalability to some extent; it still uses a huge amount of computing power and primary memory space. This work brings in a separate approach, named cluster analysis, using partitioning technique of huge transaction database. The goal of this study is to come up with an appropriate mechanism to get rid of memory problems, bring down total elapsed time and enable cluster analysis based on partitioning of entire transaction database into various suitable partitions. This partitioning approach finds out the target cluster by processing each partition in multiple iterations. Thus, this approach avoids loading of data objects at once in the computer's main memory and thereby gets rid of out of memory concerns.

***Keywords*:** Data mining, partitioning, performance, transaction database, clustering

**Corresponding author*: sampath@nitte.edu.in

Kukatlapalli Pradeep Kumar, Aynur Unal, Vinay Jha Pillai, Hari Murthy, and M. Niranjanamurthy (eds.)
Data Engineering and Data Science: Concepts and Applications, (71–84) © 2023 Scrivener Publishing LLC

4.1 Introduction

This research work brings in advanced approach which is utilized for handling of huge amount of data objects to seek the cluster of target. Cluster, in this context means putting similar data objects together [1]. Each of the cluster analysis approach has its own strong points and weaknesses [2]. Many of these methods will have to scan all of the input data objects sequentially, find out cluster of target and recompute the centroid of cluster after each data object is treated for processing. These methods create performance bottlenecks when quite a big number of iterations are applied for seeking out cluster of target of every data object. This advanced method breaks the huge set of data objects into separate multiple partitions and thereafter processes individual isolated partitions in an isolated thread of computation if multicores available in the system; otherwise each of the isolated partitions is treated for processing one after another, for determining target cluster. This method of sub-partitioning will aid in getting rid of heavy computational time requirements of clustering techniques in mining algorithms of data.

The aim in this research paper is realized in two phases: 1) Sub-partitioning on huge data set depending on availability multicores for simultaneaous computation and it will also consider the size of the dataset and primary memory space available for the purpose of sub-partitioning. 2) Identification of cluster of target on every sub-partition in an isolated workspace of computation. Comparison of total elapsed time across the abovementioned approach is done.

The various components used in systems for data mining comprise a source of data, engine for mining, data warehouse for storing information, the pattern evaluation subsystem, interface for users, and knowledge store [3]. The original source of raw data is the database, World Wide Web, data warehouse, text files, and other source files. Generally it requires a large amount of historical data for obtaining accurate results with data mining. Organizations usually keep data in datastores or data warehouses. Yet another key data source is the World Wide Web. Recently for data mining functionalities, a variety of major data mining methods were developed and applied. These are association, classification, clustering, prediction, sequential patterns, and regression as shown in Fig. 4.1.

With usage of clustering, a group of various data objects is classified as cluster of objects. One group indicates a cluster of data [4]. Data objects are subdivided into various different groups, which are determined based on the similarity of data objects. Once the classifications of data into various

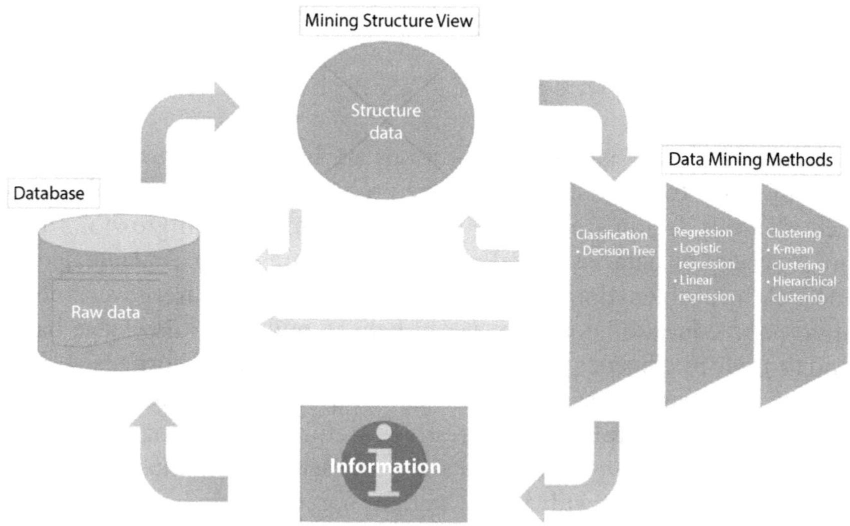

Fig. 4.1 Technical framework of data mining system.

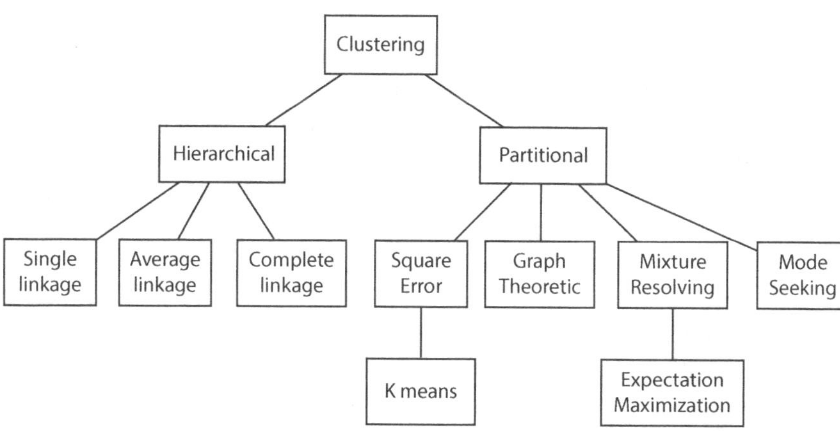

Fig. 4.2 Overall technical framework of clustering techniques.

cluster groups, all the groups are attached by a label. It leads to adaptation of the changes by performing the classification. Cluster analysis in terms of data mining provides mechanism that finds out the group of objects that look similar to each other in the group and are different from the objects present in another group. The task of cluster analysis is done with the help of various mechanisms shown in Fig. 4.2. There are different benefits of data clustering analysis such as image processing, information retrieval, pattern recognition, market study and many more. With the help of data clustering, organizations can determine new groups from its database of users and customers. Data Classification can also be performed by considering purchasing patterns.

Many data analysis-based applications are being developed and made available now to make it easier for the organizations and stakeholders to help them in deciding on future plans and organizational growth. There are a variety of data analytical applications being made available across most of the problem domains such as healthcare, agriculture, government, sports and educational system. It can help to build the knowledge base, and to apply intelligence, by processing raw data in various formats. A data analytical system is a system which takes input in the form of raw data, processes it and outputs the interesting, useful and meaningful patterns for the knowledge workers. With the proposed efficient cluster analysis system, even a dataset with a huge amount of records can be analysed efficiently through a data partitioning approach. By following this scheme, we may reduce the time consumption or memory resource consumption since it enables the carrying out of multitasking at a given point of time. Additionally, currently all the existing traditional analytics involve high usage of computer resources when a large dataset is processed. These incidents of high usage of computing resources can be overcome if we apply a partitioning-based proposed data analytical system. The key goal of this research proposal for exploring the methodologies is designing an efficient cluster analytical system which can work large datasets without any major issues while data processing. The proposed block diagram of cluster analysis is given in Fig. 4.3.

- Performance bottleneck is the major problem in data analytics with large datasets.
- Data collection is carried out with the help of datasets available over internet.
- Data partitioning-based approach is used to establish processing large datasets. The key target is to design an efficient system which is capable of dealing with a huge amount of raw data successfully wthout any performance issues.

This chapter summarizes the consequences and benefits of applying a partitioning-based cluster analysis system in data mining techniques and various application systems which can be developed along with the technical framework that can be used for realizing the aim of it. The following sections in this chapter include methodology for partitioning the dataset,

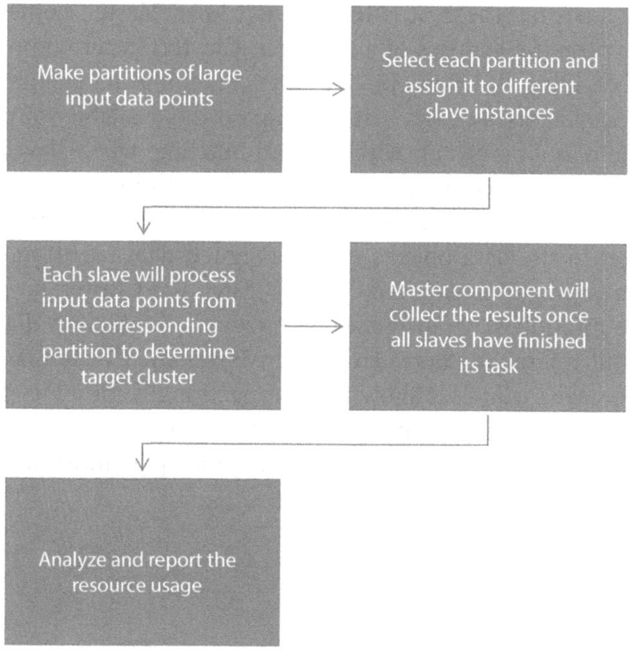

Fig. 4.3 Proposed cluster analysis system.

literature survey, algorithms developed, design and runtime interactions of various software elements, results, conclusion, and scope of future work.

4.2 Methodology Used in Proposed Cluster Analysis System

In this sub-section, various steps that are being followed and algorithms designed for the implementation are described. Steps followed to identifying cluster of target for every input object element have different sub-tasks underneath. Sub-tasks consist of dividing large input data objects into various sub-partitions, selection of individual partition and submission of it to various slave executors in separate thread of computing if multicores are available in the system. If multiprocessors are not available for simultaneous computing then each partition is processed one after another with the help of single slave instance. The slave executor will have the job of computing the cluster of target from all the data objects present in the repsective partition. Two kinds of software elements are developed for realization of this mechanism. The first is termed as ClusterMaster element and its job is to expose the features of a manager. ClusterMaster component has the job of dividing the data object points, attaching the partition to separate slave executers or a slave depending on the availabity of multicores in the system and comes up with the end outcome when slave executors have completed their computation in isolated execution space. The second element represents ClusterSlave and it does the job of finding out cluster of target in the data objects in the assigned data partition. Dynamic interaction through instances of ClusterMaster and ClusterSlave elements with usage multiple slave instances is shown via interaction diagram Fig. 4.4. Dynamic interaction between ClusterMaster and ClusterSlave element with single slave instance is shown via interaction diagram Fig. 4.5.

This cluster analysis system mainly consists of the following modules. High-level software elements and their relationship is shown in Fig. 4.6.

- Data partitioning
- Loading of partition
- Partition executor/slave
- Cluster result aggregator
- Computer resource analyser

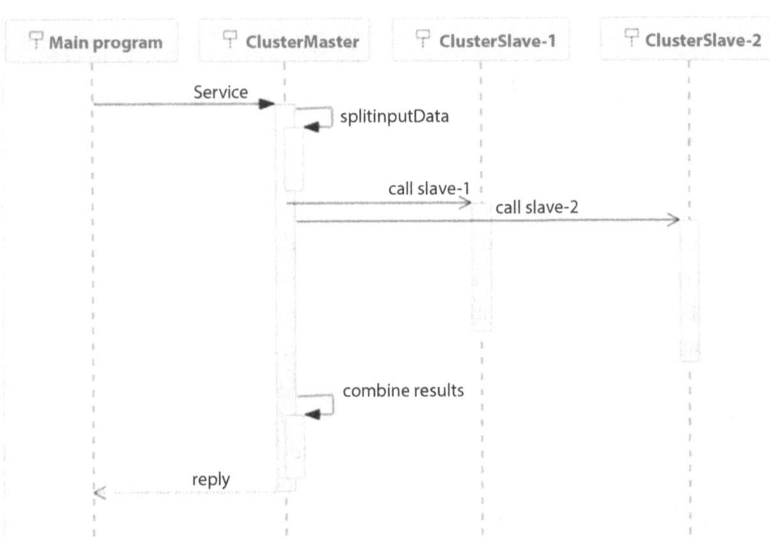

Fig. 4.4 Run-time interaction between ClusterMaster and multiple slave executors.

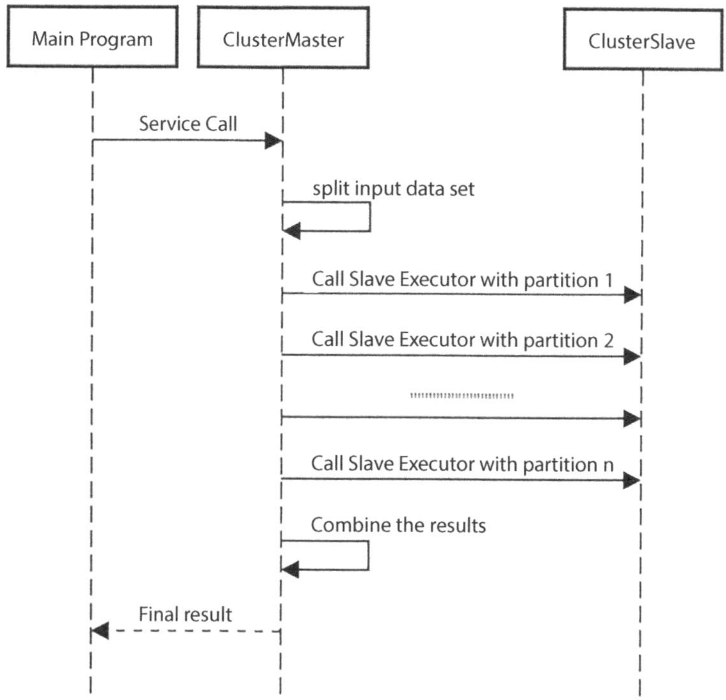

Fig. 4.5 Run-time interaction between ClusterMaster and single slave executor.

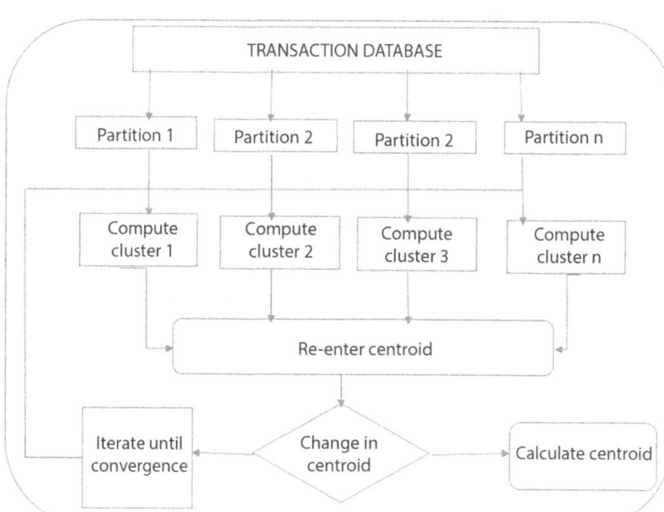

Fig. 4.6 Overview of software elements and their relationship by usage of ClusterMaster and slave executor.

4.2.1 Design of Algorithms

In this section, let us capture the algorithms being created for this cluster analysis.

Algorithm 1: PT-DTP
This is an algorithm being applied to generate partitions in N numbers on the huge data object elements.

Input: **DT**, set of data objects as input; **m**, no of computing processors in the system or available memory size in case of single processor system.
Output: **PT**, collections of partitions; PT will have partition information in the form of beginning and ending indices of devided data set.

```
   k ← |DT|    /* total number of data points */
   tn ← k/j   /* tn gives number of data points in the current workspace
of partition */
   n ← j    /* n: number of workspace of partitions */
   sidx ← 0
   For p = 1 to n
   start
   eidx ← sidx + tn
   /* sidx and eidx gives both beginning and ending indices of the current
partition */
   add { sidx, eidx } to PT
   sidx ← eidx
   End
   return PT;
```

Algorithm 2: KNNCL-PTDB
This specific algorithm is developed for executing KNN [4] (k-nearest neighbor algorithm on isolated partition associated to separate workspace of slave executor). The function KNN could be existing algorithm of clustering analysis.

```
Input: DT, set of data objects; PT, collection of isolated partitions;
Output: SCL, a collection of cluster centroids
For every partition PTq Є PT do
start
Ex← create new Executor of slave
KNN(Ex, PTq, SCL);     /* SCL represents the output having collection of clusters */
End
WaitForAllSlaveExcecutors() /* wait until all of the slave executers have completed its work */
return SCL;
```

Using a simple example, let us look at the flow of partitioned clustering method with the help of slave executors. The input data object elements are captured in Table 4.1. We are going to have two clusters for which centroids are assumed as (1.0, 1.0) and (5.0, 7.0) initially. If we assume that executing system is going to provide four different processors available for the execution of multiple slave executors. As there are twelve data object

Table 4.1 Data object points, isolated partitions and the cluster number.

Data object points for input	Partition ID	Cluster target
(1.1, 1.1)	1	1
(1.6, 2.0)		1
(3.1, 4.2)		2
(5.1 ,7.2)	2	2
(3.4, 4.9)		2
(1.6 ,1.1)		1
(3.8, 4.7)	3	2
(1.0, 1.9)		1
(5.4, 6.8)		2
(1.4, 2.6)	4	1
(1.5, 2.4)		1
(3.3, 4.3)		2

points captured in this table, four isolated data partitions having three data object points in each workspace of partition are formed. Every isolated partition workspace is provided as input data for different workspaces of ClusterSlave executors. Each of the ClusterSlave executors will compute cluster centroid in its workspace using technique of KNN algorithm.

4.3 Literature Survey on Existing Systems

This data clustering issue may be used to various clustering domain applications [5]. The capability to automation task of putting similar data items enables us to extract hidden similarities and key insights during combining a huge amount of data objects into small groups. Unlike existing traditional clustering approaches, most of the clustering algorithms are not depending on statistical distributions of input object points and so it can be useful to apply when a small amount of prior knowledge is present on a certain issue [6]. McCallum *et al.* [7] explained how the total number of iterations could be made less by partitioning an input dataset into overlapping subsets. Thereafter, iterate only input data point objects within overlapping places. However, this approach was leading to accuracy issues with the final results.

4.3.1 Experimental Results

We have performed experimentation on freely available open-source library furnished by open-source data mining library spmf [8]. This technique using multiple instances of slave's implementation for dealing with data object points is experimented using 5K, 10K and 15K data object points, respectively. With the help of resource analyser total elapsed time for all of the three categories of data objects set is depicted using a graph, Fig. 4.7. Experiments were performed on the below mentioned system platform.

- Processor used: Intel(R) Pentium(R) CPU N3540 @2.16GHZ
- System: windows 8.1, 64 bit, x-64 processor
- RAM Available: 4GB
- Processors: 4 numbers
- Programming Language: Java:JVM: 1.8

A partitioning-based approach of large data set also showed promising experimental results with regard to performance bottlenecks. An overview of overall resource behavior of our partitioning-based approach is depicted in Table 4.2.

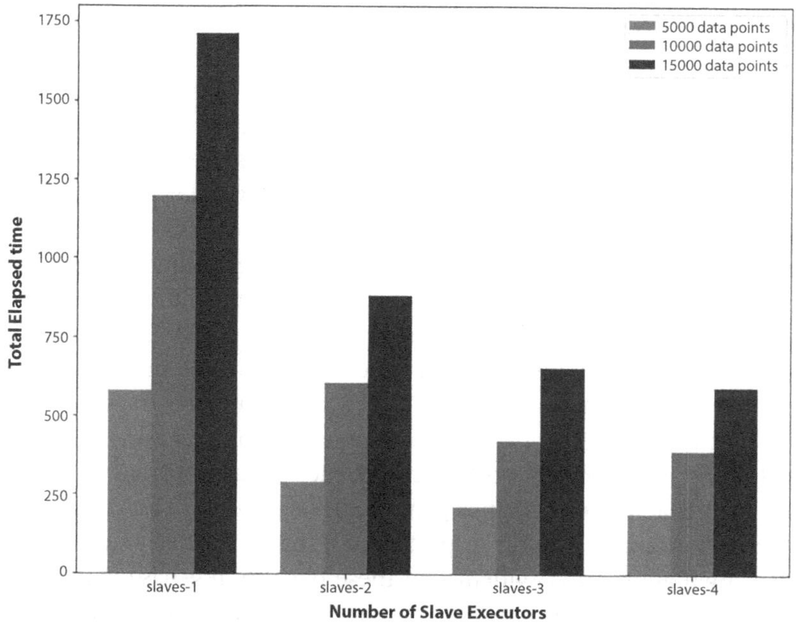

Fig. 4.7 Elapsed time with number of slave executor instances and object points with the help KNN clustering algorithm.

Table 4.2 Computing resource comparison of partitioning based cluster analysis.

Algorithm/features	Elapsed time	Memory usage	Scalability
KNN without partitioning	High	High	Low
KNN with partitioning	Less	optimal	High
KMeans without partitioning	Very High	High	Low
KMeans with partitioning	moderate	optimal	moderate

4.4 Conclusion

This section provides a comparative analysis of the partitioning-based techniques used for cluster analysis for data points. The database considered here is in the form of csv file representing various data objects. Methods to do partitioning of large data objects and algorithms for processing each workspace of partition are discussed in detail. It is concluded that most of the research studies claim that existing clustering techniques such as KNN are suffering from performance bottlenecks in terms of total elapsed time and primary memory crash when a data set with large size is involved. This approach reduces the total elapsed time by making use of multiple slave executors and eliminates memory crash with the help of usage of a single slave executor working on a data partition with smaller workspace. This partitioning and isolated workspace for each data partition-based technique also enables performance gain of 30% on average cases. By looking at performance comparison of partitioning-based cluster analysis in the table shown above, it is evident that these partitioning-based approaches provide better performance gains and accuracy. So, it is worth going for usage of these mechanisms in applications based on cluster analysis. In future, this technique can be extended to explore other remaining existing clustering techniques as well. This would enhance the efficiency and performance gains and enable the optimal computing resource usage easily in clustering techniques on a large transaction database.

References

1. Berkhin, P.: A survey of clustering data mining techniques. In: *Recent Advances in Clustering*, pp. 25–71. Springer, Berlin (2016).
2. D. Xu, Y. Tian: A comprehensive survey of clustering algorithms, *Ann Data Sci*, vol.2 (2),165-193(2015).
3. Han J. *Data mining: concepts and techniques.* San Francisco: Morgan Kaufmann Publishers Inc.; 2005. ISBN 1558609016.Google Scholar.
4. *k*-nearest neighbors algorithm, https://en.wikipedia.org/wiki/Knearest_neighbors_algorithm, last accessed 2021/06/08.
5. Gonzalez TF. Clustering to minimize the maximum intercluster distance. *Theor Comput Sci.* 1985;38:293–306. MathSciNet **Article MATH Google Scholar**.
6. Babuska R. Fuzzy clustering. http://homes.di.unimi.it/~valenti/SlideCorsi/Bioinformatica05/Fuzzy-Clustering-lecture-Babuska.pdf. Accessed 4 Jan 2016.

7. McCallum A, Nigam K, Ungar LH. Efficient clustering of high-dimensional datasets with application to reference matching. In: *Proceedings of the ACM SIGKDD International Conference on Knowledge Discovery and Data Mining*. 2000. pp. 169–78.
8. Open source library spmf, http://www.philippe-fournier-viger.com/spmf/ last accessed 2021/06/12.

5

Application of Data Science in Macromodeling of Nonlinear Dynamical Systems

Nagaraj S.[1]***, Seshachalam D.**[1] **and Jayalatha G.**[2]

[1]BMS College of Engineering, Bengaluru, India
[2]RV College of Engineering, Bengaluru, India

Abstract

Systems trained on data to predict the behaviour of the real-world physical system models are giving faster and more accurate simulation results than the nonlinear mathematical models. The reason for this is that the nonlinear structure of real-world problems with complexities cannot be handled by linearized models, and nonlinear models take long evaluation time, which includes redundant parameter evaluations. In this work, the neural network modeling of nonlinear dynamical system using supervised learning technique and proper orthogonal decomposition has been addressed.

The work is validated using two numerical models. The biological neuron spiking and ring oscillator models are considered here. They are trained with multiple temporal swatches evolved on different parametric variation and trailed out at different time instances. Neural network tries to learn such models by analysing snapshot observations from active systems. Trained neural network is able to make better prediction for validating data which is different from the training data set. The complexity of the data set is reduced by applying the POD on the original data set. This demonstrates that the neural network is successfully utilised to macro model the autonomous dynamical system.

Keywords: Data science, machine learning, neural network, dynamical system, fitzhugh-nagumo model, ring oscillator model, finite difference method

**Corresponding author*: nagaraj.research@gmail.com

Kukatlapalli Pradeep Kumar, Aynur Unal, Vinay Jha Pillai, Hari Murthy, and M. Niranjanamurthy (eds.) Data Engineering and Data Science: Concepts and Applications, (85–136) © 2023 Scrivener Publishing LLC

5.1 Introduction

Data science is a methodical initiative that builds and organizes knowledge in the form of testable explanations and predictions [1]. Data science therefore implies the systematic study of the organization, properties, analysis of data and its role in inference, including the confidence in the inference. From an engineering system modeling perspective, traditional database models are found to be inadequate for knowledge discovery due to fast access and summarization of data.

Many physical world systems are represented by highly nonlinear dynamical system models. Modeling mathematically and simulating such nonlinear dynamical systems requires higher degrees of freedom in accurately describing data flow. A nonlinear dynamical system governed by partial differential equation (PDE) needs discretization to analyse in the domain of computation. Application of finite difference (FD) scheme in discretizing the system results in a set of nonlinear algebraic equations. Employing numerical techniques like Runge-Kutta (RK) or Newton-Raphson (NR), these equations are solved. If viewed in terms of processing time and memory, these methods are observed to be inefficient. Size and complexity of time variant/invariant system are significantly reduced using various model order reduction (MOR) techniques [2]. These techniques include Passive Reduced-Order Interconnect Macromodeling Algorithm (PRIMA) method, optimal Hankel norm technique, Arnoldi, Lauguerre and proper orthogonal decomposition (POD) [3].

Nonlinear system model order reduction exposes to totally distinct confrontations which have not experienced in linearizing the model. Due to large condition numbers, a wide range of time constants embracing different orders of dimension, nonlinear systems show a highly analogue nature [4]. POD has cost effectively and efficiently solved nonlinear problems overcoming the above challenges [5]. It has successfully reduced order of the system model describing the nonlinear behaviour. This method is also known as Karhune-Loève decomposition. POD has solved many applications in various mechanical and physical problems effectively since the 1990s [6]. Recently POD has also been applied in nonlinear system macromodeling [7]. However, nonlinear evaluation of the model found disadvantage due to significant CPU cost [8]. This has been overcome by developing a macromodel of a nonlinear system using data modeling in general and artificial neural network (ANN) in particular [9–12]. Neural network models are found to be promising in reducing memory and processing time with good accuracy.

Data Science has become popular in recent years. Data science models such as ANN are applied in various fields such as dynamical system modeling, control system modeling and signal processing [13] due to their high correlation capabilities and are widely appreciated for the same. Hence, it can be confidently assumed that area of application of machine learning also comprises of pattern recognition, dynamic system control application, data classification and function approximation [14, 15]. ANN is trained using input-output relationship set before deployment. For these reasons, a high-fidelity solution must be drawn for the nonlinear system that generates "true" modes. A cost function is minimized to solve for the optimization problem which in turn is the measure of suitability of the network prediction. The machine learning framework of the current work utilizes Bayesian Regularization [16]. It is popular for its accuracy. A data science model requires huge data to build the training set. This huge data can be constructed in two ways. One way is by redoing the experiment several times and collecting the data, which is expensive. Another acceptable way is generating the data from a Mathematical model constructed with less assumptions. A large data set required to construct an ANN model initially can be produced by the latter method. Once the ANN model is built, it is validated using experimental data set.

ANN learning models predict the system behaviour in all engineering fields and provide a promising approach to overcome modeling bottleneck. Learning is for demonstrating some cutting edge formalisms. Yet there exist additionally a few formalisms for which learning is as yet a test; especially, formalisms containing timing and hybrid signals are the bottleneck for model generation. The programmed learning of models for specialized frameworks could be the major step in macromodeling. Automatic learning of the model behaviour would alleviate a designer from the troublesome activity of building a complex model for an assignment being referred to. In the past few decades, there has been a solid resurgence in the field of ANNs including specialists from numerous differing disciplines, examining the reasonableness of AI methods, with the end goal of full-scale demonstration of complex dynamical systems utilizing a neural system [17–21].

By observing the advantages of POD in reducing the nonlinear model order and ANN in macromodeling of nonlinear algebraic equation, POD-ANN combination is generated to gain the benefit from both. Hence the POD-ANN reduced order macromodel of nonlinear complex systems has been developed as shown in Figure 5.1 [12, 17, 18].

The computational units of brain called neurons make over trillions of synaptic connections and hence form intricate networks. Each neuron

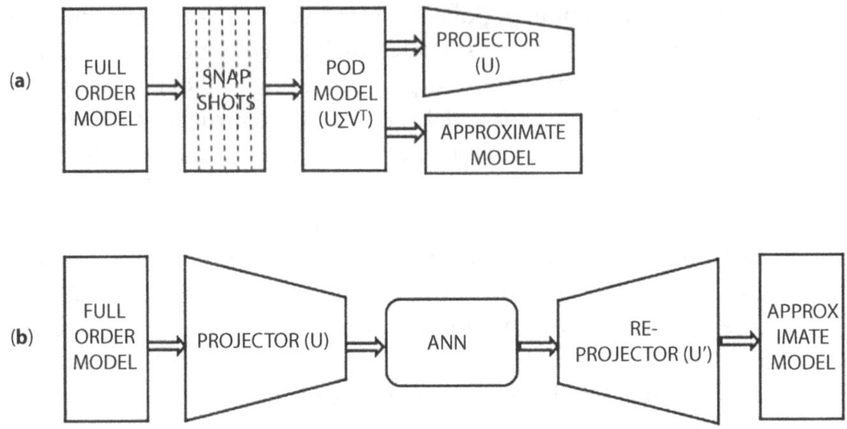

Figure 5.1 Block diagram of model. (a) Block diagram for constructing projector of the model. (b) Block diagram for constructing approximate model using projector.

encodes its internal state as membrane potential and spike (action potential). This is characterized as an atomic event where the membrane and action potentials' quick decay and rise with 5 to 10 ms occurrence. For inter-neuronal interactions, the significance of electric propagation on membrane of neurons is important. Going on, the neurotransmitter release is triggered by the propagation of action potential to the axon. This intern changes the post-synaptic neuron's membrane potential for synaptic interactions. This leads to interactions which are inhibitory or excitatory. These interactions are present as an underlying entity beneath the brain's many basic operations such as differentiation of the neuronal representation, amplification and synchronization [22]. Here, the FitzHugh-Nagumo system is found to be one of the normally used systems which explains the transmission conditions of node [23].

This parametrized PDE has bistable nonlinearity. This means that the nonlinear system has two stable states. This condition exists due to a single positive feedback loop in which under the proper circumstances, P motif activates Q and Q motif activates P or the equivalent double-negative feedback loop (P hinders Q and Q hinders P) create a bistable all-or-none switch [24]. This system is reduced using linearization, POD and POD-ANN methods to speed up the computation. The results of every method have been compared.

Voltage-controlled oscillators (VCOs) find a wide range of use in commercial applications because of their integrated nature. In various electronic systems, cross-coupled and ring oscillators are two essential categories of VCOs. In applications like disk-drive read channels [25] and clock recovery circuits [26], ring oscillators are employed. Similarly, in frequency synthesizers and function generators, cross-coupled oscillators are used.

For the development of microelectronics, the last few decades have been an exciting period. This watershed moment brings us to the very edge of implementing a complete system on a single chip at scales previously unimagined. Yet, as with most watershed moments, the implementation described above is not without a bottleneck. The main performance bottleneck is the challenge of wire placement which masks active device blinding speed. Modeling increasingly complicated effects is a major problem. Added to that, the complexity of computation needs to be sufficiently small for any proposed model in the design cycle which allows for many iterations.

It has been observed that an ANN model can be constructed when we have sufficient data to train the model. If the model is constructed with less data, then accuracy and reliability of the model is assured. In physical systems, data collection is very challenging and complex. In such cases, constructing the predictive model will be a great challenge. This chapter introduces the reader to ANN modeling of physical systems when less data is available and an appropriate mathematical model is known. Here, preliminary work of using ANN to assist modeling of nonlinear autonomous dynamic system has been presented. Data-driven supervised machine learning framework is utilised to learn the nonlinear behaviour of the autonomous dynamic system.

The present work is ordered as follows. Sections (5.2) and (5.3) narrate the very nature of Nonlinear autonomous dynamical system. Data science life cycle is explained in section (5.4). ANN modeling is explained in section (5.5). The method of extracting linearity of the nonlinear behaviour of FitzHugh-Nagumo system, Ring Oscillator system and VLSI Interconnect systems are explained in sections (5.6), (5.7) and (5.8), respectively. Section (5.9) details a macro-model using machine learning. In section (5.10), methodology followed for ANN modeling is listed. Finally numerical results obtained and conclusions of this work have been brought out in sections (5.11) and (5.12), respectively.

5.2 Nonlinear Autonomous Dynamical System

A nonlinear dynamic system has been described using the following state space model,

$$\dot{v}(t) = Av(t) + f(v(t)) + Bu(t),$$
$$y(t) = C^T v(t), \quad (5.1)$$

where $v(t) \in R^N$ is a state vector, $f: R^N \rightarrow R^N$ is a function of nonlinear vector values, A is $N \times N$ dimensioned system matrix, B is $N \times M$ dimensioned input matrix, C is $N \times K$ dimensioned output matrix, $y: R \rightarrow R^K$ is the output signal and $u: R \rightarrow R^M$ is an input signal.

When force input signal to the system is zero, the system is called autonomous system and is described by,

$$\dot{v}(t) = Av(t) + f(v(t)),$$
$$y(t) = C^T v(t). \quad (5.2)$$

5.3 Nonlinear System - MOR

In the current section, systems which are nonlinear [27] appearing in many applications of engineering are considered. Dynamical systems which are nonlinear state-space form are considered as follows:

$$\frac{d}{dt} f(x(t)) = g(x(t)) + Cu(t),$$
$$y(t) = L^T x(t). \quad (5.3)$$

Here $x \in \mathbb{R}^n$ is a "State vector" (Vector of unknowns), as in electrical or control engineering. "State variables" are $x(t)$. $C \in \mathbb{R}^{nxm}$, the input matrix and $L \in \mathbb{R}^{nxq}$ is the output matrix. Also $f, g: \mathbb{R}^n \rightarrow \mathbb{R}^n$. Presently, several works on MOR techniques for systems with nonlinearity are available. Quadratic method is one such technique [28]. It is also one of the simplest among them. The closest and most accurate comparison to the quadratic method is bilinearization technique [29, 30].

Variation analysis based methods [31–37], result in the occurrence of smaller errors as compared to the two previous methods. When strong nonlinearities are dealt, for nonlinear function, piece-wise linear approximation technique is preferred [38]. The above techniques provide results by extending moment matching technique or considering Balanced Truncation (BT)- MOR technique. Here using transfer function, accuracy of MOR is measured.

The abovementioned MOR techniques for nonlinear systems either do nonlinearity approximation with a lower degree polynomial [28–30, 33, 36] or lower degree polynomial weighted sum [38] or transforming system (3) to quadratic bilinear system [31, 32, 34, 35]. For reduced model, the function, $f(V\ z(t))$ computed directly, is replaced by bilinear approximation or polynomial evaluation. Here V is the transformation matrix and $z(t)$ is reduced order state variable.

5.3.1 Proper Orthogonal Decomposition

To reduce nonlinear systems, MOR techniques based on time domain and snapshot like reduced basis method and POD [39], evaluate $f(V\ z(t))$ directly. Projection bases are obtained from frequency or time domain snapshots in POD or Karhunen-Loève expansion [40–46]. More so using time-domain snapshots, reduced basis generation is inexpensive. Frequency domain snapshot generation is expensive. However, results in adequate bases and model extraction cost is reduced by spatial symmetries [46]. Very importantly, employing POD technique and connecting BT makes it possible to determine truncated approximate balanced realizations of original linear systems efficiently [46–48].

Galerkin projection-based POD finds innumerable applications, such as aerodynamics [49], computational fluid dynamics [50, 51], optimal control [52] and compressible flow [53]. To improve accuracy and efficiency of particular applications, modifications and extensions of POD are proposed. Chief among the modifications and extensions are related to handling gappy (reduced) data sets [54] for reconstruction of unsteady flow from limited data. Other important modifications and extensions include obtaining the placement location of optimal sensor where gappy POD [55] is proposed. For parabolic PDEs that are parameterized, extended POD with reduced bases approximation framework is proposed in [56] in combination with greedy sampling procedure. In aeronautic application for change in parameter sensitive solutions, Grassmann manifold tangent space interpolation techniques are proposed [57, 58] to construct robust and accurate reduced system of POD efficiently.

Construction of POD basis is effected by selection of snapshot in approximating solution of original space. They are specific to a few applications. In applications of optimal control, a way of avoiding dependence on selection of snapshot is suggested in [59]. [60] proposes adaptive sampling constrained to a model of choosing snapshots for parametric spaces of high dimensional large-scale systems. A compact POD with a goal-oriented framework for static systems having applications of optimizations employing snapshots of state vectors with respect to input parameters of the system is proposed by [61].

5.4 Data Science Life Cycle

Data modeling is constructed through its life cycle as illustrated in Figure 5.2.

5.4.1 Problem Identification

Problem identification is the most important step in data science life cycle. The problem may be to predict outcomes, show correlations, identify anomalies, identify patterns, or categorizing or grouping of data. The problem needs to be specific and conclusive. A machine learning algorithm is decided depending on the type of problem.

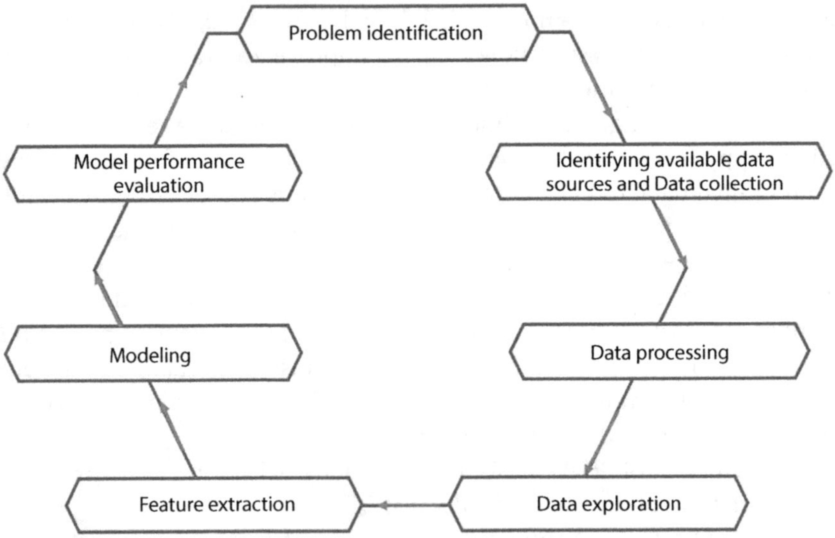

Figure 5.2 Data science life cycle.

5.4.2 Identifying Available Data Sources and Data Collection

Identifying data sources and collecting data is the most challenging task in data science. Some of the challenges with data collection are cost of data collection, mode of collection (manual, semi-automation, automation), data format, accuracy and reliability of data, duration of time in collection of data, data pre-processing.

5.4.3 Data Processing

Data set collected from different sources usually have unstructured data, corrupted data, incomplete data, duplicate data, incompletely formatted data. When data set collected from different sources get merged, may have duplications and mislabelled. Hence data need to be processed and cleaned, before being used for further analysis. In data processing, the data set is modified by appropriate missing data and removing irrelevant data.

5.4.4 Data Exploration

Exploratory Data Analysis (EDA) is the initial stage in data exploration. In this process, target and input data are differentiated depending on the requirement and manipulating available data sources. This is achieved by taking a broad look at patterns, trends, outliers, unexpected results and so on in existing data, using visual and quantitative.

5.4.5 Feature Extraction

Feature extraction is the process of obtaining dominant group of data set from raw data set by dimensionality reduction. To observe the dominant input variables, which have maximum influence on the behaviour of target variables, dimensionality reduction or order reduction is performed through any reduction techniques. Since most of the system shows non-linear behaviour, proper orthogonal decomposition is found to be a better technique among present methods.

5.4.6 Modeling

Based on principal component analysis, variables are identified. Using ANN, considering these variables, a system model is developed. A back propagating multilayer perceptron network is used which comprises one of each hidden, input and output layer. Each variable that describes attributes which are tested will correspond to input layer nodes. Also, class count is the same as neuron count in the output layer. Depending on training data amount and task complexity, neurons and hidden layer count is decided. Every neuron in output and hidden layer gets connected to all previous layer nodes in association with numerical weight, regulating signal magnitude passing between the two neurons. Employing supervised learning technique, neural network is trained.

5.4.7 Model Performance Evaluation

To indicate variation in the model, mean-square error (MSE) is employed, verifying model ability.

5.5 Artificial Neural Network in Modeling

Artificial neural networks (ANNs) [62] similar to their biological counterpart are systems for computation containing neurons that are information processing units, interconnected by weighted synapses. Specifically, the capacity to learn from experience is an attractive ANN feature [63]. There are applications that are difficult to be dealt with through computer algorithms employing rule-based programming. Chief among such applications are speech recognition, image identification and cluster detection. It is here that ANN is found to be most useful [64]. Its successful application to traditional contexts can be seen in the approximation of continuous function, which is a valid programming approach alternative in investigating uncertainties, nonlinearities and complexities found in higher order systems [65]. ANN learning is accomplished through a process of training, which exposes a network to a collection of samples and adjusting the weights accordingly, to provide good responses in situations of similarity.

5.5.1 Machine Learning

Machine learning (ML) deals with scientifically studying statistical models and algorithms used by computer systems in performing a specific task without the use of explicit instructions. Rather, ML relies on inference and patterns. ML can be best understood as a subset of Artificial Intelligence (AI). Algorithms in ML possess sample data called "training data". Such data are used to mathematically construct models. These models are used in making decisions or predictions for performing tasks without explicit programming [66]. ML algorithms and applications also include system modeling, where conventional algorithms fail to develop task performance effectively.

Different types of models are researched and used for/in ML systems. By exploring further, it is observed that biological neural network constituting brains of animals inspired systems, *viz.*, connectionist systems or ANNs are computing systems [67]. These systems "learn" to perform tasks without any task-specific rule programming.

5.5.2 Biological Neuron Model

A vertebrate's information processing system, also known as nervous system (NS), is broadly divided into central NS (CNS) and peripheral NS (PNS). While the CNS contains spinal cord and brain, the PNS comprises information transmitters called nerves. The nerves act as an interface between the organs and brain for collection and transmission of information. Going further, externally stimulated cells of sensor organs produce sensory signals in the form of electrical signals. These signals are transmitted to the CNS via sensory nerves. The information in the form of signal is managed, stored and processed to generate suitable motor signals by the effectors to concerned muscles for suitable action [67].

The direct involvement of CNS and PNS is found in the processing of information. Neurons, which are a large amount of modified cells, accomplish this at the cellular level as shown in Figure 5.3. Continuous communication between the processing elements is maintained by traversing through connections of thick net in the form of electric signals. For example, each neuron on an average is linked to 10^3-10^4 neurons in a human being. As explained in further discussions, neurons are characterised by

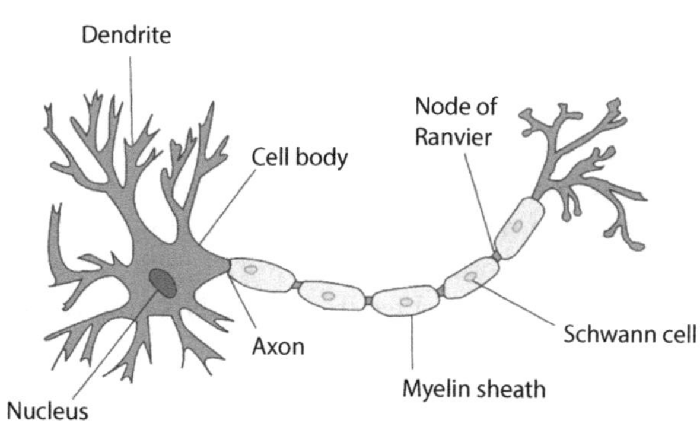

Figure 5.3 Illustration of a biological neuron [63].

a simple structure equipped for rapidly collecting incoming signals and generating an output pulse if accumulated input signal exceeds the action potential (threshold). In other words, if the function of a neuron is developed in terms of typical nonlinear input-output mapping, neurons act as switches [63].

Neural networks in biology familiarize themselves through the procedure of learning with specific class problems. Exposing networks to collective situations during the process of learning allows the possibility of deriving a set of tools that can provide reasonable solutions for similar incidents. Using the same logic, a generalisation of cognitive systems should be possible. Furthermore, against noisy inputs that are external errors, successfully trained neural networks show fault tolerance (at a discrete level). Cognitive capabilities as a whole are affected slightly in the case of a small group of neurons or a neuron that is dead or damaged. In such cases, risks of faulty transmission are taken care of by other processing nodes [63].

5.5.3 Artificial Neural Networks

Motivated by the processing system of biological information explained so far, neural network or simply ANN is a model which is computationally capable of learning from the data it has observed. This provides an alternate to the paradigm of algorithmic programming [65]. Quite similar to its biological counterpart, ANNs comprise artificial neurons that are an assortment of processing units and weighted guided synaptic network between the neurons. Following the direction dictated by synapses, data travels through the connections among neurons. With synapses as the oriented edges and neurons as nodes, for all practical purposes, ANN is an oriented graph which has been conjured as a network for a specific situation by adjusting the weights through training process [68].

5.5.4 Network Topologies

A network *topology* is defined by the design of neurons, i.e., its interconnection within a network. Various network architectures are available in literature, with a certain number of them available for custom-made specific tasks or applications. Two popular network topologies are investigated in the study: recurrent ANN and feedforward ANN are presented.

Feedforward ANN. In a feedforward ANN, arranging neurons (*perceptron*) [69] into layers, one for *input layer* of source neurons M_I, *hidden layers K*, computing neurons H_k ($k = 1, \ldots ,K$) and computing neurons M_O for *output layer*. Neurons have an inherent property in that neurons in a layer are connected to the next layer of neurons moving to the output layer but not in the same layer. Then $\mathbf{p} \in \mathbb{R}^{M_I}$, an *activation pattern* is applied from the source nodes to the network in the initial layer. In the first hidden layer neurons, the input signal is provided. Next, the input pattern is given to a successive layer from output signals for every hidden layer. The travel of information to the last layer, i.e., the output layer of network, constitutes the overall output components $\mathbf{q} \in \mathbb{R}^{M_O}$ in response to input pattern $\mathbf{p}$ of the network. A map between the *output space* $\mathbb{R}^{M_O}$ and *input space* $\mathbb{R}^{M_I}$ is established from the feedforward network, creating an effective network architecture. Such an architecture is particularly suitable for approximations and classification of continuous functions.

Based on the layers with trainable weights or equivalently total hidden layers, feedforward ANN is classified as Single-layer perceptrons (SLPs) and Multi-layer perceptrons (MLPs). In SLPs, hidden layer is absent and hence input layer is directly connected to the output layer. Feedforward networks can be classified according to the number of hidden layers or on the equivalent number of layers of trainable weights. SLPs consist of the input and output layer, without any hidden layer. As neurons in input layer have no information processing operations, the name SLP is used considering only the output layer. Hence, the SLPs range of application is limited due to their simple structure. For instance, binary activation function with unique output neuron and binary input vector is considered as a classifier, whereas an input space (unit hypercube) is split by the means of a hyper plane. Therefore, SLPs allow properly represented data to be uniquely and *linearly separable* [63]. Increase in space dimension, however, results in decrease of share of linearly separable sets, which make SLPs attractive.

It is observed that practically SLPs or two layer MLPs are sufficient for modeling most of the applications. In the above-discussed binary classifier, one hidden layer is sufficient for representing convex polygon. Additionally, two hidden layers are used to represent any arbitrary set. It must be noted that increasing the hidden layers for representation will not improve the network capability. However, practical advice or rules applicable to the number of training samples, number of hidden layers and neurons have had limited discussion in the literature.

Finally, it has been stated that synapses connect neuron pairs in contiguous layers. Recent development in the variants connect neurons within the same layer in *lateral connections* and skip layers in *shortcut connections*. However, perceptron's original definition is maintained in this work.

Recurrent ANN. Here, each neuron binds with different neurons, forbidding autosynapses (neuron output is the input of same neuron in subsequent time step). If every neuron is linked with the rest of the neurons, then the ANN is said to be *completely linked.* As a result, input and output neurons cannot be distinguished from one another. Thus the initial *network state* is the activation state set of all neurons in the network and representing network input. Similarly, final network state represents the output. Hence recurrent ANN communicates with its surrounding environment through the states of the neurons. Some of the examples of recurrent ANN are electrically charged particles within a magnetic field behaviour inspired by Hopfield networks [70], cluster detection, hyperspectral data classification using self-organizing maps [71].

5.5.4.1 NARX Neural Network

In the present work, chaotic time series are dealt with proposed architectural approaches which are based on "Nonlinear Autoregressive models with eXogenous input (NARX model)" [67, 72, 73]. Modeling time series and systems with nonlinearity, this model is well suited and powerful. Dynamical NARX neural network has important applications in control systems. In light of the same a few advantages of NARX with gradient descent algorithm are noted: (1) More effective learning happens in NARX networks with gradient descent than in other ANN (2) Convergence is faster with better generalization compared to other ANNs. Compared to recurrence neural networks that are conventional, NARX networks are better to discover long time dependences. Thus, the benefits of back-propagation-through-time (BPTT) algorithm enjoys when they are calculated using gradients and why long-term dependences are benefitted by output delays are explained in [72, 73].

5.5.5 ANN Modeling Using Mathematical Models

Order of the Systems governed by nonlinear differential equations is significantly reduced by POD techniques [5]. Evaluating nonlinear polynomial portion has a disadvantage in the form of remarkable CPU cost [74]. ANN-based methods have been proven to be an effective alternative approach in the literature [9–11]. These methods are shown to be efficient at approximating static systems. Modeling of complex dynamic systems using ANN is achieved by using complex ANN architecture supported by robust training process [12].

Here ANNs are referred for approximating parameters linking of parameterized PDEs with corresponding coefficients in the solution of reduced basis. Thus, research depends on collecting samples obtained through

POD applied on high-fidelity model in accomplishing this task. In present work, autonomous nonlinear dissipative system described by equation (2) has been explained through Neuron spiking model using FitzHugh-Nagumo system, Ring Oscillator system and VLSI Interconnect system using Telegraph equation.

5.6 Neuron Spiking Model Using FitzHugh-Nagumo (F-N) System

The activation and deactivation dynamics of spiking neuron is constructed using Hodgkin-Huxley (HH) model. The FitzHugh-Nagumo system is a simplified version of the HH model. The model with voltage υ and recovery of voltage ω is given by:

$$\begin{aligned} \varepsilon\upsilon_t(x,t) &= \varepsilon^2\upsilon_{xx}(x,t) + f\upsilon(x,t) - \omega(x,t) + c, \\ \omega_t(x,t) &= b\upsilon(x,t) - \gamma\omega(x,t) + c. \end{aligned} \tag{5.4}$$

Function $f(\upsilon) = \upsilon(\upsilon - 0.1)(1 - \upsilon)$ with initial condition: $\upsilon(x,0) = 0, \omega(x,0) = 0, x \in [0,L]$, and boundary condition: $\varepsilon\upsilon_t(0,t) = -i_0(t)$, $\varepsilon\upsilon_t(L,t) = 0, t \geq 0$, considering the parameters $c = 0.05, \gamma = 2, b = 0.5, \varepsilon = 0.015, L = 1$. The stimulus is $i_0(t) = 50000t^3 exp(-15t)$. To solve the PDE equation (5.4), it needs to be discretised using any of the discretising methods. Here, the central FD method is used spatially with the forward Euler time integration to discretise the above equation.

5.6.1 Linearization of F-N System

F-N is an *Autonomous dynamic system* because of its implicit time dependency and *Nonlinearity*. Linear approximation of the system over the equilibrium point is constructed to get the low-cost solution in considerably less time. It is well known from the Hartman-Grobman theorem [75] that a nonlinear system close to its equilibrium point is a "topological conjugate", i.e., geometrically similar to a linear system. Thus, the F-N system (5.4) at equilibrium state is written as:

$$\begin{aligned} E\upsilon' &= \varepsilon^2\upsilon_{xx} + f(\upsilon) - \omega + c, \\ \omega' &= b\upsilon - \gamma\omega + c. \end{aligned} \tag{5.5}$$

The above equation (5.5) is renamed as:

$$F(V,W) = \varepsilon\upsilon' = \varepsilon^2\upsilon_{xx} + f(\upsilon) - \omega + c,$$
$$G(V,W) = \omega' = b\upsilon - \gamma\omega + c. \tag{5.6}$$

By solving the equations $F(V, W) = 0$ and $G(V, W) = 0$ simultaneously, the steady state solution is written as (V_0, W_0). Considering a new point $V = V_0 + p$, $W = W_0 + q$, p and q emerge as the smallest positive numbers for which linear approximation is nearly equal to nonlinear system.

By the *Taylor series expansion of two variables*, the expression (5.6) is written as:

$$V' = p\frac{\partial F(V_0,W_0)}{\partial V} + q\frac{\partial F(V_0,W_0)}{\partial W},$$
$$W' = p\frac{\partial G(V_0,W_0)}{\partial V} + q\frac{\partial G(V_0,W_0)}{\partial W}.$$

As p and q are considered very small, the higher order terms $O(h^2,k^2,hk)$ are neglected. The linear equation in matrix form is written as:

$$\mathbf{V}^0 = \mathbf{A}\mathbf{V}. \tag{5.7}$$

Where $\mathbf{V} = (V, W)^T$ and the matrix $\mathbf{A}$ is given by:

$$A = \begin{pmatrix} \frac{\partial F(V_0,W_0)}{\partial V} & \frac{\partial F(V_0,W_0)}{\partial W} \\ \frac{\partial G(V_0,W_0)}{\partial V} & \frac{\partial G(V_0,W_0)}{\partial W} \end{pmatrix}$$

The values of **A** are given in section 3.6.3. The cost of evaluation of the system is still reduced using linear reduced order models developed using any linear system MOR techniques.

5.6.2 Reduced Order Model of Linear System

For reducing the linear system, the Arnoldi method is usually used to get better linear approximation. The Linear system MOR is derived by transforming the system (5.7) to:

$$\mathbf{A}^{-1}\mathrm{V}' = \mathbf{V}, \tag{5.8}$$

then reducing equation (5.8) by the system Arnoldi as explained in [76] and finally the reduced system is transformed back to be with the same form as equation (5.7).

It is observed that the linear system MOR is 200 times smaller than the original system and approximates the original linear system accurately. Most of the linear system MOR techniques including the Arnoldi are based on moment matching, which always match the steady state performance perfectly. Hence, linear system MOR techniques are used for nonlinear system analysis whose characteristics are actually close to linear systems in specific conditions.

5.6.3 Finite Difference Discretization of F-N System

Let the discretization of a spacial domain with step size Δx be $x_i = i\Delta x$ with $x_0 = 0$ and $x_{n+1} = L$ for $i = 0,1,...,n+1$. The discretization of time domain with step size Δt, for $j = 0,1,...,t$ is given by $t_j = j\Delta t$. At the mesh point (x_i,t_j) of discretised system, let the solution be $\omega(x_i,t_j)$ and $\upsilon(x_i,t_j)$ for $j = 0,1,...,t$ and $i = 0,1,\ldots,n+1$.

$$\varepsilon\left(\frac{\upsilon_i^{j+1}-\upsilon_i^j}{\Delta t}\right)=\varepsilon^2\left(\frac{\upsilon_{i-1}^j-2\upsilon_i^j+\upsilon_{i+1}^j}{\Delta x^2}\right)+f(\upsilon_i^j)-\omega_i^j+c,\frac{\omega_i^{j+1}-\omega_i^j}{\Delta t}=b\upsilon_i^j-\gamma\omega_i^j+c. \tag{5.9}$$

Initial conditions are: $\forall\ i = 1,....,n + 1$, $\upsilon_i^0 = 0$ and $\omega_i^0 = 0$.

Conditions at the boundary are:

$\dfrac{\upsilon_1^j-\upsilon_0^j}{\Delta x} = -i_0(t_j) \Rightarrow \upsilon_0^j = \upsilon_1^j + \Delta x i_0(t_j)$ and $\dfrac{\upsilon_{n+1}^j-\upsilon_N^j}{\Delta x} = 0 \Rightarrow \upsilon_{n+1}^j = \upsilon_N^j$, for all $j = 1,....,t$. That is,

$$\frac{\upsilon_0^j - 2\upsilon_1^j + \upsilon_2^j}{\Delta x^2} = \frac{(\upsilon_1^j + \Delta x i_0(t_j)) - 2\upsilon_1^j + \upsilon_2^j}{\Delta x^2} = \frac{-\upsilon_1^j + \upsilon_2^j}{\Delta x^2} + \frac{i_0(t_j)}{\Delta x},$$
$$\frac{\upsilon_{n-1}^j - 2\upsilon_n^j + \upsilon_{n+1}^j}{\Delta x^2} = \frac{(\upsilon_n^j + \Delta x i_0(t_j)) - 2\upsilon_n^j + \upsilon_{n+1}^j}{\Delta x^2} = \frac{-\upsilon_n^j + \upsilon_{n+1}^j}{\Delta x^2} + \frac{i_0(t_j)}{\Delta x}. \tag{5.10}$$

Let $\mathbf{v}^j = \left[\upsilon_0^j, \upsilon_1^j, \ldots \upsilon_n^j\right]^T \in \mathbb{R}^n$, $\mathbf{w}^j = \left[\omega_0^j, \omega_1^j, \ldots \omega_n^j\right]^T \in \mathbb{R}^n$ and $\mathbf{y}^j = \begin{bmatrix} \hat{\mathbf{v}}^j \\ \hat{\mathbf{w}}^j \end{bmatrix} \in \mathbb{R}^{2n}$ Then for $j = 0,1,2,\cdots,t$ and $\mathbf{y}^0 = 0$, the FD discritized full order system is given by:

$$\mathbf{y}^{j+1} - \mathbf{y}^j = \frac{1}{\mathbf{Q}}[-\mathbf{A}\mathbf{y}^j + \mathrm{h}(t_j) + \mathbf{F}(\mathbf{y}^j)], \tag{5.11}$$

$$\mathbf{Q} = \frac{1}{\Delta t}\begin{bmatrix} \varepsilon\mathbf{I} & 0 \\ 0 & \varepsilon\mathbf{I} \end{bmatrix} \in \mathbb{R}^{2n\times 2n},\ \mathbf{A} = \begin{bmatrix} \mathbf{L} & \mathbf{I} \\ -b\mathbf{I} & \gamma\mathbf{I} \end{bmatrix} \in \mathbb{R}^{2n\times 2n},\ \mathbf{L} = \frac{\varepsilon^2}{\Delta x^2}\begin{pmatrix} 1 & -1 & & & \\ -1 & 2 & -1 & & \\ & \ddots & \ddots & \ddots & \\ & & -1 & 2 & -1 \\ & & & -1 & 1 \end{pmatrix} \in \mathbb{R}^{n\times n},$$

$$\mathbf{I} = \text{Identity matrix} \in \mathbb{R}^{n\times n} \quad \mathbf{h}(t) = \frac{\varepsilon^2}{\Delta x}\begin{bmatrix} \mathbf{h}_0(t) \\ 0 \end{bmatrix} + c \in \mathbb{R}^{2n}, \text{with } \mathbf{h}_0(t) = \begin{bmatrix} i_0(t) \\ 0 \\ \vdots \\ 0 \end{bmatrix} \in \mathbb{R}^n,$$

$$\mathbf{c} = c\begin{bmatrix} 1 \\ 1 \\ \vdots \\ 1 \end{bmatrix} \in \mathbb{R}^{2n},\ \mathbf{F}(\mathbf{y}^j) = \begin{bmatrix} f(\mathbf{v}^j) \\ 0 \end{bmatrix} \in \mathbb{R}^{2n},\ \text{with } f(\mathbf{v}^j) = \begin{bmatrix} f(\upsilon_1^j) \\ \vdots \\ f(\upsilon_n^j) \end{bmatrix} \in \mathbb{R}^n, j = 0,1,2,\ldots.$$

5.6.4 MOR of F-N System Using POD-Galerkin Method

From the *snapshot* set, the MOR model is constructed using POD basis. Here the *snapshot* is defined as a numerical solution of equation (5.4) at time 't'.

Let $\mathbb{V}$ and $\mathbb{W}$ be the *snapshot* matrices defined at times $t_1, \ldots, t_{n_s}$. Here, n_s is defined as the total number of snapshots. $\mathbb{V}$ and $\mathbb{W}$ are defined by:

$$\mathbb{V} = \begin{bmatrix} | & & | \\ \mathbf{v}^1 & \cdots & \mathbf{v}^{n_s} \\ | & & | \end{bmatrix} \in \mathbb{R}^{n \times n_s}, \mathbb{W} = \begin{bmatrix} | & & | \\ \mathbf{w}^1 & \cdots & \mathbf{w}^{n_s} \\ | & & | \end{bmatrix} \in \mathbb{R}^{n \times n_s}. \tag{5.12}$$

When POD is applied to the snapshot matrix $\mathbb{V}$, the matrix is decomposed as:

$$\mathbb{V} = \mathbf{U}_\upsilon \Sigma \Upsilon^T.$$

The POD basis matrix for $\mathbb{V}$ is denoted by $\mathbf{U}_\upsilon \in R^{n\times k}$, where k is the number of largest singular values considered for POD basis and $k \leq min\{rank(\mathbb{V}),rank(\mathbb{W})\}$. Similarly, denoting the basis matrix of POD for $\mathbb{W}$ by $\mathbf{U}_\omega \in \mathbb{R}^{nxk}$.

The system and solution of MOR for discretized FD are acquired by projecting on the POD basis matrix **U**, which is defined by:

$$\mathbf{U} = \begin{bmatrix} \mathbf{U}_\upsilon & 0 \\ 0 & \mathbf{U}_\omega \end{bmatrix} \in \mathbb{R}^{2n\times 2k}. \tag{5.13}$$

The Galerkin projection is applied to equation (5.11) and $\mathbf{y}^j$ of the full order system is substituted with $\mathbf{U}\mathbf{y}^j$. b Equation (5.11) is rewritten in reduced order form with $\mathbf{y}^0 = 0$, as: b

$$\hat{\mathbf{y}}^{j+1} - \hat{\mathbf{y}}^j = \frac{1}{\mathbf{U}^T\mathbf{Q}\mathbf{U}}[-\mathbf{U}^T\mathbf{A}\mathbf{U}\hat{\mathbf{y}}^j + \mathbf{U}^T\mathbf{h}(t_j) + \mathbf{U}^T\mathbf{F}(\mathbf{U}\hat{\mathbf{y}}^j)]. \tag{5.14}$$

Hence, the reduced order model of F-N system using POD-MOR with = 0, is given by:

$$\hat{\mathbf{y}}^{j+1} - \hat{\mathbf{y}}^j = \frac{1}{\hat{\mathbf{Q}}}[-\hat{\mathbf{A}}\hat{\mathbf{y}}^j + \hat{\mathbf{h}}(t_j) + \hat{\mathbf{F}}(\hat{\mathbf{y}}^j)]. \tag{5.15}$$

Where

$$Q_b = U^T QU = \frac{1}{4t}\begin{bmatrix} ^{\varepsilon}1_k & 0 \\ 0 & ^{\varepsilon}1_k \end{bmatrix} \quad 2R^{2k-2k}, \quad 1_k = \text{Identify matrix } 2R^{k-k},$$

$$A_b = U^T AU = \frac{1}{4t}\begin{bmatrix} L & 1_k \\ -bl_k & yl_k \end{bmatrix} \quad 2R^{2k-2k}, \quad h^b(t_j) = U^T h(t_j), F^b(by^j) = U^T F(Uby^j).$$

5.7 Ring Oscillator Model

A typical ring oscillator shown in Figure 5.4 consists of a feedback loop connected to digital-inverters, which are odd in number. Digital inverters having combination of Low Pass Filter (LPF) and nonlinearity of cascaded *tanh*(.) function in the proposed model are shown in Figure 5.5. Nonlinear term in the model is expressed as $V_n = V_0\, tanh(V_i/V_s)$. Here, the internal slope adjustment is V_s, input voltage is V_i and the saturation voltage is V_0 [77]. Thus, the behaviour of the inverter is modeled as:

$$V_0 + TV'_0 = -V_{sat} tan(V_i/V_s).$$

In state-space, the model would be formulated as follows:

$$\dot{V}_0 = -1/T * V_0 - V_{sat} * V_i / T * tan(V_i / V_s),$$

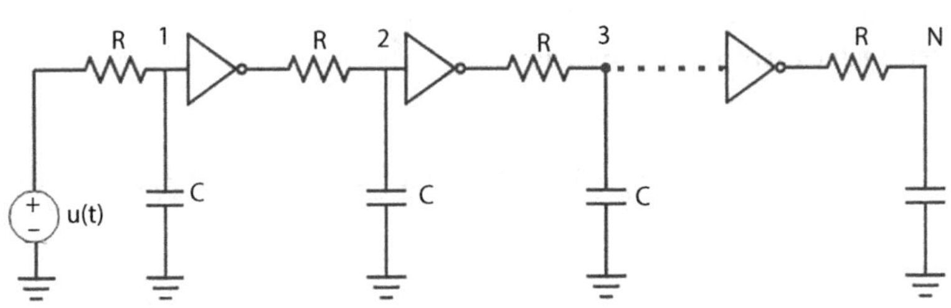

Figure 5.4 Inverter chain circuit.

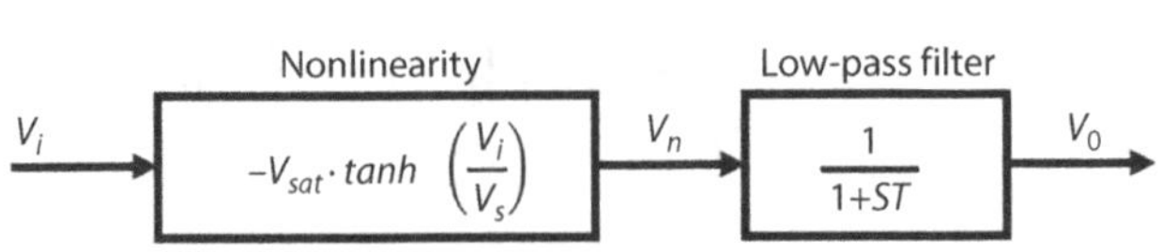

Figure 5.5 Low pass filter with nonlinearity.

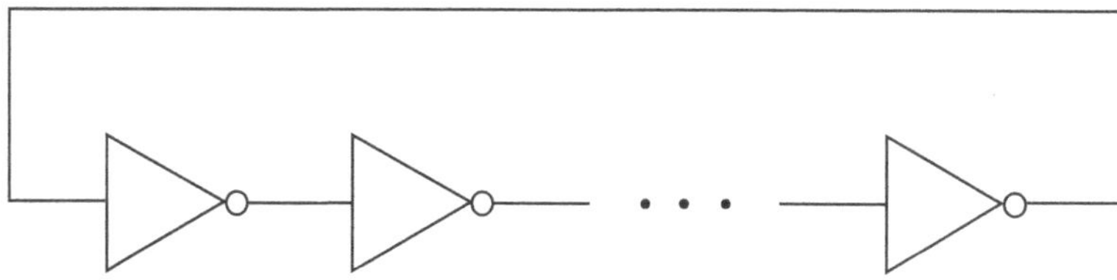

Figure 5.6 n-Stage feedback model for ring oscillator [7, 8].

The above Figure 5.6 represent the n-Stage feedback model for ring oscillator.

$$\dot{x} = -\eta x - \beta tanh(\alpha x),$$

where $\eta = 1/T, \alpha = 1/V_s, \beta = V_{sat} * V_i/T$ as shown in Figures 5.7 and 5.8, expressing system model in the feedback loop form as follows:

$$\dot{x}_1 = -\eta_1 x_1 - \beta_1 tanh(\alpha_1 x_n),$$

$$\dot{x}_2 = -\eta_2 x_2 - \beta_2 tanh(\alpha_2 x_1),$$

$$\dot{x}_3 = -\eta_3 x_3 - \beta_3 tanh(\alpha_3 x_2),$$

$$\dot{x}_n = -\eta_n x_n - \beta_n tanh(\alpha_n x_{n-1}). \tag{5.16}$$

All n blocks have a negative gain. Therefore, for an odd number n, interconnection is a negative feedback system [78]. For limit cycle existence, the necessary condition is global stability, which is essential under systems of negative feedback [79]. Further, transformation with n as odd number, $D = diag(d_1, \cdots, d_n)$ with $d_i = (-1)^{i+1}$, a compact form is obtained. Here n identical oscillator characteristic models are considered. Denoting coupling between the oscillator models by input $u(t)$, $\alpha_1 = \alpha_2 = \cdots = 5$ as internal slope factor inverse, $T = RC$ as time constant of circuit, $R_1 = R_2 = \cdots = 1k\ \Omega$

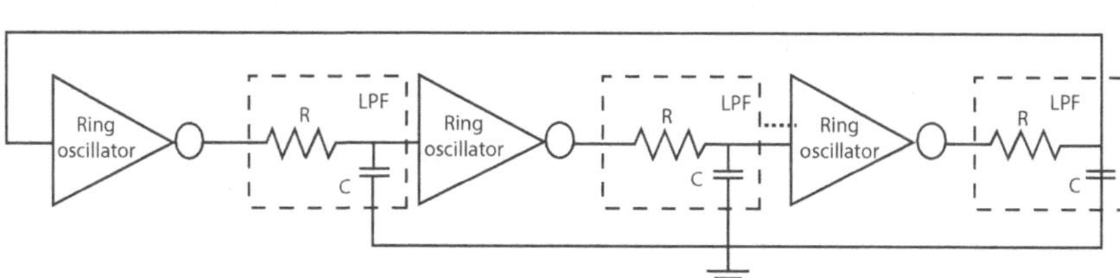

Figure 5.7 Model of ring oscillator with LPF.

and $C_1 = C_2 = \cdots = 2\mu F$, would mean that ring oscillator stages are similar. Formulating oscillator model as:

$$\dot{x}(t) = -\beta \tanh(\alpha x(t)) - \eta x(t) + u(t). \tag{5.17}$$

Nonlinear ring oscillator model simulation in capturing slow variation transient part is challenging as it takes a longer period of time. Further, it is stated in a linearized form applied to cyclic systems that solutions at sustained point or periodic orbit is described by $\omega-$ limit [78].

5.7.1 Model Order Reduction of Ring Oscillator Circuit

A nascent field in ROM research is the MOR technique used for nonlinear systems. Here, it makes sense in minimizing the simulation time and improving the criteria of $n-$stage ring oscillator design for high computational costs. Some numerical computation resources will take care nonlinearity, tanh(.) which shows unpredictable behaviour. The formulation of the standard model is given below:

$$\begin{aligned} \dot{x}(t) &= f(x) + \mathrm{A}x(t) + \mathrm{B}u(t), \\ y(t) &= \mathrm{C}x(t). \end{aligned} \tag{5.18}$$

In equation (5.18), $f(x) = -\beta * \tanh(\alpha x(t))$ is the nonlinear term [80]. The error approximation and accuracy of the adopted methodology for MOR are explained in subsequent sections.

5.7.2 Ring Oscillator Circuit Approximation Using Linear System MOR

The evaluation of the linearized ring oscillator equation has substantially reduced the computational cost. Additional reductions in such models lead to a noticeable reduction in the computational costs. One of the implicit moment matching methods, called "Arnoldi method", [76] has been used in this work, to reduce the linear approximation. Using the projection basis V [76], a reduced order model of linear system is constructed as:

$$X = V\,Z, \tag{5.19}$$

where $Z(t) = [z_1(t),...,z_{r-1}(t)]^T$, r is the reduced order and projection basis $V \in \mathbb{R}^{Nxr}$.

Substituting equation (5.19) in equation (5.17), a reduced order linear model is given by:

$$\frac{dz}{dt} = V^T J_F VZ + V^T (fX_0 - J_F X_0) \tag{5.20}$$

It is to be noted here that the reduced order linear model utilizes least computation cost, compromising the accuracy of the model. Thus the optimum model which gives good accuracy with less computation time is given by the combination of machine learnt neural network model and a reduced order linear model.

5.7.3 POD-ANN Macromodel of Ring Oscillator Circuit

The Hilbert space basis dimension is reduced by retaining original system accuracy employing POD. Fundamentally, it is similar to the SVD of a Euclidean space. The ring oscillator model attributes contained by the premise components are extricated proficiently by POD. In fact it indicates a snapshot matrix $\{x_1......x_n\} \subset \mathbb{R}^n$ in Euclidean space. By noting the system response called snapshots that are samples of trajectories, for different instances of time t_i, the snapshot matrix is optimally constructed.

Taking y=span $\{x_1......x_n\} \subset \mathbb{R}^n$ and system order as $N = \dim(y)$, space y is approximated optimally by the linear span of $\{\phi\}_{i=1}^{q} \subset \mathbb{R}^n$, orthonormal vector set of dimension q of reduced basis (for $q \ll N$). By the problem of minimization, optimality is taken into consideration as below:

$$min \sum_{j=1}^{N} \left\langle \left\| V(x,t_i) - \sum_{j=1}^{N} \phi_k(t_i)\alpha_k(x) \right\|^2 \right\rangle \quad i = 1,....N, \tag{5.21}$$

where L_2 - norm is $||\cdot||^2$ and total observation average is $\langle\cdot\rangle$.

Employing SVD to $V = [x^1,...x^{ns}] \in \mathbb{R}^{nxn_s}$, snapshot matrix, U, the left singular vector is realized. By solving the nonlinear model (5.21) such that,

$$U\Sigma W^T = V, \tag{5.22}$$

where $U = [u_1,...u_q] \in \mathbb{R}^{nxq}$ is right orthogonal vector, $W = [w_1,...w_q] \in \mathbb{R}^{n_s xq}$ is left singular orthonormal basis vector, Σ =diag$(\sigma_1,....,\sigma_q) \in \mathbb{R}^{qxq}$ is with $\sigma_1 > \sigma_2 .... > \sigma_q > 0$, $W \in \mathbb{R}^{n_s xq}$ and $rank(V) \leq (q,n_s)$. Optimal solution u_i is obtained by minimising the POD basis, as presented in equation (5.21), where $i = 1....q$. By approximating the snapshots using the above method, minimum 2-norm error is given by:

$$\sum_{i=1}^{n_s} \left\| y_i - \sum_{j=1}^{K} (Y_i^T \upsilon_j)\upsilon_j \right\|_2^2 = \sum_{j=k+1}^{r} \sigma_j^2. \tag{5.23}$$

In [81] on POD basis further discussion is available. POD together with snapshot ensemble in conjunction with POD-ANN provides comparatively better approximation.

It is to be noted here that, although expressing equation (5.15) is reduced on the (POD) basis, the nonlinear term computation complexity still depends on dimension n of the full FD system. This nonlinear computation in each iteration substantially increases the computation cost. Hence POD-ANN method is introduced which gives more accurate results with low computation costs.

5.8 Nonlinear VLSI Interconnect Model Using Telegraph Equation

An infinitesimal piece of interconnect system is shown in Figure 5.8. Insulating the cable with imperfection, a capacitance and conductance with respect to leakage current are assumed. Hence, let v be the voltage applied at a distance x from one end of the cable at time instant t. Representing the current flow at x as i for time t, the current and voltage at time t at $x + dx$ are denoted as $i + \Delta i$ and $v + \Delta v$ respectively. The telegraph equation describing the interconnect model is discretized into infinitesimal parts of length dx. Parameters of unit length of resistance (R), capacitance (C), inductance (L) and conductance (G) are assumed. Here, Ohm's law across RLC is applied, resulting in:

$$\upsilon_R = iR, \quad \upsilon_L = L\frac{di}{dt}, \quad \upsilon_C = \frac{1}{C}\int i_c dt. \tag{5.24}$$

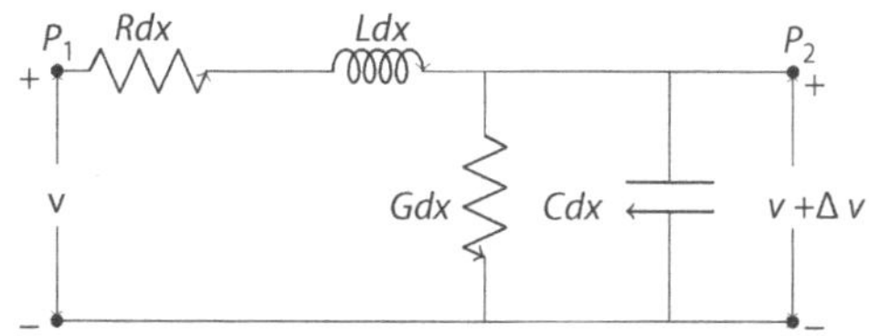

Figure 5.8 Circuit diagram of interconnect model.

The sum of voltage at point P_1 and voltage drop across points P_1-P_2 is equal to the voltage at point P_2 and is written as:

$$(\upsilon + \Delta\upsilon) - \upsilon = -\left[iRdx + Ldx\frac{\partial i}{\partial t}\right] \tag{5.25}$$

5.8.1 Macromodeling of VLSI Interconnect

Partially differentiating equation (5.25) with respect to *x* and also observing *dx* tending to zero, the equation is reduced to:

$$\frac{\partial \upsilon}{\partial x} = -\left[L\frac{\partial i}{\partial t} + Ri\right]. \tag{5.26}$$

It is also observed that the current leakage to the ground is equal to the current passing through points P_1 and P_2. Hence leakage current is given by:

$$(i + \Delta i) - i = -i_c dx - (Gdx)v. \tag{5.27}$$

The current to the ground through the capacitor from equation (5.24) is reduced to:

$$i_c = C\frac{\partial \upsilon}{\partial t}. \tag{5.28}$$

Differentiating equations (5.25) and (5.28) with respect to *t* and *x*, followed by eliminating derivatives of *v*, gives:

$$\frac{\partial^2 i}{\partial x^2} = (RG)i + (LG + RC)\frac{\partial i}{\partial t} + LC\frac{\partial^2 i}{\partial t^2}. \tag{5.29}$$

Similarly, differentiating equations (5.25) and (5.28) with respect to t, x followed by eliminating derivatives of i, gives:

$$\frac{\partial^2 \upsilon}{\partial x^2} = (RG)\upsilon + (LG + RC)\frac{\partial \upsilon}{\partial t} + LC\frac{\partial^2 \upsilon}{\partial t^2}. \tag{5.30}$$

Equations (5.29) and (5.30) are addressed as single dimensional second order hyperbolic Telegraph equations. Considering the nonlinear terms [82], the Telegraph equation is presented as:

$$LC\frac{\partial^2 \upsilon}{\partial x^2} = \frac{\partial^2 \upsilon}{\partial t^2} + 2\frac{\partial \upsilon}{\partial t} + \upsilon^2 - e^{2x-4t} + e^{x-2t}. \tag{5.31}$$

5.8.2 Discretisation of Interconnect Model

For full order numerical evaluation of system (31), the system is semi-discretized spacially, using FDM as $v_i = i\Delta x$ and semi discretized in time as $v^j = j\Delta t$. Hence,

$$\frac{\partial^2 \upsilon}{\partial x^2} = \frac{\upsilon_{i+1}^j - 2\upsilon_i^j + \upsilon_{i-1}^j}{(\Delta x)^2}, \quad \frac{\partial^2 \upsilon}{\partial t^2} = \frac{\upsilon_i^{j+1} - 2\upsilon_i^j + \upsilon_i^{j-1}}{(\Delta t)^2}, \quad \frac{\partial \upsilon}{\partial t} = \frac{\upsilon_i^{j+1} - \upsilon_i^{j-1}}{2(\Delta t)},$$
$$v^2 = (v_i^{j+1})^2, e^{2x-4t} = e^{2(i+\Delta x)-4(j+\Delta t)}, e^{x-2t} = e^{(i+\Delta x)-2(j+\Delta t)}. \tag{5.32}$$

Substituting equation (5.32) in telegraph equation (5.31) gives:

$$LC\frac{\upsilon_{i+1}^j - 2\upsilon_i^j + \upsilon_{i-1}^j}{(\Delta x)^2} = \frac{\upsilon_i^{j+1} - 2\upsilon_i^j + \upsilon_i^{j-1}}{(\Delta t)^2} + 2\left(\frac{\upsilon_i^{j+1} - \upsilon_i^{j-1}}{2(\Delta t)}\right) - (\upsilon_i^{j+1})^2 + e^{2(i+\Delta x)-4(j+\Delta t)} + e^{(i+\Delta x)-2(j+\Delta t)}. \tag{5.33}$$

The system (5.33) is a nonlinear full order inter-connect model based on Telegraph equation, which is solved by integrating the time and space variables.

5.8.3 Linearization of VLSI Interconnect Model

The full order system (5.33) is written in the form:

$$F(V)=LC\frac{\upsilon_{i+1}^{j}-2\upsilon_{i}^{j}+\upsilon_{i-1}^{j}}{(\Delta x)^{2}}-\frac{\upsilon_{i}^{j+1}-2\upsilon_{i}^{j}+\upsilon_{i}^{j-1}}{(\Delta t)^{2}}-2\left(\frac{\upsilon_{i}^{j+1}-\upsilon_{i}^{j-1}}{2(\Delta t)}\right)+(\upsilon_{i}^{j+1})^{2}-e^{2(i+\Delta x)-4(j+\Delta t)}-e^{(i+\Delta x)-2(j+\Delta t)}$$

(5.34)

(5.34) Equation (5.34) is linearized to achieve faster approximation. The linear model using the Taylor series expansion of the system is given by:

$$L(V)=V^{(0)}-\frac{J_{F}^{(0)}(V-V^{(0)})}{F(V^{(0)})}+\mathcal{O}(\Delta t^{2}), \tag{5.35}$$

where $J_F^{(0)}$ is the 0th time level Jacobian matrix of F evaluated at $V^{(0)}$ and is given by:

$$J_{F}^{(0)}=\begin{pmatrix}\left(\frac{\partial f_{1}}{\partial \upsilon_{1}}\right)^{(0)} & \left(\frac{\partial f_{2}}{\partial \upsilon_{2}}\right)^{(0)} & \cdots & \left(\frac{\partial f_{1}}{\partial \upsilon}\right)^{(0)}\\ \vdots & \vdots & & \vdots\\ \left(\frac{\partial f_{N-1}}{\partial \upsilon_{1}}\right)^{(0)} & \left(\frac{\partial f_{N-1}}{\partial \upsilon_{2}}\right)^{(0)} & \cdots & \left(\frac{\partial f_{N-1}}{\partial \upsilon_{N-1}}\right)^{(0)}\end{pmatrix}.$$

Hence the Jacobian matrix of the system (5.34) is given by:

$$J_{F}^{(0)}=\left(\frac{1}{(\Delta t)^{2}}+\frac{1}{\Delta t}+\frac{LC}{(\Delta x)^{2}}\right)I+2VI.$$

Where I is the identity matrix and $V = (v_1, v_2 \ldots v_{n-1})$. Hence the linear model is given by:

$$L(V)=V^{(0)}-\frac{\left(\left(\frac{1}{(\Delta t)^{2}}+\frac{1}{\Delta t}+\frac{LC}{(\Delta x)^{2}}\right)I+2VI\right)(V-V^{(0)})}{F(V^{(0)})}. \tag{5.36}$$

5.8.4 Reduced Order Linear Model of VLSI Interconnect

The computational cost of VLSI Interconnect model using Telegraph equation is noticeably reduced by working on the order of linear model using the Arnoldi method as discussed in earlier works [76]. Using the projection basis W [76], the reduced order model of linear system (5.36) is constructed as

$$V = WZ, \tag{5.37}$$

where $Z(t) = [z_1(t),...,z_{r-1}(t)]^T$, r is the reduced order and projection basis $W \in R^{N\times r}$. Substituting equation (5.37) in equation (5.36), the reduced order linear model is given by:

$$\frac{dz}{dt} = W^T J_F WZ + W^T(fV_0 - J_F V_0). \tag{5.38}$$

It must be noted that the reduced order linear model involves the least computation cost, compromising the accuracy of the model. To derive the optimum model which offers good accuracy with less computation time is given by a combination of machine learnt neural network model and reduced order linear model.

5.9 Macromodel Using Machine Learning

The elementary structure of the basic feed forward ANN as shown in Figure 5.9 is composed of M layers, each having number of predefined neuron cells. Every ANN layer is amalgamated with transfer function and each neuron with bias [14]. The process is described by the equation,

$$\mathbf{X}_m = G_2(\mathbf{W}_m \mathbf{X}_m -_1 + \mathbf{B}_m). \tag{5.39}$$

Here m^{th} layer biasing parameter matrix is represented by B_m. The m and $m - 1$ layer linking weight matrix is represented by W_m and input for the mth layer is given by X_{m-1}. Hence initial input layer is given by

$$\mathbf{X}^0 = [x_1^0, x_2^0, \ldots, x_{n_s}^0], \tag{5.40}$$

and initial weight matrix is given by

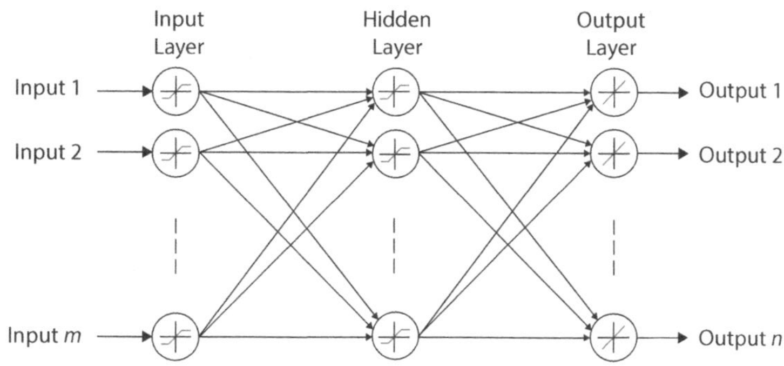

Figure 5.9 Neural network abstraction.

$$\mathbf{W}^1 = \begin{bmatrix} \mathbf{w}_1^{1T} \\ \mathbf{w}_2^{1T}, \\ \vdots, \\ \mathbf{w}_{n_q}^{1\ T} \end{bmatrix} \tag{5.41}$$

Every unit cell with transfer function G_m at a particular layer m operates on input and bias realising an output. This output is further "fed forward" across the network. The neurons of first layer connected with input data set pass the data without computation. The elements of succeeding layers called hidden layers (see Figure 5.10) perform computation using inputs and respective biases. The output layer is the final layer which generally consists of linear activation function with a bias.

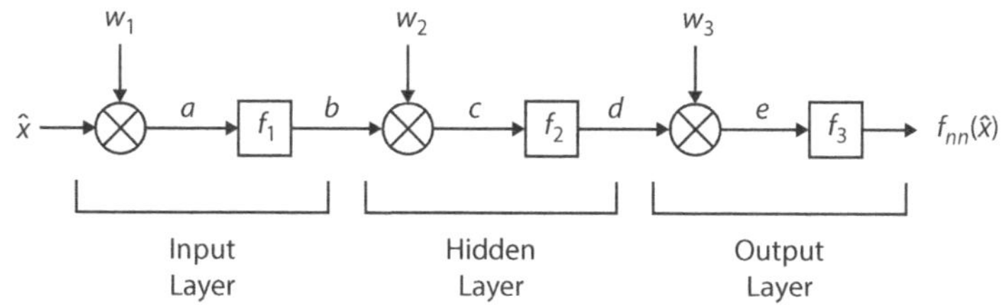

Figure 5.10 Internal structure of feed forward neural network.

5.9.1 Activation Function

In general, any differential function can be considered as activation function of layer other than input layer [83]. Usually monotonically increasing, bounded and differentiable sigmoidal functions are used as activation function [84]. Here each neuron performs computation on input and bias using activation function. For improving the overall behaviour of ANN by obtaining the benefit of saturation behaviour, the Tan-Sigmoid activation is used in present investigation which is represented as

$$G_m(s) = \frac{2}{1+e^{(-2s)}} - 1 \tag{5.42}$$

5.9.2 Bayesian Regularization

Biases and linear weights that are used to predict the target are called best fit parameters. To regulate the set of best fit parameters, the ANN is trained, minimizing the error between the output and target. A single layer feed forward ANN training objective function is specified as

$$F_D = \sum_{i=1}^{N} \left\| \mathbf{t}_i - \mathbf{x}_i^2 \right\|_2, \tag{5.43}$$

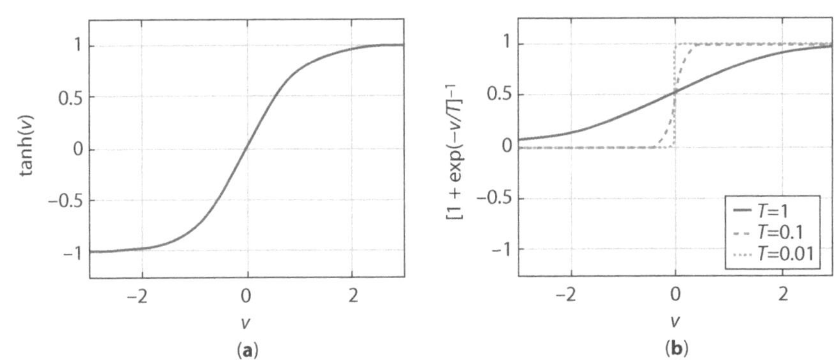

Figure 5.11 Activation functions. (a) Hyperbolic tangent activation function. (b) Logistic activation function for three different T parameters; note that the logistic function looks like Heaviside function, as T decreases.

which is 2-norm error between targets and outputs. To prevent the overfitting of noise data in this squared error based training, the popular extreme learning machine (ELM) and Bayesian regularisation (BR) approach is used. Mathematically, the BR framework objective function is given by

$$E = \beta F_D + \alpha F_W, \tag{5.44}$$

where α and β are dynamically adjustable weight coefficients and F_W is the sum of squared weights which is mathematically represented as,

$$F_W = \| \mathbf{W}^1 \|_2 + \| \mathbf{W}^2 \|_2, \tag{5.45}$$

for single hidden layer. In Bayesian regularisation, the network errors are minimized by penalizing the sum of the squared weights [16, 85]. The above Figure 5.11 shows the activation functions.

5.9.3 Optimization

Gradient-based optimization is used to set the weight co-efficient through back propagation. To obtain the gradient-based optimization, the combination for Gauss-Newton's and steepest descent based fast search technique called Levenberg-Marquardt (LM) technique is utilized ([86] and [87]). The initialization of layer weights is according to the Nguyen-Windrow methodology [15] as,

$$\gamma = P^{Tot} - 2\alpha tr(\mathbf{H})^{-1}, \tag{5.46}$$

where P^{Tot} denotes total number of parameters, $\mathbf{H}$ denotes the Hessian and is approximately given by

$$\mathbf{H} \approx 2\beta \mathbf{J}^T\mathbf{J} + 2\alpha \mathbf{I}, \tag{5.47}$$

where $\mathbf{J}$ is called training set error Jacobian matrix got from the chain rule of backpropagation [84].

5.10 MOR of Dynamical Systems Using POD-ANN

A neural network is modeled using the "learning through experience" technique. Here supervised learning methodology is utilized by providing appropriate inputs and corresponding outputs. The network gets trained with supervised machine learning algorithm and self-validates based on untrained set of data. The logical steps stated below briefs the methodology followed:

1. Construct the reduced order nonlinear model using POD.
2. Construct the reduced order linear model for the neuron spiking model using Arnoldi algorithm.
3. Construct the neural network with feed forward network model considering one input, one output and one hidden layer. Also in hidden layer consider two neurons, in input and output layers one neuron each.
4. For different points, get the output data for reduced order nonlinear and linear neuron spiking models and store the same as output and input data sets, respectively.
5. Training ANN from 80% of data set and same is validated with 20% of data set, with supervised learning method.
6. Check for the accuracy of the output providing new input data. If not satisfied then increase the neurons in hidden layer.
7. When the output found satisfactory, then deploy the neural network model for real-time execution.

5.10.1 Accuracy and Performance Index

This section focuses on accuracy of the system's response approximated by neural network reduced models. Let full order system response be $\mathbf{y}$ and POD-ANN reduced model obtained with a POD basis of dimension l be $\mathbf{y}^{(l)} = \Phi_l \tilde{y}$. Then with reference to the full-order-system's response $\mathbf{y}$, approximate response $\bar{y}$ with relative error $L_2(\Omega)$ is given by:

$$\frac{\|\mathbf{y}-\bar{\mathbf{y}}\|_{L_2(\Omega)}}{\|\mathbf{y}\|_{L_2(\Omega)}} := \bar{e}, \tag{5.48}$$

where $\Omega = [0,T] \times [0,L]$, in time and space respectively. $\bar{y} = \mathbf{y}^{(l)}$ is the approximate response for POD-ANN in (5.48).

Performance of the presented models is also evaluated by mean squared error (MSE). This is also referred as the risk function corresponding to full order solution. For computing MSE, neural network approximation of reduced order model and numerical evaluation of full order model are considered. Difference between the above attributes is calculated to find the error. Mean of the square of the above error is calculated to get the MSE as given by,

$$MSE = \frac{1}{N}\sum_{i=1}^{N}(\hat{y}_i - y_i)^2. \tag{5.49}$$

5.11 Numerical Results

This chapter focuses on building a neural network model for nonlinear dynamic physical systems, when much less data is available for training. Here snap shot data is produced by the appropriate mathematical model. Data set generated is cleaned and reshaped.

This chapter also focuses on how computational time and complexity is reduced using reduced order modeling technique like proper orthogonal decomposition (POD). It is observed that by reducing the order of the model, the evaluation complexity of nonlinear term of original problem remains unchanged. It is also observed that the neural network developed using POD model is found to be much more efficient and cost effective. This has been proved using three case studies – one is the biological neuron model, another is the ring oscillator model, and a third is the VLSI interconnect model.

5.11.1 F-N System

The performance of novel methods designed and applied on neuron spiking model using F-N system is discussed in the present section. This work requires maintaining accuracy of full order FD model in the reduced model as well. The initial conditions for the F-N system are:

$$x \in [0,L], \quad \omega(x,0) = 0, \quad \upsilon(x,0) = 0. \tag{5.50}$$

The boundary conditions are functions of time in exponential form, which serve as model input and are defined as:

$$\varepsilon\upsilon_t(0,t) = -i_0(t), \quad \varepsilon\upsilon_t(L,t) = 0, \quad t \geq 0, \tag{5.51}$$

Table 5.1 Comparison of three models.

Case study	Mathematical method	Soft program for realising the model	Remarks on merits/novelty of work
Neuron spiking model using FitzHugh-Nagumo system	POD - ANN	POD algorithm and supervised machine learning algorithm in Python programming environment	POD-ANN reduced order model for parameterized PDE, having bistable nonlinearity
Ring Oscillator model using Inverter chain circuit	POD - ANN	POD algorithm and supervised machine learning algorithm in Python programming environment	POD-ANN reduced order model for unpredictable tanh(.) nonlinearity
VLSI Interconnect model using Telegraph equation	POD - ANN	POD algorithm and supervised machine learning algorithm in Python programming environment	POD-ANN reduced order model for spacially discretized second order hyperbolic equation

considering the parameters $c = 0.05$, $\gamma = 2, b = 0.5$, $\varepsilon = 0.015$, $L = 1$. The stimulus is $i_0(t) = 50000t^3 exp(-15t)$.

FD scheme is applied on the nonlinear equation (5.4), with 1024 grid points in space and 1000 grid points in time. The numerical full order solution of system (5.11) and phase time plot with initial condition (5.50) are presented in Figure 5.13 for $\epsilon = 0.015$. Figure 5.12 gives singular values obtained from POD in descending order. The reduced order $r = 5$ is selected and POD is applied on the full order FD model to obtain reduced order data used as output to ANN. The linear part of the model is also considered and POD is applied separately on the same. The reduced order model of the linear model is separately constructed and is used as an input

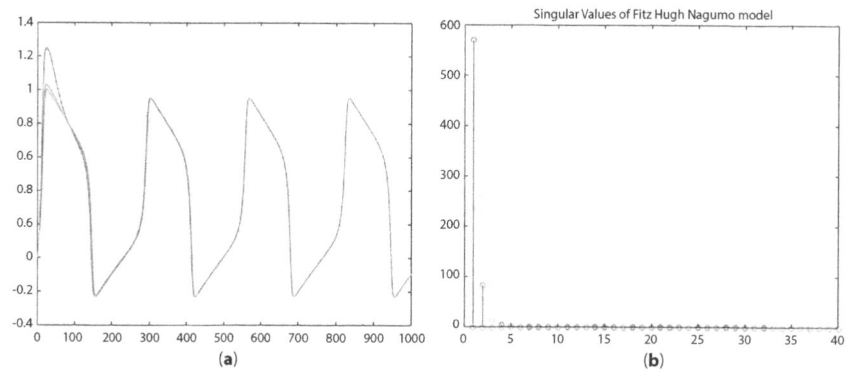

Figure 5.12 Phase time plot and singular value plot of Fitz-Hugh Nagumo model. (a) Phase time plot of full order Fitz-Hugh Nagumo model. (b) Singular values of Fitz-Hugh Nagumo model.

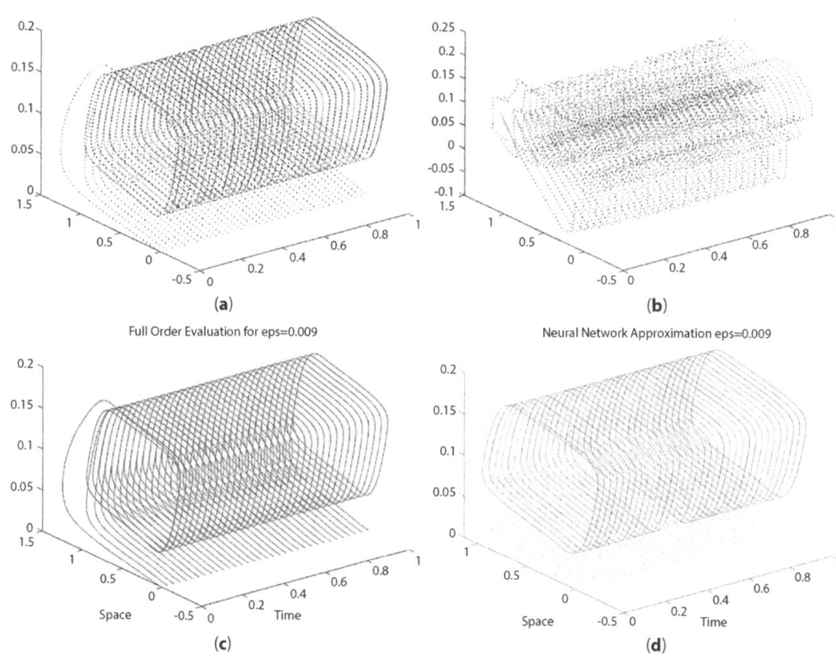

Figure 5.13 Linear solution, full order solution and Neural network approximation of Fitz-Hugh Nagumo model. (a) Full order solution of Fitz-Hugh Nagumo model for ϵ =0.015. (b) Linear solution of Fitz-Hugh Nagumo model for ϵ =0.09. (c) Full order solution of Fitz-Hugh Nagumo model for ϵ = 0.09. (d) Neural network approximation of Fitz-Hugh Nagumo model for ϵ = 0.09.

to ANN. This procedure has been repeated for 100 different values. These 100 data sample sets are used to train the ANN model. The trained POD-ANN model is verified for $\epsilon = 0.09$. Linear solution, full order solution and ANN approximation of F-N model for $\epsilon = 0.09$ are shown in Figure 5.13.

5.11.2 Ring Oscillator Model

In this section, the expediency of less computation cost of linear model with POD-ANN method for the nonlinear ring oscillator circuit [88] is considered as shown in Figure 5.4 given by equations (5.16) and (5.17). Numerical outcomes obtained by applying POD-ANN technique with POD for ring oscillator circuit model are presented in section 5.7.1. Nonlinear ring oscillator consists of inactive linear segments with resistance and capacitance forming low pass filter. These components are assumed as lumped parameters. $N = 100$ is considered to be a framework for full order.

5.11.3 Reduced Order POD Approximation of Ring Oscillator

POD is applied on full order model. From the graph shown in Figure 5.14, it is observed from inspection that dominant ten singular values influence almost 98% of the oscillator behaviour. Hence, the order of the reduced system $q = 10$ is considered. Orthonormal matrix U of SVD equation (5.22)

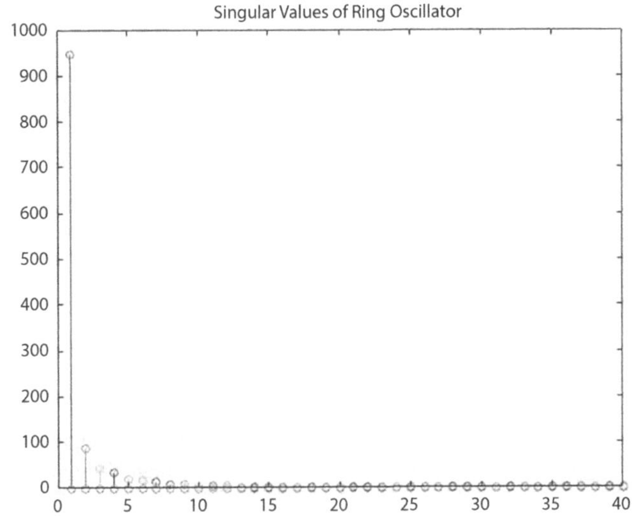

Figure 5.14 Singular values in descending order.

is used as projection matrix which is multiplied by the resultant output to get the full order approximation with order $N = 100$.

5.11.3.1 *Study of POD-ANN Approximation of Ring Oscillator for Variation in Amplitude of Input Signal and for Different Input Signals*

The POD-ANN reduced order model has successfully approximated the output for several input signals having variations in amplitude. POD-ANN with one hidden, one output and one input is used for modeling. The input and output layers have one neuron each and the hidden layer has two neurons. To train the POD-ANN for approximation, supervised training was given with several basic input waves, viz., sinusoidal, triangular and step inputs as shown in Figure 5.15. POD-ANN model obtained after training is verified for different input cases as mentioned in equation (5.52).

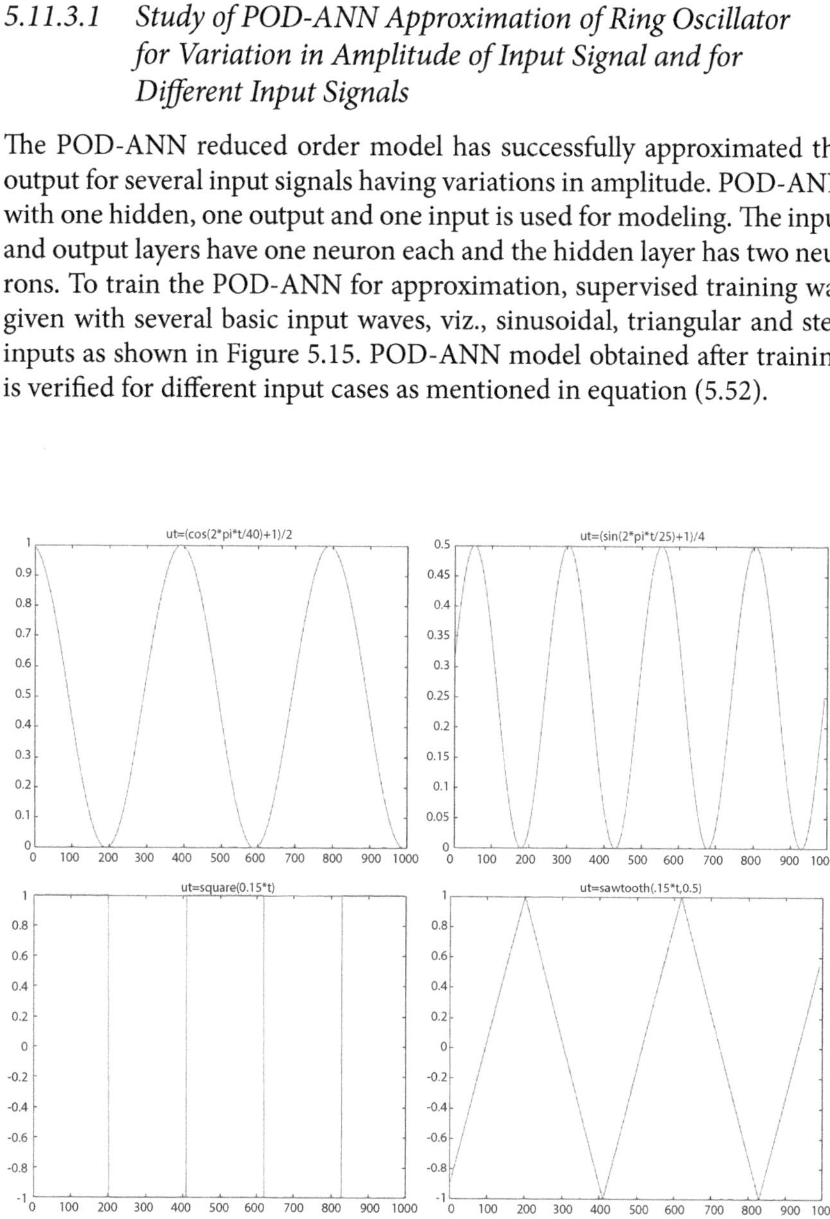

Figure 5.15 Different input wave forms used to train the POD-ANN to approximate the ring oscillator circuit behaviour.

$$
\begin{aligned}
input_1 &= sawtooth(.35*t,0.25)+(sin(2*pi*t/25)+1)/4,\\
input_2 &= 1/2*(1+cos(2*pi*t/10))+square(.15*t),\\
input_3 &= 1/2*(1+cos(2*pi*t/10))+(cos(2*pi*t/50)+1)/2,\\
input_4 &= 1/2*(1+cos(2*pi*t/10))+(sin(2*pi*t/25)+ 1)/6.
\end{aligned} \tag{5.52}
$$

The solution obtained from the same was compared with the original and POD reduced model solutions. Comparative results are shown in Figure 5.16. To observe the accuracy of the results, MSE of POD-ANN model and POD model is tabulated in Table 5.2 and time taken for full order evaluation and ANN approximation is tabulated in Table 5.3.

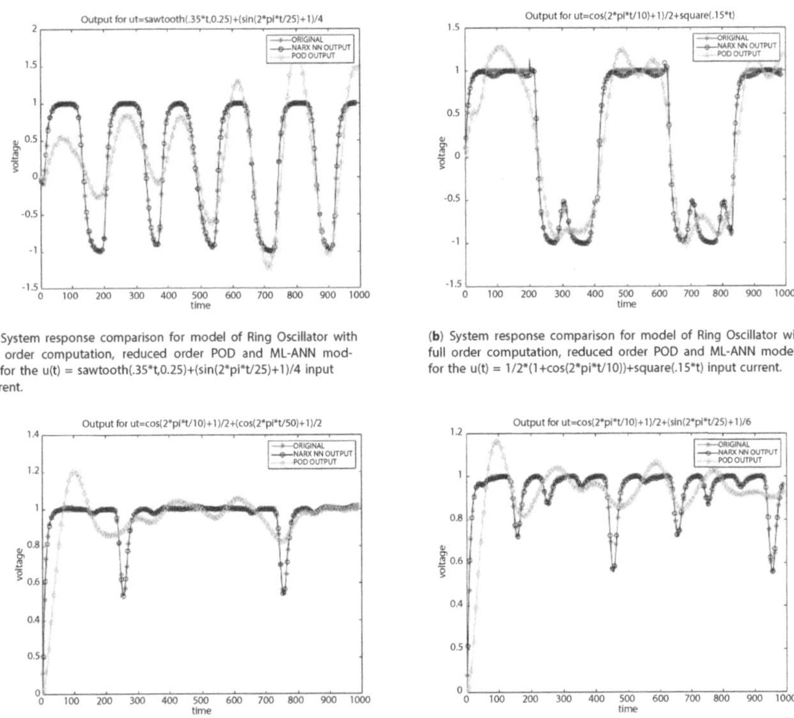

(a) System response comparison for model of Ring Oscillator with full order computation, reduced order POD and ML-ANN models for the u(t) = sawtooth(.35*t,0.25)+(sin(2*pi*t/25)+1)/4 input current.

(b) System response comparison for model of Ring Oscillator with full order computation, reduced order POD and ML-ANN models for the u(t) = 1/2*(1+cos(2*pi*t/10))+square(.15*t) input current.

(c) System response comparison for model of Ring Oscillator with full order computation, reduced order POD and ML-ANN models for the u(t) = 1/2*(1+cos(2*pi*t/10))+(cos(2*pi*t/50)+1)/2 input current.

(d) System response comparison for model of Ring Oscillator with full order computation, reduced order POD and ML-ANN models for the u(t) = 1/2*(1+cos(2*pi*t/10))+(sin(2*pi*t/25)+1)/6 input current.

Figure 5.16 Ring oscillator response for different inputs.

Table 5.2 Performance of POD and neural network approximation of nonlinear ring oscillator circuit for different input signals.

Input signal	MSE (POD)	MSE (NN)
sawtooth(.35*t,0.25) +(sin(2*pi*t/25)+1)/4	0.1681	3.4429e-05
1/2*(1+cos(2*pi*t/10)) +square(.15*t)	0.0567	6.4114e-04
1/2*(1+cos(2*pi*t/10)) +(cos(2*pi*t/50)+1)/2	0.0196	4.5679e-05
1/2*(1+cos(2*pi*t/10)) +(sin(2*pi*t/25)+1)/6	0.0199	1.5494e-05

Table 5.3 Time taken for full order computation and neural network approximation of nonlinear ring oscillator circuit for different input signals (in seconds).

Input signal	Full order execution time	ANN execution time
sawtooth(.35*t,0.25) +(sin(2*pi*t/25)+1)/4	0.5260	0.1006
1/2*(1+cos(2*pi*t/10)) +square(.15*t)	1.9110	0.1089
1/2*(1+cos(2*pi*t/10)) +(cos(2*pi*t/50)+1)/2	1.7754	0.1207
1/2*(1+cos(2*pi*t/10)) +(sin(2*pi*t/25)+1)/6	0.3166	0.0973

Note: For time recording, Intel Core i5, 2.5 GHz processor, DDR3 RAM with 16 GB, 1600 MHz in MacBook Pro has been used.

5.11.3.2 *POD-ANN Approximation of Ring Oscillator for Variation in Frequency*

In order to investigate the performance of the POD-ANN model of Ring Oscillator for variation in frequency, POD-ANN is trained with 2kHz, 20kHz, 100kHz and 200kHz frequencies. 100 recordings are considered from every frequency for the ensemble of snapshot and POD algorithm is applied on the same. The POD-ANN is trained using the snapshot ensemble data. The POD-ANN was verified for different input frequencies and the results are presented in Figure 5.17. The comparative MSE results are tabulated in Table 5.4 and also time taken for full order evaluation and ANN approximation is tabulated in Table 5.5. Training samples, test inputs and testing technique are identical to what is followed in F-N system. Obtaining extrapolation results with interpolation, comparison with

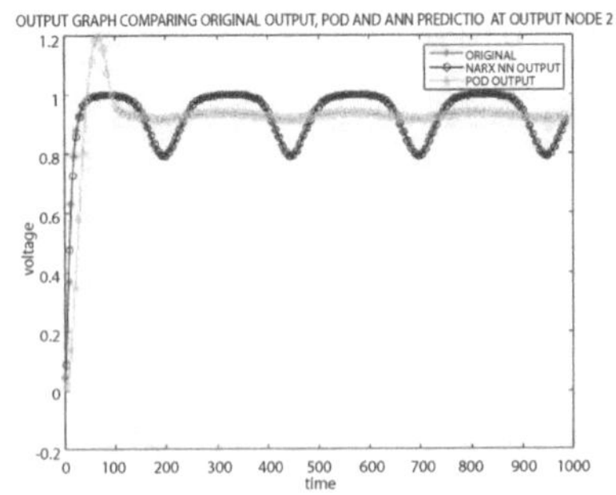

(a) System response comparison for model of Ring Oscillator with full order computation, reduced order POD and ML-ANN models for the ut = (cos(2*pi*t/freq)+1)/10, frequency=16k Hz, input current.

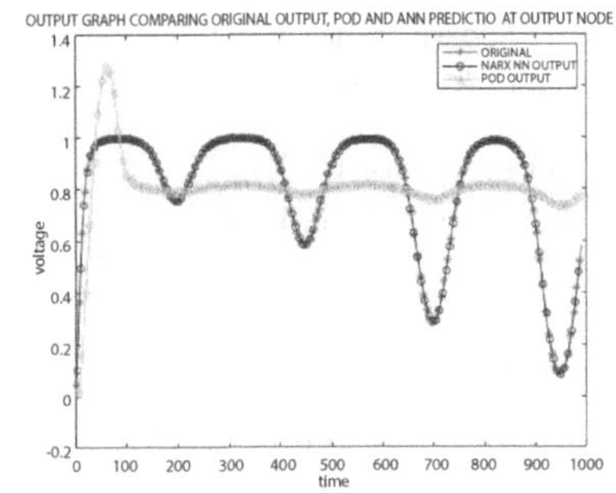

(b) System response comparison for model of Ring Oscillator with full order computation, reduced order POD and ML-ANN models for the ut = (cos(2*pi*t/freq)+1)/10, frequency=200 Hz, input current.

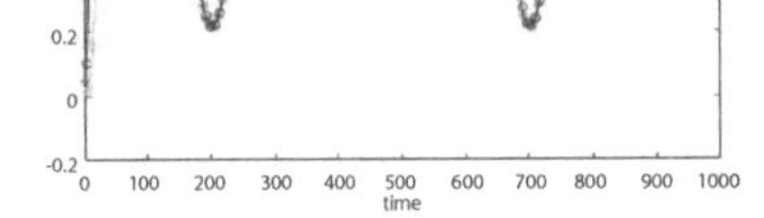

(c) System response comparison for model of Ring Oscillator with full order computation, reduced order POD and ML-ANN models for the ut = (cos(2*pi*t/freq)+1)/10, frequency=50 Hz, input current.

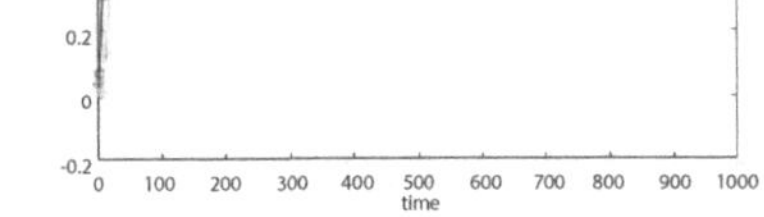

(d) System response comparison for model of Ring Oscillator with full order computation, reduced order POD and ML-ANN models for the ut = (cos(2*pi*t/freq)+1)/10, frequency=4 Hz, input current.

Figure 5.17 System response of ring oscillator for different frequency of input signals.

Table 5.4 Performance of POD and neural network approximation of ring oscillator circuit for different frequency signals.

Signal frequency	MSE (POD)	MSE (NN)	Signal frequency	MSE (POD)	MSE (NN)
4 Hz	0.0305	9.0011e-05	1k Hz	0.0110	1.4789e-05
50 Hz	0.0466	2.1068e-05	16k Hz	0.0104	3.9048e-05
200 Hz	0.0528	1.8706e-05	300k Hz	0.0104	5.2137e-05

Table 5.5 Execution Time taken for full order model evaluation and neural network approximation of ring oscillator circuit for different frequency signals (in seconds).

Signal frequency	Full order execution time	ANN execution time
4 Hz	1.5759	0.0776
50 Hz	1.5917	0.0627
200 Hz	1.5758	0.0630
1k Hz	1.6352	0.0845
16k Hz	1.5983	0.0630
300k Hz	1.5818	0.0574

Note: For time recording, Intel Core i5, 2.5 GHz processor, DDR3 RAM with 16 GB, 1600 MHz in MacBook Pro has been used.

original data is made. From several trials, it is also noted that the developed POD-ANN model interpolates as well as extrapolates accurately. This is one of the novel inferences made from the present study.

5.11.4 POD-ANN Approximation of VLSI Interconnect

In this section, the performance of the devised novel method, applied on VLSI Interconnect model using Telegraph equation, is discussed. The intention of the present work is to ascertain the preserving of the accuracy of the FD model of full order by reduction in model order.

Initial conditions to Telegraph equation are:

$$v_t(x,0) = -2e^x, \quad v(x,0) = e^x. \tag{5.53}$$

The boundary conditions to the model serving as input which are functions of time in exponential form is defined as:

$$v_{Nx+1}(t) = e^{2-2\Delta t}, \quad v_0(t) = e^{-2\Delta t}. \tag{5.54}$$

It is seen that the FD scheme is applied on the nonlinear equation (5.29), with 50 grid points in space and time. Further, singular values from POD are derived in descending order as shown in Figure 5.18.

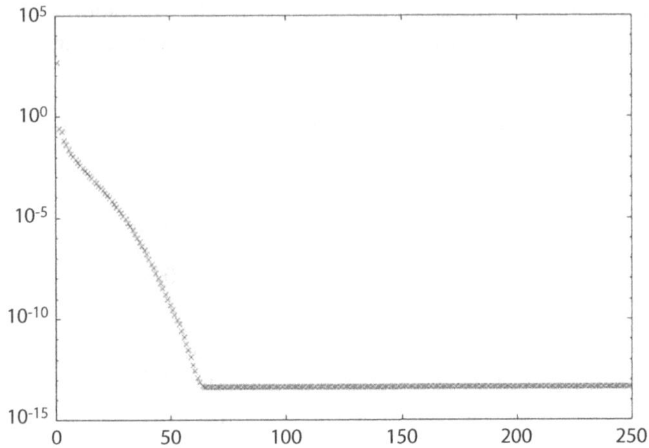

Figure 5.18 Singular values of telegraph equation for $LC = 1.2sec$.

The reduced order $r = 10$ is selected and POD is applied on the full order FD model to get reduced order data used as output to ANN. The linear part of the model is also considered and POD is applied separately on the same. The reduced order model of the linear model which is separately constructed and is used as an input to ANN. This procedure has been repeated for 200 different LC constants. These 200 data sample sets have been used to train the ANN model. The trained POD-ANN model is verified for LC constants $c = 0.05, 1.2$ and 1.9 as depicted in Figure 5.19. Also, the feedforward neural network with 10 neurons on a single hidden layer is used for the ANN model.

The solution obtained from POD-ANN is observed to be a close match with the exact and FDM solution as evident from Figure 5.20. Note that at any instant of time, the voltage grows exponentially with distance. Additionally, there is an exponential decay in $v(x,t)$ at a particular distance.

5.12 Conclusion

It is observed from Figure 5.13 that the ANN data science model has successfully traced the nonlinear behaviour from the linear model. Such successful tracing provides variations in the system parameter. The performance of the model is verified with the full order FD model using the mean square error and L_2 norm. It is observed from Table 5.6 that the errors are within the acceptable limit and thus the model shows the better

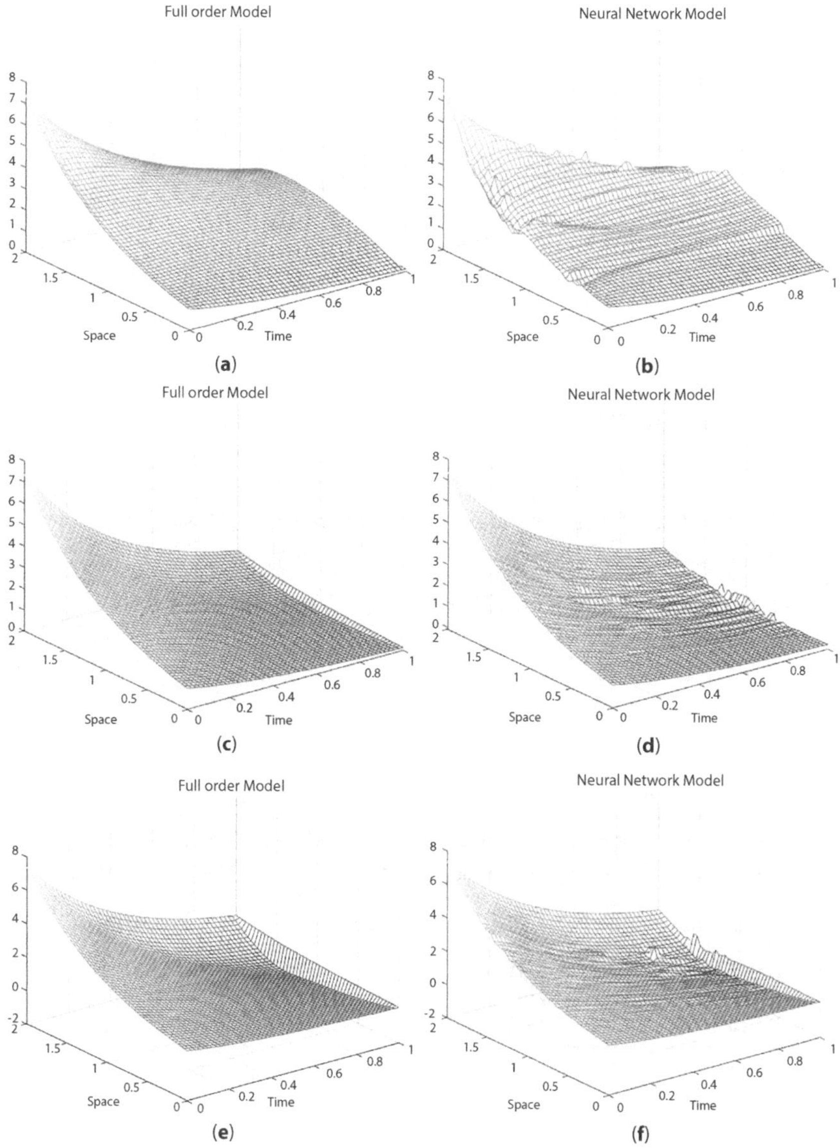

Figure 5.19 Full order and ANN model solution of Telegraph equation for LC constants = 0.05,1.2 and 1.9 Seconds, respectively.

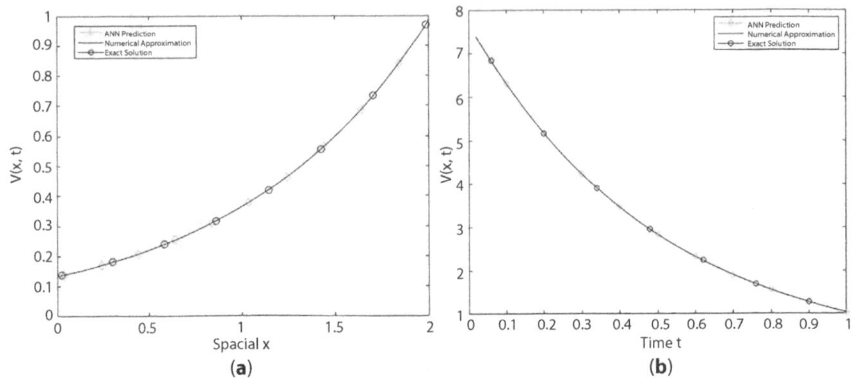

Figure 5.20 Original solution and ANN approximation of amplitude in relation with space and time.

Table 5.6 Performance index - Mean Square Error (MSE) and relative error $L_2(\Omega)$ for Fitzhugh Nagumo model.

Parametric value (v)	MSE (ANN)	Relative error $L_2(\Omega)$
0.015	0.0597	0.4417
0.019	0.0777	0.4104
0.022	0.0858	0.3889

Table 5.7 Execution time* (in seconds) taken for full order model evaluation and ANN approximation of for Fitz-Hugh Nagumo model.

Parametric value (v)	Full order execution	ANN execution
0.015	1020	0.8189
0.019	1053	0.7347
0.022	1065	0.7609

Note: For recording time, Intel Core i5 processor, 2.5 GHz, DDR3 RAM 1600 MHz,16 GB, MacBook Pro is used.

performance. It is also observed from Table 5.7, that a lesser amount of time has been taken by POD-ANN model solution compared to full order model solutions.

It is observed from the plots that the trained network is customized to successfully delineate the nonlinear behaviour from the linear model, given variation in the system parameter. The performance of the model is verified with the full order FD model using the $L_2(\Omega)$ error and mean square error. It is observed from Table 5.8 that the errors are in acceptable limit of 1%. Further, the model shows an improved level of performance. It is also observed from Table 5.9, that the highly limited time of only 5 – 10% of full order execution time has been taken by POD-ANN model solution compared to the full order model solution.

Table 5.8 Performance of neural network approximation of nonlinear interconnect circuit for different *LC* constants.

LC (Sec)	MSE	L_2 (Ω)
0.05	0.0488	0.3216
1.2	0.0018	0.0509
1.9	0.0035	0.0579

Table 5.9 Execution time taken by neural network approximation and full order model of nonlinear interconnect circuit for different *LC* Constants (in seconds).

LC	Full order execution	ANN execution
0.05	0.4508	0.0105
1.2	0.4405	0.0116
1.9	0.4813	0.0106

Note: For time recording, Intel Core i5, 2.5 GHz processor, DDR3 RAM with 16 GB, 1600 MHz in MacBook Pro has been used.

References

1. S. E. Bibri, "The Sciences Underlying Smart Sustainable Urbanism: Unprecedented Paradigmatic and Scholarly Shifts in Light of Big Data Science and Analytics," *Smart Cities*, vol. 2, no. 2, pp. 179–213, 2019.
2. Z. Lu, K. Lu, K. Zhang, H. Zhang, X. Gu, Y. Jin, S. Zhao, C. Fu, and Y. Yang, "A Review of Model Order Reduction Methods for Large-Scale Structure Systems," *Shock and Vibration*, vol. 2021, p. 6631180, 2021.
3. W. Schilders, H. Vorst, van der, and J. Rommes, *Model order reduction: theory, research aspects and applications*. Mathematics in industry, Germany: Springer, 2008.
4. D. S. Glass, X. Jin, and I. H. Riedel-Kruse, "Nonlinear delay differential equations and their application to modeling biological network motifs," *Nature Communications*, vol. 12, no. 1, p. 1788, 2021.
5. T. Henneron and S. Clenet, "Model order reduction of non-linear magnetostatic problems based on POD and DEI methods," *IEEE Transactions on Magnetics*, vol. 50, no. 2, pp. 33–36, 2014.
6. G. Kerschen, J.-C. Golinval, A. F. Vakakis, and L. A. Bergman, "The method of proper orthogonal decomposition for dynamical characterization and order reduction of mechanical systems: an overview," *Nonlinear Dynamics*, vol. 41, no. 1-3, pp. 147–169, 2005.
7. M. Kanaan and R. Khazaka, "Model order reduction for nonlinear macromodeling of RF circuits," in *New Circuits and Systems Conference (NEWCAS), 2011 IEEE 9th International*, pp. 217–220, IEEE, 2011.
8. M. Striebel and J. Rommes, "Model order reduction of nonlinear systems: status, open issues, and applications," 06 2018.
9. Y. Fang, M. C. Yagoub, F. Wang, and Q.-J. Zhang, "A new macromodeling approach for nonlinear microwave circuits based on recurrent neural networks," *IEEE Transactions on Microwave Theory and Techniques*, vol. 48, no. 12, pp. 2335–2344, 2000.
10. Q.-J. Zhang, K. C. Gupta, and V. K. Devabhaktuni, "Artificial neural networks for rf and microwave design-from theory to practice," *IEEE Transactions on Microwave Theory and Techniques*, vol. 51, no. 4, pp. 1339–1350, 2003.
11. H. Kabir, M. Yu, and Q. Zhang, "Recent advances of neural network-based EM-CAD," *International Journal of RF and Microwave Computer-Aided Engineering*, vol. 20, no. 5, pp. 502–511, 2010.
12. M. Kanaan and R. Khazaka, "Nonlinear time-domain macromodeling using proper orthogonal decomposition and feedforward neural networks," in *30th Canadian Conference on Electrical and Computer Engineering (CCECE)*, pp. 1–4, IEEE, 2017.
13. Nrgaard, Magnus, O. E. Ravn, N. K. Poulsen, and L. K. Hansen, *Neural Networks for Modelling and Control of Dynamic Systems: A Practitioner's Handbook*. Berlin, Heidelberg: Springer-Verlag, 1st ed., 2000.

14. H. B. Demuth, M. H. Beale, O. De Jess, and M. T. Hagan, *Neural Network Design*. USA: Martin Hagan, 2nd ed., 2014.
15. B. Widrow, D. E. Rumelhart, and M. A. Lehr, "Neural networks: Applications in industry, business and science," *Commun. ACM*, vol. 37, pp. 93–105, 1994.
16. F. Dan Foresee and M. T. Hagan, "Gauss-newton approximation to bayesian learning," in *Proceedings of International Conference on Neural Networks (ICNN'97)*, vol. 3, pp. 1930–1935 vol. 3, June 1997.
17. S. Nagaraj, D. Seshachalam, and S. Hucharaddi, "Model order reduction of nonlinear circuit using proper orthogonal decomposition and nonlinear autoregressive with exogenous input (narx) neural network," in *2018 16th ACM/IEEE International Conference on Formal Methods and Models for System Design (MEMOCODE)*, pp. 1–4, Oct 2018.
18. S. Nagaraj and D. Seshachalam, "Neural network macromodeling of nonlinear electrical circuit for variable frequency inputs using karhunen-loeve decomposition," in *Third IEEE Interna- tional Conference on Computational Systems and Information Technology for Sustainable Solutions (CSITSS-2018), Bengaluru, India*, 2018.
19. T. Nguyen and J. E. Schutt-Aine, "A pseudo-supervised machine learning approach to broadband lti macromodeling," in *2018 IEEE International Symposium on Electromagnetic Compatibility and 2018 IEEE Asia-Pacific Symposium on Electromagnetic Compatibility (EMC/APEMC)*, pp. 1018–1021, May 2018.
20. C. A. L. Bailer-Jones, D. J. C. MacKay, and P. J. Withers, "A recurrent neural network for modelling dynamical systems," *Network: Computation in Neural Systems*, vol. 9, no. 4, pp. 531–547, 1998. PMID: 10221578.
21. R. C. B. Rego and F. M. U. d. Araújo, "Nonlinear control system with reinforcement learning and neural networks based Lyapunov functions," *IEEE Latin America Transactions*, vol. 19, no. 8, pp. 1253–1260, 2021.
22. G. M. Shepherd, ed., *The Synaptic Organization of the Brain,* 5th Edition. Oxford University Press, 2004.
23. E. M. Izhikevich, "Which model to use for cortical spiking neurons?," *IEEE Transactions on Neural Networks*, vol. 15, no. 5, pp. 1063–1070, 2004.
24. O. Brandman, J. E. Ferrell, R. Li, and T. Meyer, "Interlinked fast and slow positive feedback loops drive reliable cell decisions," *Science*, vol. 310, no. 5747, pp. 496–498, 2005.
25. M. Negahban, R. Behrasi, G. Tsang, H. Abouhossein, and G. Bouchaya, *A two-chip CMOS read channel for hard-disk drives. ISSCC Dig. Tech. Works*, pp. 216-217, 1993.
26. L. DeVito, J. Newton, R. Croughwell, J. Bulzacchelli, and F. Benkley, A 52 and 155 MHz clock-recovery PLL. *ISSCC Dig. Tech. Works*, 1991.
27. O. V. Rudenko and C. M. Hedberg, "Strong and weak nonlinear dynamics: models, classification, examples," *Acoustical Physics*, vol. 59, no. 6, pp. 644–650, 2013.

28. Y. Chen, "Model order reduction for nonlinear systems," Master's thesis, Massachusetts Institute of Technology, 1999.
29. Z. Bai and D. Skoogh, "A projection method for model reduction of bilinear dynamical systems," *Linear Algebra Appl.*, vol. 415, no. 23, pp. 406–425, 2006.
30. J. Phillips, "Projection frameworks for model reduction of weakly nonlinear systems," *In Proc. Design Automation Conference*, pp. 184–189, 2000.
31. P. Benner and T. Breiten, "Krylov-subspace based model reduction of nonlinear circuit models using bilinear and quadratic-linear approximations," *Progress in industrial Mathematics at ECMI 2010, Mathematics in Industry*, vol. 17, no. 1, pp. 53–159, 2012.
32. T. Breiten and P. Benner, "Two-sided moment matching methods for nonlinear model reduction," *Preprint MPIMD/12-12, Max Planck Institute Magdeburg Preprints*, 2012.
33. L. Feng, X. Zeng, C. Chiang, D. Zhou, and Q. Fang, "Direct nonlinear order reduction with variational analysis," In *Proc. Design, Automation and Test in Europe*, pp. 1316–1321, 2004.
34. C. Gu, "QLMOR: A new projection-based approach for nonlinear model order reduction," *In Proc. International Conference on Computer-Aided Design*, pp. 389–396, 2009.
35. ChenjieGu., "QLMOR: A projection-based nonlinear model order reduction approach using quadratic-linear representation of nonlinear systems," *IEEE Trans. Comput.-Aided Design*, vol. 30, no. 9, pp. 1307–1320, 2011.
36. J. Phillips, "Projection-based approaches for model reduction of weakly nonlinear time-varying systems," *IEEE Trans. Comput.-Aided Des. Integr. Circuits Syst.*, vol. 22, no. 2, pp. 171–187, 2003.
37. W. Rugh, *Nonlinear System Theory*, Johns Hopkins University Press, Baltimore, 1981.
38. M. Rewienski and J. White, "A trajectory piecewise-linear approach to model order reduction and fast simulation of nonlinear circuits and micromachined devices," *IEEE Trans. Comput.-Aided Des. Integr. Circuits Syst.*, vol. 22, no. 2, pp. 155–170, 2003.
39. M. Grepl, Y. Maday, N. Nguyen, and A. Patera, "Efficient reduced-basis treatment of nonaffine and nonlinear partial differential equations," *Mathematical Modelling and Numerical Analysis (ESAIM: M2AN)*, vol. 41, pp. 575– 605, 05 2007.
40. G. Berkooz, P. Holmes, and J. Lumley, "The Proper Orthogonal Decomposition in the Analysis of Turbulent Flows," *Annual Review of Fluid Mechanics*, vol. 25, pp. 539–575, 1993.
41. K. Hall, J. Thomas, and E. Dowell, Reduced-order modelling of unsteady small-disturbance flows using a frequency domain proper orthogonal decomposition technique.
42. T. Kim, "Frequency-Domain Karhunen-Loeve Method and Its Application to Linear Dynamic Systems," *AIAA Journal*, vol. 36, no. 11, pp. 2117–2123, 1998.

43. L. Sirovich, "Turbulence and the Dynamics of Coherent Structures. Part 1: Coherent Structures," *Quarterly of Applied Mathematics*, vol. 45, no. 3, pp. 561-71, 1987.
44. K. E. Willcox, J. D. Paduano, J. Peraire, and K. C. Hall, "Low order aerodynamic models for aeroelastic control of turbo machines," *Proceedings of the American Institute of Aeronautics and Astronautics Structures, Structural Dynamics, and Materials Conference*, vol. 3, pp. 1–11, 1999.
45. K. Willcox, J. Peraire, and J. White, "An Arnoldi approach for generation of reduced-orde models for turbo machinery," *Computers and Fluids*, vol. 31, no. 3, pp. 369–89, 2002.
46. K. Willcox and J. Peraire, "Balanced Model Reduction via the Proper Orthogonal Decomposition," *AIAA Journal, vol. 40, no. 11, pp. 2323-30,,* 2002.
47. S. Lall, J. E. Marsden, and S. Glavaski, "Empirical model reduction of controlled nonlinear systems," *Proceedings of the 14th International Federation of Automatic Control World Congress*, vol. 6, pp. 473-8, 1999.
48. J. R. Li, "Model Reduction of Large Linear Systems via Low Rank System Gramians," Ph.D. dissertation, Massachusetts Institute of Technology, Cambridge, MA, 2000.
49. Jinghong Chen and Sung Mo Kang, "An algorithm for automatic model-order reduction of nonlinear MEMS devices," in *Proc. IEEE Int. Symp. Circuits and Systems (ISCAS)*, vol. 2, pp. 445–448, 2000.
50. R. Kao and M. Horowitz, "Timing analysis for piecewise linear Rsim," *IEEE Transactions on Computer-Aided Design of Integrated Circuits and Systems*, vol. 13, no. 12, pp. 1498–1512, 1994.
51. A. Megrteski, "Lecture notes for Multivariable Control Systems," Massachusetts Institute of Technology, 2002.
52. M. Kamon, N. A. Marques, L. M. Silveira, and J. White, "Automatic generation of accurate circuit models of 3-D interconnect," *and Manufacturing Technology: Part B IEEE Transactions on Components, Packaging*, vol. 21, no. 3, pp. 225–240, 1998.
53. J. E. Mehner, L. D. Gabbay, and S. D. Senturia, "Computer-Aided Generation of Nonlinear Reduced-Order Dynamic Macromodels - II: Stress-Stiffened Case," *Journal of Microelectromechanical Systems*, vol. 9, no. 2, pp. 270–278, 2000.
54. R. F. Remis and P. M. van den Berg, "A Modified Lanczos Algorithm for the Computation of Transient Electromagnetic Wavefields," *IEEE Transactions on Microwave Theory and Techniques*, vol. 45, no. 12, pp. 2139–2149, 1997.
55. E. Gad, A. Dounavis, M. Nakhla, and R. Achar, "Passive model order reduction of multi port distributed interconnects," *Proceedings of the 37th Design Automation Conference*, pp. 526–531, 2000.
56. L. D. Gabbay, J. E. Mehner, and S. D. Senturia, "Computer-Aided Generation of NonlinearReduced-Order Dynamic Macromodels - I: Non-Stress-Stiffened

Case," *Journal of Microelectromechanical Systems*, vol. 9, no. 2, pp. 262–269, 2000.

57. A. Ariffin and N. Munro, "Representation of a class of nonlinear systems with parameter- dependent linear approximations," *Proceedings of the UKACC International Conference on Control*, vol. 1, pp. 537–542, 1998.
58. E. Acar, L. T. Pileggi, and S. R. Nassif, "A linear-centric simulation framework for parametric fluctuations," in *Design, Automation and Test in Europe Conference and Exhibition*, pp. 568–575, 2002.
59. F. Karray and T. A. W. Dwyer, "On the nonlinear system identification of a class of bilinear dynamical models," in *Proc. 30th IEEE Conf. Decision and Control*, vol. 1, pp. 778–782, 1991.
60. Jinghong Chen and Sung-Mo Kang, "Model-order reduction of nonlinear MEMS devices through arclength-based Karhunen-Loeve decomposition," in *Proc. ISCAS 2001. The 2001 IEEE Int. Symp. Circuits and Systems* (Cat. No.01CH37196), vol. 3, pp. 457–460, May 2001.
61. H. Aling, R. L. Kosut, A. Emami-Naeini, and J. L. Ebert, "Nonlinear model reduction with application to rapid thermal processing," in *Proc. 35th IEEE Conf. Decision and Control*, vol. 4, pp. 4305–4310, Dec. 1996.
62. S. Haykin, "Neural Networks: A comprehensive foundation," *UpperSaddleRiver, NJ: Prentice Hall.*, 2004.
63. K. David, *A Brief Introduction to Neural Networks*. 2007.
64. The MathWorks, Inc. (2016). Machine learning challenges: Choosing the best model and avoiding overfitting. Retrieved from https://it.mathworks.com/campaigns/products/offer/common-machine-learning-challenges.html.
65. M. A. Nielsen, "Neural Networks and Deep Learning," 2018.
66. O. I. Abiodun, A. Jantan, A. E. Omolara, K. V. Dada, A. M. Umar, O. U. Linus, H. Arshad, A. A. Kazaure, U. Gana, and M. U. Kiru, "Comprehensive review of artificial neural network applications to pattern recognition," *IEEE Access*, vol. 7, pp. 158820–158846, 2019.
67. Y. Liu, Y. Sun, B. Xue, M. Zhang, G. G. Yen, and K. C. Tan, "A survey on evolutionary neural architecture search," *IEEE Transactions on Neural Networks and Learning Systems*, pp. 1–21, 2021.
68. Z. Wu, S. Pan, F. Chen, G. Long, C. Zhang, and P. S. Yu, "A comprehensive survey on graph neural networks," *IEEE Transactions on Neural Networks and Learning Systems*, vol. 32, no. 1, pp. 4–24, 2021.
69. T. Moldwin and I. Segev, "Perceptron Learning and Classification in a Modeled Cortical Pyramidal Cell," *Frontiers in Computational Neuroscience*, vol. 14, p. 33, 2020.
70. Z. Uykan, "On the Working Principle of the Hopfield Neural Networks and its Equivalence to the GADIA in Optimization," *IEEE Transactions on Neural Networks and Learning Systems*, vol. 31, no. 9, pp. 3294–3304, 2020.

71. F. M. Riese, S. Keller, and S. Hinz, "Supervised and Semi-Supervised Self-Organizing Maps for Regression and Classification Focusing on Hyperspectral Data," *Remote Sensing*, vol. 12, pp. 1–23, dec 2019.
72. Q. Liu, W. Chen, H. Hu, Q. Zhu, and Z. Xie, "An Optimal NARX Neural Network Identification Model for a Magnetorheological Damper with Force-Distortion Behaviour," *Frontiers in Materials*, vol. 7, p. 10, 2020.
73. Y. Raptodimos and I. Lazakis, "Application of NARX neural network for predicting marine engine performance parameters," *Ships and Offshore Structures*, vol. 15, no. 4, pp. 443–452, 2020.
74. M. Striebel and J. Rommes, "Model order reduction of nonlinear systems: status, open issues, and applications," 06 2018.
75. P. Hartman, "A lemma in the theory of structural stability of differential equations," *Proceedings of the American Mathematical Society*, vol. 11, no. 4, pp. 610–620, 1960.
76. L. Miguel Silveira, M. Kamon, and J. White, "Efficient reduced-order modeling of frequency-dependent coupling inductances associated with 3-D interconnect structures and Manufacturing Technology: Part B," *IEEE Transactions on Components, Packaging*, vol. 19, pp. 283–288, May 1996.
77. J. Roychowdhury, *Numerical simulation and modelling of electronic and biochemical systems*. Now Publishers Inc, 2009.
78. X. Ge, M. Arcak, and K. N. Salama, "Nonlinear Analysis of Ring Oscillator and Cross-Coupled Oscillator Circuits," *Dynamics of Continuous, Discrete and Impulsive Systems Series B: Applications and Algorithms*, 2010.
79. H. K. Khalil and J. W. Grizzle, *Nonlinear systems*. Upper Saddle River, NJ: Prentice Hall, 2002.
80. J. N. Kutz, *Data-driven modeling and scientific computation: methods for complex systems and big data*. Oxford University Press, 2013.
81. S. Nagaraj, D. Seshachalam, and S. Hucharaddi, "Model Order Reduction of Nonlinear Circuit using Proper Orthogonal Decomposition and Nonlinear Autoregressive with eXogenous input (NARX) Neural Network," in *Proc. 16th ACM/IEEE Int. Conf. Formal Methods and Models for System Design (MEMOCODE)*, pp. 1–4, Oct. 2018.
82. R. K. C. Vineet K. Srivastava, Mukesh K. Awasthi and M. Tamsir, "The Telegraph Equation and Its Solution by Reduced Differential Transform Method," *Modelling and Simulation in Engineering, Article ID 746351*, p. 6, 2013.
83. G. Zhang, B. E. Patuwo, and M. Y. Hu, "Forecasting with artificial neural networks:: The state of the art," *International Journal of Forecasting*, vol. 14, no. 1, pp. 35–62, 1998.
84. I. Goodfellow, Y. Bengio, and A. Courville, ""Deep Learning"," *MIT press*, 2016.
85. D. J. MacKay, "Bayesian interpolation," *Neural Computation*, vol. 4, pp. 415–447, 1991.

86. K. Levenberg, "A method for the solution of certain nonlinear problems in least squares," *Quarterly of Applied Mathematics*, vol. 2, no. 2, pp. 164–168, 1944.
87. D. Marquardt, "An algorithm for least-squares estimation of nonlinear parameters," *Journal of the Society for Industrial and Applied Mathematics*, vol. 11, no. 2, pp. 431–441, 1963.
88. C. Gu, Model order reduction of nonlinear dynamical systems. PhD thesis, University of California, Berkeley, 2011.

6

Comparative Analysis of Various Ensemble Approaches for Web Page Classification

J. Dutta*, Yong Woon Kim and Dalia Dominic

Centre for Digital Innovation, CHRIST (Deemed to be University), Mysore Road, Kumbalgodu, Bangalore, India

Abstract

The amount of data available on web pages is enormous, and extracting the relevant information and classifying them is an important task. Web page classification finds applications in web content filtering, maintaining and expanding web directories, building efficient crawlers, etc. Machine Learning methods known for their well-established classification approaches have proved to be effective in web page classification. The present work uses ensemble methods like Bagging Meta Estimator, Random Forest, Adaptive boosting, Gradient Tree boosting, Extreme Gradient boosting and stacking to improve single classifier's results. One dataset is manually created to classify web pages into IoT projects and non-IoT projects. Another publicly available dataset is used to classify publications- and conference-related web pages. The advantage of the Ensemble methods over single classifiers has been validated, and various parameters to tune the Ensemble classifiers have been presented and analysed, with accuracy being the metric for performance. Features like learning rate, number of estimators, and maximum number of features have been tuned besides other parameters, and a comparison has been presented.

***Keywords*:** Machine learning, web scraping, web page classification, ensemble learning models

**Corresponding author*: jyotirmoy.dutta@christuniversity.in

Kukatlapalli Pradeep Kumar, Aynur Unal, Vinay Jha Pillai, Hari Murthy, and M. Niranjanamurthy (eds.) Data Engineering and Data Science: Concepts and Applications, (137–172) © 2023 Scrivener Publishing LLC

6.1 Introduction

There is a vast amount of data available on web pages. An important task is the extraction of appropriate information from web pages and their subsequent classification. According to the definition, web page classification pertains to the process of assigning predefined categories to web pages. Web page classification has various applications in web content filtering, contextual advertising, construction of knowledge bases, maintaining or expanding web directories, building domain-specific search engines, and improving search result quality [1, 2].

Manual classification is not a practical solution, considering the vast amount of data; therefore, different approaches have been used traditionally. Statistical techniques in which the web pages are classified based on term-frequency analysis [3], combining feature selection with a web page classification algorithm [4], and classifying web content using textual and visual features [5] are some methods used for web page classification. Besides these, the use of Machine Learning approaches are found to be effective and provides an alternative way of web page classification [1, 6–12].

Many works have reported the improvement of results by using Ensemble models [13–16]. Ensemble methods are mainly used for reducing generalisation errors [17]. Essentially, the Ensemble classifier works by combining multiple models to produce better results. In contrast to other learning approaches, which create a single learner based on training data, Ensemble methods construct a set of learners and combine them. These methods tend to be more effective if the base learners are accurate and diverse [18].

There are many approaches using machine learning ensemble models, and in this context, some survey requirements regarding the ensemble approach for web page classification was needed. The present work explores the effect of ensemble methods on web page classification. Ensemble methods like Bagging Meta Estimator, Random Forest, Gradient Tree boosting, Adaptive boosting, Extreme Gradient boosting and stacking have been used for two data sets. One manually collected dataset has 3,215 URLs, and the other has 9,481 publicly available data [19]. The first data set includes URLs from two categories: IoT project and Non-IoT project. IoT project contains 1,390 project URLs that relate to web pages with IoT-related projects. Non-IoT project has 1,825 URLs of web pages that are not IoT related, like, for example, pages with food recipes. The publicly available dataset includes URLs from two categories: Publications and

Conferences. Publications contains 6,329 websites related to content on publications and Conferences contains 3,152 web pages related to information on conferences.

Individual classifiers like K Nearest Neighbour (KNN), Support Vector Machines (SVM), Decision tree (DT), Naive Bayes (NB), and Stochastic Gradient Descent (SGD) were used to classify these labelled data. After this, Ensemble methods were used to improve the performance, and the results were analysed and compared. The Ensemble methods show improved results and consolidate that using the Ensemble classifier improves the accuracy compared to that obtained by a single classifier. Another outcome is to tune various parameters to get the highest accuracy. The methods were implemented using the Python scikit-learn library on a machine with 16 GB of RAM, GTX 1080, and Ubuntu.

Compared to other reported works, the significant contribution of this work is a comprehensive overview of three ensemble methods, bagging, boosting and stacking with five different classifiers, i.e., KNN, SVM, DT, NB and SGD, as the base classifier. Tuning various parameters to obtain the best accuracy has also been presented.

The chapter is organised as follows. Section 6.2 contains a survey of related literature, and section 6.3 describes the approach and methods used in this work. A brief review of all Ensemble methods used in this work is presented in section 6.4. Results and comparisons of various single and Ensemble classifiers are given in section 6.5. The conclusion of the work is presented in section 6.6.

6.2 Literature Survey

Numerous works have used Machine Learning approaches to classify web pages. The existing Machine Learning algorithms used for web page classification and the available related literature have been described by Lassri *et al.* [1]. In work reported by Kwon *et al.*, an automatic web page classifier based on an adaptation of the KNN approach has been proposed [6]. SVM classifiers were suggested by Sun *et al.* to classify web pages based on their text and context feature sets [7]. The work by Masetic *et al.* evaluates the performance of a decision tree classifier which proved to be effective in detecting malicious websites [8]. Another study by Woogue *et al.* experimented on seven algorithms that categorised web pages based on the educational domain [11]. Few works have presented the use of the Naive Bayes method. In the work by Patil *et al.*, the NB approach has been used to classify websites based on their home page content [12].

Wang *et al.* report using WebDoc, an automatic document classification system implemented to classify web documents based on the Library of Congress classification scheme. WebDoc creates a knowledge base using the training data and then classifies documents based on the knowledge base [20]. In a work by Tsukada *et al.*, the authors have constructed decision trees that identify the relevant category for each web page. This method achieved good accuracy when classifying web-page categories on Yahoo! Japan [21]. In the paper by Awad, the author has used SVM, KNN and Generalized Instance Set to perform a behaviour comparison on the web page classification problem.

It was determined that the least storage requirement and least computation cost were incurred during the SVM with only a few negative documents for building centroids [26].

Multiple reported works have used Ensemble methods to improve the results obtained by single classifiers. Wan *et al.* used the four well-known classification methods: bagging, boosting, stacking, and Random Forest, to improve the results by combining their outputs [16]. In work reported by Han *et al.*, an Ensemble learning method to combine various classification models such as logistic regression, SVM, Decision Tree, Naive Bayes, KNN, and Random Forest has been used [15].

In work by Onan, a comparative analysis of correlation, consistency, information gain, and chi-square-based feature selection methods, as well as bagging, dagging, boosting, and random subspace, has been presented. The author used four different base learners: KNN, NB, C4.5, and FURIA [22].

In the work by Xia *et al.*, Ensemble methods used for sentiment classification are examined, and two different feature selection methods were used. NB, maximum entropy and SVM were used as base learners. For examining the effectiveness of Ensemble learning, the fix-combination, weight-combination, and meta-classifier combination methods were used [23].

Wang *et al.* examined the performance of Ensemble learning for sentiment classification. Bagging, boosting and random subspace Ensemble methods were used along with five different base learners, including NB, maximum entropy, DT, SVM, and KNN. Results showed that ensemble learning improves the predictive performance of base learners in sentiment classification [24]. Adaboost and SGD were used to filter spam emails by Elshoush *et al.*, and the results show these as the best choices [25]. Table 6.1 summarises selected reported works.

Table 6.1 Summary of selected reported works for web-page classification.

Reported work	Classifiers used	Ensemble models	Reported results	Evaluation dataset
[5]	SVM, KNN, NB, ANN	No	This paper has surveyed some existing classification techniques and approaches. The goal of classification is to generate more certain, precise and accurate system results.	Universal data set single label dataset: http://web.ist.utl.pt/acardoso/datasets/ multi-label data sets: (http://sci2s.ugr.es/keel/multilabel.php)
[6]	KNN	No	The performance in the validation test was improved using feature selection in KNN approach.	Hanmir (http://www.hanmir.com)
[7]	SVM	No	0.6 values for the F1 measure	WebKB
[9]	SVM, Maxent, ELM	No	The accuracy of detecting malicious web-pages was improved by 98.24%	Benchmark data set

(*Continued*)

Table 6.1 Summary of selected reported works for web-page classification. (*Continued*)

Reported work	Classifiers used	Ensemble models	Reported results	Evaluation dataset
[11]	Logistic Regression, Multinomial NB, Linear SVM, DT, Random Forest, KNN, Multilayer Perceptron	No	Linear models such as Logistic Regression and Linear SVM showed to have good accuracy when used for text classification.	Manually collected 1200 URLs related to the educational domain. Mostly collected from Wikipedia
[15]	LR, SVM, NB, DT, KNN, Random Forest	Stacking	Classification achieved a 67.32 % accuracy rate on the test dataset.	Kaggle, and the FIRE2019
[21]	DT	No	Web pages classified by this method into top-level categories on Yahoo! Japan are classified with high accuracy.	Randomly downloaded 200 Web pages from each of the top 14 categories in Yahoo! Japan, and the total number of Web pages for the experiments is 1,000.

(*Continued*)

Table 6.1 Summary of selected reported works for web-page classification. (*Continued*)

Reported work	Classifiers used	Ensemble models	Reported results	Evaluation dataset
[22]	NB, KNN, C4.5, FURIA	AdaBoost, bagging, dagging, random subspace algorithm	Combining AdaBoost and Naive Bayes algorithms with consistency-based feature selection provides the best average predictive performance (88.1%).	DMOZ-50 Open Directory Project dataset
[26]	SVM, KNN, GIS	No	SVM with few negative documents builds the centroids with the smallest storage requirements and the lowest online computation costs.	Aljazeera News 21578
[27]	NB, SMO, J48	No	Naive Bayes classifier perform better than SMO and J48 and precision greater than 0.9.	Indian news websites corpus (1025)

(*Continued*)

Table 6.1 Summary of selected reported works for web-page classification. (*Continued*)

Reported work	Classifiers used	Ensemble models	Reported results	Evaluation dataset
[28]	MLP, NB, RBF, SVM	No	Compared with Naive Bayes, RBF, and SVM, MLP can achieve the highest accuracy of 97.60%.	Data was collected from 3 online store websites, including Jabong, Global Reebok, and Asos.

6.3 Material and Methods

The first part of the chapter involves extracting information from the web using web scraping. Web scraping is an automated method to extract content from websites. The second part of the chapter consists of converting the scraped data to vector formats, training it on several classifiers, and improving classification accuracy using Ensemble learning methods.

For extracting the contents from the websites, Beautiful Soup web scraping tool was used. This tool was used to parse the page and search for specific elements for a manually curated dataset, i.e., DS1. It has been used to extract the raw contents from all 3,215 web pages. Parsed content includes 3,215 project URLs, as well as their labels and contents. The labels are divided into IoT project and Non–IoT project. There are 1,390 rows labelled IoT project and 1,825 rows labelled Non-IoT project. A public dataset, DS2, with 9,481 URLs with contents and their labels, was available in the public domain and used in this work [19]. The data in DS2 is classified into two categories, i.e., Publications and Conferences with 6,329 rows labelled as Publications and 3,152 rows labelled as Conferences.

Preprocessing helps to convert the data into clean and formatted data by removing unwanted information, improving the data quality, and increasing the model's accuracy and efficiency. Stopword removal, Lemmatization technique and Regular expression are the three preprocessing techniques used in the present work.

Stop words are the most common words found in any natural language which carries less information. Stop word removal helps to focus only on the essential words. The Lemmatization process is another way of preprocessing for finding the basic form of a word, and it converts all the words to their noun forms. Regular expression, has been used to remove some symbols. Table 6.2 below shows one example of these preprocessing steps.

At first, the dataset is divided into a training dataset (85%) and a testing dataset (15%). 'Term Frequency — Inverse Document Frequency' (TF-IDF) has been used for converting the text format to vector format as a next step. TF-IDF is a technique for quantifying a word in documents [29]. TF-IDF provides relevant keywords, and using those keywords, some specific documents can be categorised. Mathematically, the TF-IDF score is calculated as shown in equation (6.1) [30].

$$tfidf(t,d,D) = tf(t,d).idf(t,D) \tag{6.1}$$

$$\text{where, } tf(t,d) = \log\,(1 + freq(t,d))$$

$$idf(t,D) = \log\,(\mathrm{N}/(count(d \in D{:}t \in d))$$

Table 6.2 Original content after removing regular expressions, Stop words and lemmatization.

Original content	Applying regular expressions	Removing stop words	Performing lemmatization (Final content)
Home Assistant will take care of creating graphs and automate processes triggered by your readings. This tutorial has finished, now it's up to you to add features and customise it for your own purposes!	Home assistant will take care of creating graphs and automate processes triggered by your readings this tutorial has finished now it up to you to add features and customise it for your own purposes	Home assistant take care creating graphs automate processes triggered readings tutorial finished add features customise purposes	Home assistant take care create graph automate process trigger read tutorial finish add feature customise purpose

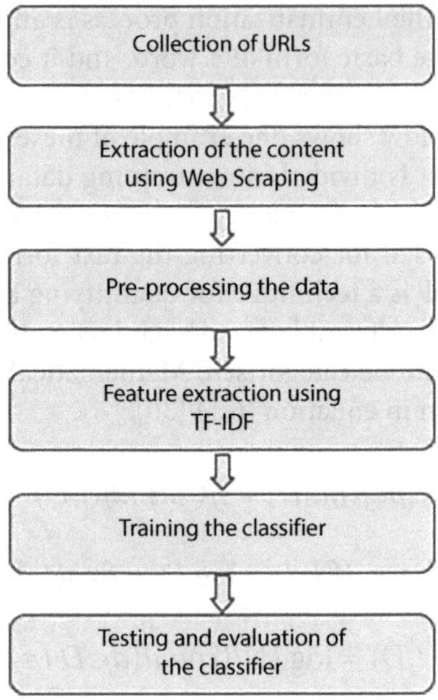

Figure 6.1 Work flow.

Here, 't' is the word in the document 'd' from the document set 'D'.

Initially, five individual supervised Machine Learning algorithms were used to train and test the dataset, namely NB, SVM, KNN, SGD and DT. After that, Ensemble methods bagging, boosting and stacking were used to improve the performance and were compared based on the testing accuracy. The main steps are collecting the relevant URLs, extracting the content, pre-processing the data, and extracting the features with TF-IDF. The next step is to train the individual classifiers and then train the ensemble classifiers with parameter tuning. The final step is testing and evaluating the classifier. Figure 6.1 illustrates the conceptual framework of this study.

6.4 Ensemble Classifiers

This section briefly revisits all the Ensemble methods used in this work, namely Meta-estimator and Random Forest under bagging, Adaptive boosting, Gradient Tree boosting, and Extreme Gradient boosting for boosting ensemble classifier and stacking ensemble classifier.

6.4.1 Bagging

Bagging is an Ensemble method introduced by Leo Breiman in 1994 to reduce overfitting [31]. The purpose of bagging is to combine the predictions of several base learners to produce more accurate results. The idea is to combine the predictions of several base learners to create more accurate output. The bootstrap procedure is applied to a high-variance machine learning algorithm, generally a decision tree. It is used when the goal is to decrease the variance of a decision tree classifier. This process makes several subsets of data randomly selected from the training sample with replacement. Each of the subset is used to train their decision trees. Different trees provide better predictions than a single decision tree classifier, and give better accuracy.

6.4.1.1 Bagging Meta Estimator

Bagging Meta Estimator creates random subsets from the original dataset, and these subsets of the dataset contain all features. These smaller sets are fitted with a base estimator, and finally, the results of each model are combined. The main parameters to adjust when using these methods are the number of trees and the size of the random subsets of features used to split a node.

6.4.1.2 Random Forest

Random Forest is another type of bagging ensemble which uses decision trees as the base estimator. It is an extension to the Bagging Meta Estimator algorithm, but the set of features are selected randomly, which decides the best split at each decision tree node. An individual tree in the Random Forest is fitted on each subset, and the final prediction is obtained by averaging the predictions of all decision trees.

6.4.2 Boosting

Boosting is an ensemble method used to create a collection of predictors. It is a sequential process in which each subsequent model is intended to correct the errors in the previous model. Three types of boosting, such as AdaBoost, Gradient Tree boosting, and XGBoost, have been considered.

6.4.2.1 AdaBoost

Adaptive Boosting or AdaBoost is a boosting technique that builds one robust classifier by combining multiple weak classifiers. It was first proposed by Yoav Freund and Robert Schapire [32].

6.4.2.2 Gradient Tree Boosting

Gradient Tree boosting method (GBM) is another Ensemble algorithm that can be used to solve regression and classification problems. It uses the boosting method to combine weak learners into strong learners. The regression trees are used as the base learner, and each succeeding tree in the series builds on the errors determined by the previous tree.

6.4.2.3 XGBoost

Extreme Gradient boosting (XGBoost) is an improved version of gradient boosting. XGBoost improves the speed and efficiency of the model, and it supports parallelisation by creating decision trees. XGBoost is reported to be ten times faster than other Gradient Boosting techniques and has high predictive power. A variety of regularisation are included to reduce overfitting and enhance overall performance.

6.4.3 Stacking

Unlike bagging, stacking uses different models, which fit the same dataset instead of using samples of the training dataset. Unlike boosting, in stacking, instead of a series of models that correct prior predictions, a single model is used to learn how to combine the predictions from the contributing models.

6.5 Results

The Ensemble methods are implemented for the manually curated dataset (DS1) and the publicly available dataset (DS2). Various parameters have been tuned to find out the best accuracy and analyse the effect of each parameter on accuracy. A comparison of single and Ensemble classifiers has been provided at the end.

6.5.1 Bagging Meta Estimator

Initially, the base estimator has been kept as NB, SVM, KNN, DT and SGD. It has been observed in Table 6.3 that the accuracy of DT and SVM is the best at 99.79%, keeping all other parameters the same. However, since SVM is more time-consuming, DT was taken as the base estimator for further parameter tuning. The best results have been highlighted in Table 6.3.

Once the Decision Tree is fixed as the base estimator, the next step is to tune the number of trees to observe the accuracy. The parameter n_estimators that represents the number of trees has been varied from 2 to 120. As observed in Table 6.4, for the manually collected dataset, DS1, the value of 20 gives the best accuracy of 99.58% taking 1.95 seconds, and after that, it saturates and shows no further improvements. Similarly, for the public dataset, DS2, the best result of 94.37% is observed with the number of estimators set at 90 with training time being 45.27 seconds.

The graph shown in Figure 6.2 represents these trends. As observed, the improvement in accuracy saturates after a certain number of trees and remains constant after that.

Figure 6.3 indicates that training time increases as the number of trees increases.

The parameter max_sample (representing the maximum number of samples to train each base estimator) with a value of 0.5 (50%) gives the best accuracy at 99.58% for DS1. The value of 0.4 (40%) gives the best accuracy of 94.23% for DS2. max_features are equal to the total features (n_features) for regression problems and the square root of the total number of features for classification tasks. max_features were varied from 0.1 to 1.0, which means using 10% to 100% of the features. It is observed in

Table 6.3 Bagging meta estimator with various base estimators for DS1.

Base estimator	Accuracy (%)	F1 score	Precision	Recall
DT	**99.79**	1	1	1
NB	90.26	0.90	0.90	0.91
SVM	**99.79**	1	1	1
KNN	96.06	0.96	0.96	0.96
SGD	99.17	0.99	0.99	0.99

Table 6.4 Number of trees versus accuracy and number of trees versus time.

No. of trees	Accuracy (%)	Training time (seconds)	Accuracy (%)	Training time (seconds)
	DS1	DS1	DS2	DS2
2	99.17	0.277	93.39	1.077
10	99.37	1.078	93.88	5.303
20	**99.58**	**1.957**	94.02	9.937
30	99.58	2.813	94.09	14.923
40	99.58	3.748	94.09	20.144
50	99.58	4.73	94.02	25.283
60	99.58	5.612	94.09	29.924
70	99.58	6.537	94.30	34.745
80	99.58	7.6	94.30	40.249
90	99.58	8.503	**94.37**	**45.271**
100	99.58	9.397	94.23	50.432
110	99.58	10.593	94.30	57.675
120	99.58	11.317	94.16	60.964

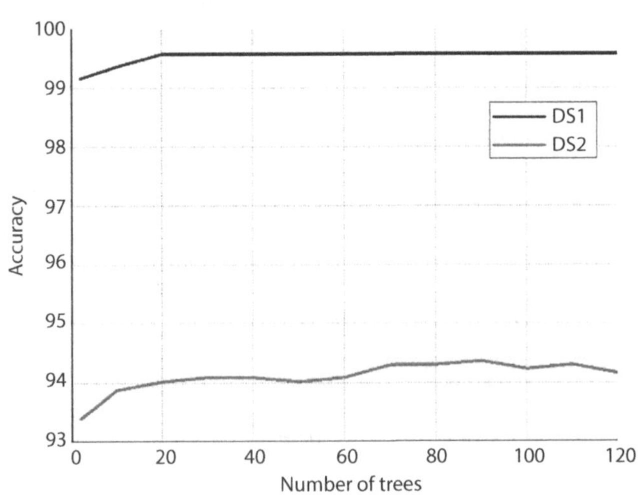

Figure 6.2 Number of trees versus accuracy for bagging meta estimator.

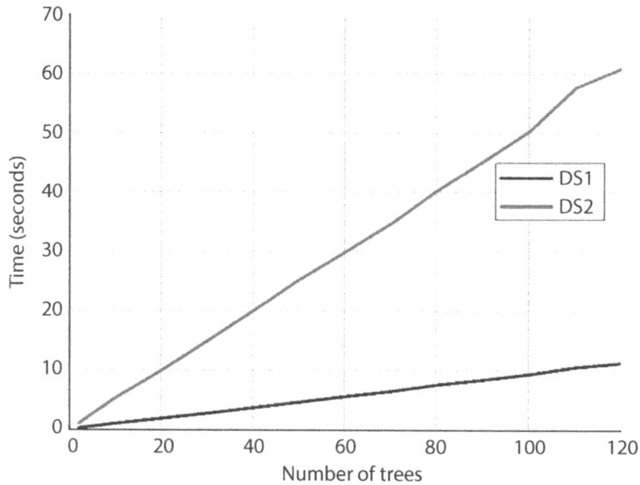

Figure 6.3 Number of trees versus training time in bagging meta estimator.

Table 6.5 Size of the random subset versus accuracy in bagging meta estimator.

Size of the random subset (%)	Accuracy (%)	Accuracy (%)
	DS1	DS2
10	96.27	85.52
20	96.89	91.00
30	**99.79**	92.83
40	99.37	93.67
50	99.58	94.16
60	99.37	**95.22**
70	99.17	94.37
80	99.17	94.51
90	99.37	94.02
100	99.37	93.88

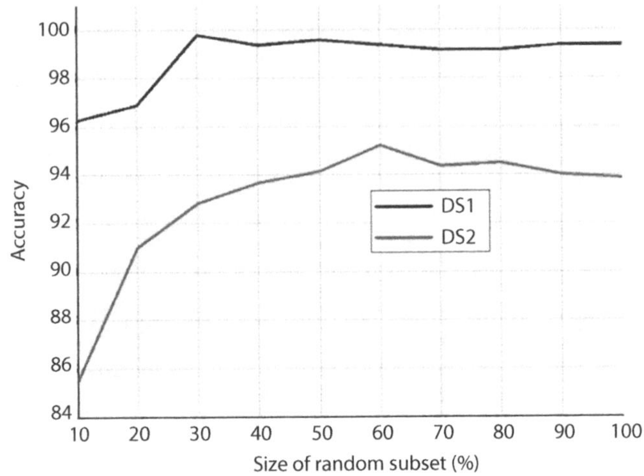

Figure 6.4 Accuracy versus size of random subset in bagging meta estimator.

Table 6.5 that a value of 0.3 for DS1 and 0.6 for DS2 is sufficient to give good accuracy.

These values are represented in the graphs as shown in Figure 6.4. It is observed that once a particular level of accuracy is reached, it remains constant.

Table 6.6 represents the best parameters for Bagging Meta Estimator implementation for DS1. For computation, the parameter bootstrap was kept 'True', meaning that the samples are drawn with replacement and bootstrap features were kept as 'False', meaning that the features were drawn without replacement.

6.5.2 Random Forest

For Random Forest, Table 6.7 represents n_estimators, representing the number of trees versus accuracy and time for both datasets. It is observed that accuracy becomes constant after a certain number of trees. For both DS1 and DS2, the accuracy becomes constant after the n_estimators value of 40. The trends are represented in the graph below in Figure 6.5. As expected, as the number of trees increases, it leads to more time consumption in the training time almost linearly.

Table 6.6 Best parameter of bagging meta estimator for DS1.

Bagging meta estimator	Parameter grid	Best parameter	Accuracy (%)	F1 score	Precision	Recall
n_estimators	{10,2,4,8,10,20,30,50,70, 100,120}	20	99.58	1	1	1
max_samples	{1.0,0.9,0.7,0.5,0.3,0.2}	0.5	99.58	1	1	1
max_features	{1.0,0.9,0.7,0.6,0.3,0.2}	0.3	99.79	1	1	1
Bootstrap	{True,False}	True	99.37	99	99	99
bootstrap_ features	{False,True}	False	99.37	99	99	99

Table 6.7 Number of trees versus accuracy and number of trees versus time for random forest.

No. of trees	Accuracy (%)	Training time (sec)	Accuracy (%)	Training time (sec)
	DS1	DS1	DS2	DS2
10	99.58	0.128	94.30	0.467
20	99.58	0.199	94.72	0.88
30	99.37	0.286	94.82	1.419
40	**99.79**	0.342	**95.01**	1.974
50	99.79	0.43	95.01	2.32
80	99.79	0.67	95.01	3.436
100	99.79	0.796	94.94	4.425
150	99.79	1.174	94.94	6.849
200	99.79	1.626	94.86	8.821

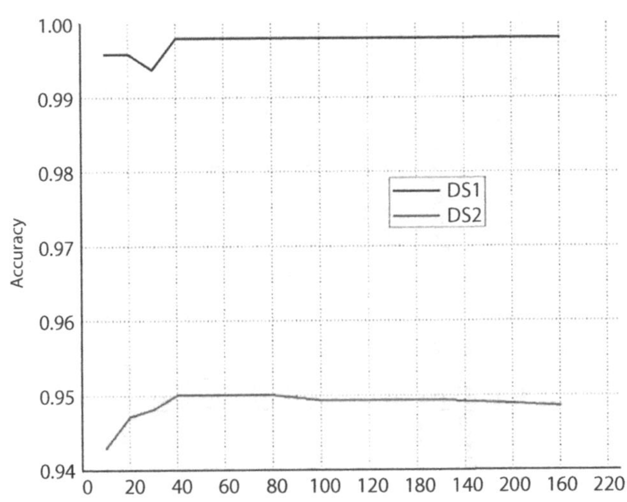

Figure 6.5 Number of trees versus accuracy for bagging random forest.

Most of the time, the performance of a model will not change whether we use Gini or Entropy as the criteria. In this work also the same is observed. For both cases, the accuracy is 99.79%. However, in terms of computation, entropy takes more time. Gini criteria take 0.784 seconds, and entropy criteria take 0.957 seconds.

Gini criteria have been used in all calculations. max_features, i.e., the maximum number of features required to train each base estimator has been set at sqrt, and max_depth is the depth of each tree in the forest. With more splits in the tree, the data is captured in more detail. If max_depth is set to 'None', each tree will expand until every leaf is pure. When all of the data on a leaf is from the same class, the leaf is a pure leaf. Table 6.8 represents the effect of changing the minimum number of samples required at leaf node, min_samples_split. When we increase this parameter, the forest becomes more constrained as each tree needs to consider more samples at each node. When we require all the samples at each node, it is observed that the model is underfitting and cannot learn enough about the data. The best value for both DS1 and DS2 is two and is plotted in Figure 6.6.

min_samples_leaf parameter is similar to min_samples_splits; however, this describes the minimum number of samples at the leaves, the base of the tree. A value of 1 gives the best result for DS1, and a value of 4 gives the best results for DS2.

Table 6.8 min_samples_split versus accuracy in bagging random forest.

min_samples_split	**Accuracy (%)**	**Accuracy (%)**
	DS1	DS2
2	**99.79**	**94.94**
3	99.79	94.86
5	99.58	94..65
10	99.58	94.72
30	99.37	94.72
50	99.37	94.65
100	99.37	94.51
150	99.37	94.30

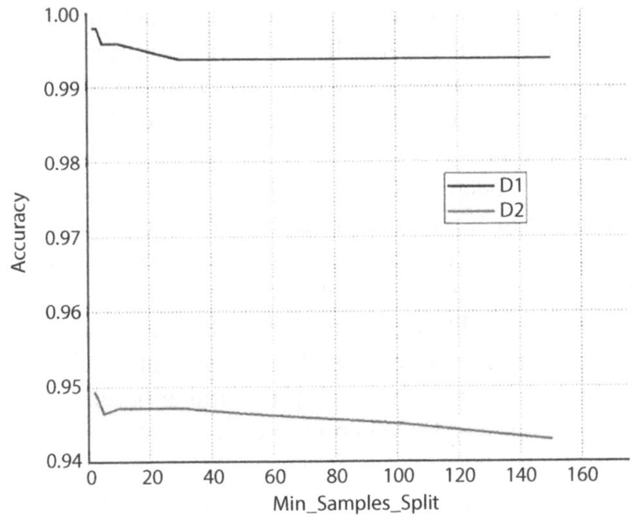

Figure 6.6 min_samples_split versus accuracy for bagging random forest.

The parameter max_leaf_ nodes is implemented using 2, 5, 10, 30 and 50. The value of 'None', the default parameter in the Scikit library, gives the best result. 'None' indicates an unlimited number of nodes.

Table 6.9 gives the best parameters for the implementation of DS1. n_estimator is kept at 100, min_samples_split at 2, min_samples_leaf is 1 with the criteria set to Gini. max_features set at sqrt of the n features. max_depth and max_leaf_nodes are kept as 'None'.

6.5.3 AdaBoost

The AdaBoost algorithm is implemented using both the DS1 dataset as well as with DS2 dataset. The first parameter of the study was to decide the n_estimators. Values of 5, 20, 30, 50, 100, 150, 200, 220, 230, 245, 250, 260, 300 were used on the dataset. The value of 245 comes as the best parameter for DS1 with an accuracy of 99.79% and 50 for DS2 with an accuracy of 94.94%. The other parameter studied was the learning rate. It was varied from 0.1,0.2,0.3,0.4,0.5,0.6,0.7,0.8,0.9 and 1.0. For DS1, a learning rate of 0.5 provides maximum accuracy of 99.79%, and for DS2, the learning rate of 1 gives the maximum accuracy of 94.94%.

Table 6.10 represents the learning rates with respective accuracies for DS1 and DS2, and the rates are plotted in Figure 6.7. As observed, the accuracy becomes constant after a particular level which is more evident in DS2.

Table 6.9 Random forest best parameters for DS1.

Random forest	Parameter grid	Best parameter	Accuracy (%)	F1 score	Precision	Recall
n-estimators	{10,20,30,40,50,80,100, 150,200}	100	99.79	1	1	1
Criterion	{gini,entropy}	Gini	99.79	1	1	1
max_features	{auto,sqrt,log2,None,300}	Sqrt	99.79	1	1	1
max_depth	{None,2,3,4,8,20}	None	99.79	1	1	1
min_ samples_ split	{2,3,5,10,30,50,100,150}	2	99.79	1	1	1
min_ samples_ leaf	{1,2,10,20,50,100}	1	99.79	1	1	1
max_leaf_ nodes	{None,2,5,10,30,50}	None	99.79	1	1	1

Table 6.10 Learning rate versus accuracy in AdaBoost algorithm.

Learning_rate	Accuracy (%)	Accuracy (%)
	DS1	DS2
0.001	98.96	68.37
0.01	98.96	70.06
0.1	98.96	84.89
0.2	99.17	91.35
0.3	99.58	93.32
0.4	99.37	93.67
0.5	**99.79**	94.58
0.6	99.58	94.23
0.7	99.58	94.65
0.8	99.58	94.44
0.9	99.79	94.72
1	99.58	**94.94**

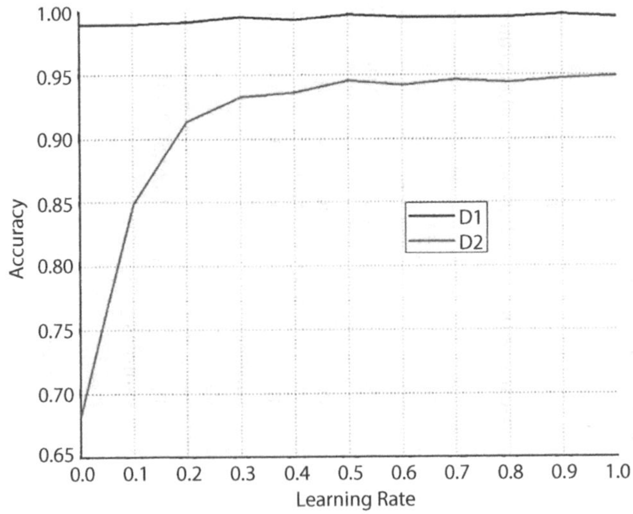

Figure 6.7 Learning rate versus accuracy for AdaBoost algorithm.

Table 6.11 Best parameters of Adaboost for DS1.

Adaboost parameter	Parameter grid	Best parameter	Accuracy (%)	F1 score	Precision	Recall
n_estimators	{50,5,20,30,100,150, 200,220,230,245, 250,260,300}	245	99.79	1	1	1
learning_rate	{1.0,0.1,0.01,0.001, 0.2,0.3,0.4,0.5,0.6, 0.7,0.8,0.9,}	0.9	99.79	1	1	1
Algorithm	SAMME.R,SAMME	SAMME.R	99.58	1	1	1

Table 6.11 tabulates the best parameters for the Adaboost algorithm for the dataset DS1.

6.5.4 Gradient Tree Boosting

Number of estimators was varied with 10, 20, 30, 50, 70, 90, 110, 120, 150, 300 and the maximum accuracy of 99.58% was obtained at 100 for DS1, and 99.22% was obtained at 300 for DS2. Learning rate of 0.1 gives the best accuracy for DS1 and 0.2 for DS2 respectively when the rate was kept at 0.1, 0.2, 0.3, 0.4, 0.5, 0.6, 0.7, 0.8, 0.9 and 1.0. Another parameter of importance is min_samples_split, which varies from 2, 5, 10, 20 and 50. The best accuracy of 99.58% was obtained at 2 for DS1 and 94.72% at 2 for DS2. The parameter min_samples_leaf varied from 1, 10, 20, 30, 50. The best accuracy of 99.58% is obtained at 1 for DS1, and 94.79% is obtained at 10 for DS2.

Table 6.12 represents the max_depth with respective accuracies for DS1 and DS2 and are plotted in Figure 6.8. The maximum accuracy of 99.58% was obtained at 3 for DS1 and 95.01% at 5 for DS2.

Table 6.13 represents the max_leaf_nodes with respective accuracies for DS1 and DS2. This trend is plotted in Figure 6.9.

Table 6.14 tabulates all the best parameters of Gradient Tree boosting for the dataset DS1. n_estimator of 100 with a learning rate of 0.1,

Table 6.12 max_depth versus accuracy for gradient tree boosting.

max_depth	**Accuracy (%)**	**Accuracy (%)**
	DS1	DS2
3	**0.9958**	0.9472
2	0.9937	0.9402
5	0.9937	**0.9501**
10	0.9937	0.9486
20	0.9958	0.9444
30	0.9958	0.9423

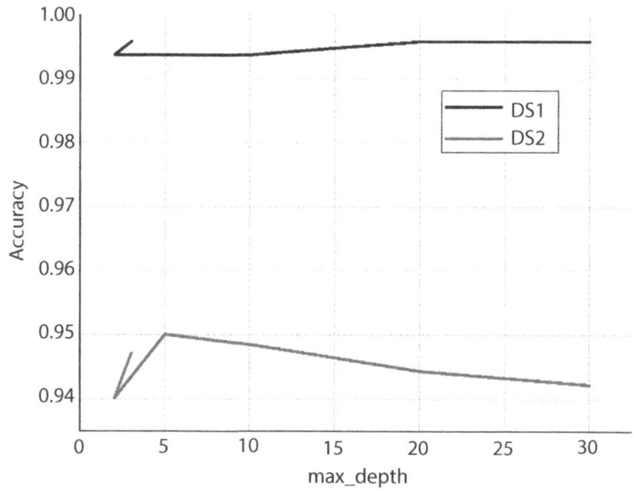

Figure 6.8 max_depth versus accuracy for gradient tree boosting.

Table 6.13 max_leaf_nodes versus accuracy for gradient tree boosting.

max_leaf_nodes	**Accuracy (%)**	**Accuracy (%)**
	DS1	DS2
None	**99.58**	**94.65**
2	99.17	90.65
3	99.37	93.6
5	99.58	94.3
10	99.58	94.65
15	99.58	94.65

min_samples_leaf of 1 ,min_samples_split of 2, min_weight_fraction_leaf kept at 0.0, max_depth at 3, max_leaf_nodes and max_features set as 'None' gives the best results. min_weight_fraction_leaf is similar to min_samples_leaf, but instead of representing an integer value, it represents a fraction of the number of observations. The default value is 0.0.

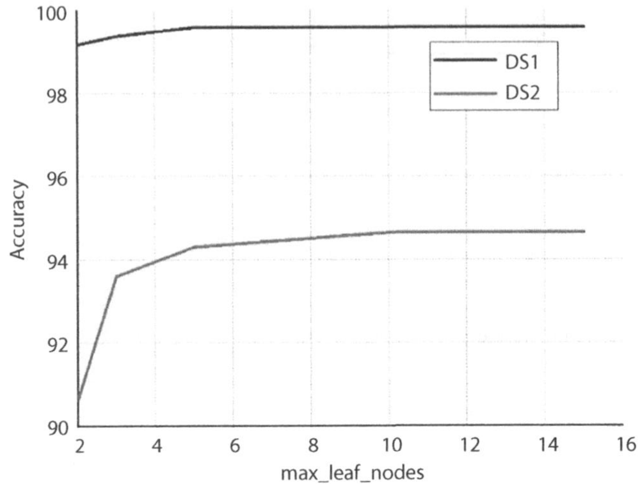

Figure 6.9 max_leaf_nodes versus accuracy for gradient tree boosting.

6.5.5 XGBoost

For XGBoost, learning_rate, min_child_weight, max_depth, subsample and the colsample_bytree were tuned. Learning rate was tuned from 0.1, 0.2, 0.3, 0.4, 0.5, 0.6, 0.7, 0.8, 0.9, 1.0. The best accuracy 99.79% observed at 0.3.

Table 6.15 represents the min_child_weight with respective accuracies for DS1 and DS2 and is plotted in Figure 6.10. The best accuracy of 99.79% was observed at 1 for DS1 and 95.43% at 2 for DS2.

Table 6.16 represents the subsample with respective accuracies for DS1 and DS2, plotted in Figure 6.11. The best accuracy of 99.79% was observed at 1 for DS1 and 95.57% was observed at 0.9 for DS2, when the rate was varied from 0.01, 0.1, 0.2, 0.3, 0.4, 0.5, 0.6, 0.7, 0.8, 0.9 and 1.0.

Table 6.17 tabulates all the best parameters of XGBoost for DS1. Learning_rate of 0.3, min_child_weight of 1, max_depth at 6, subsample kept at 1, colsample_bytree at 1, gives the best results.

Table 6.14 Gradient tree boosting best parameters for DS1.

Gradient boosting	Parameter grid	Best parameter	Accuracy (%)	F1 score	Precision	Recall
n-estimators	{100,10,20,30,50,70,90,120, 150,300}	100	99.58	1	1	1
learning_rate	{,0.1,0.2,0.3,0.4,0.5,0.6,0.7,0.8, 0.9,1.0}	0.1	99.58	1	1	1
min_samples_split	{2,5,10,20,50}	2	99.58	1	1	1
min_samples_leaf	{1,10,20,30,50}	1	99.58	1	1	1
min_weight_ fraction_leaf	(0.0,0.1,0.2,0.3,0.4}	0.0	99.58	1	1	1
max_depth	{3,5,10,20,30}	3	99.58	1	1	1
max_leaf_nodes	{None,2,3,5,10,15}	None	99.58	1	1	1
max_features	{None,auto,sqrt,log2,}	None	99.58	1	1	1

Table 6.15 min_child_weight versus accuracy in XGBoost.

min_child_weight	Accuracy (%)	Accuracy (%)
	DS1	DS2
1	**99.79**	95.29
2	99.79	**95.43**
5	99.37	94.94
10	99.37	95.01
20	98.96	92.62
30	98.96	87.56

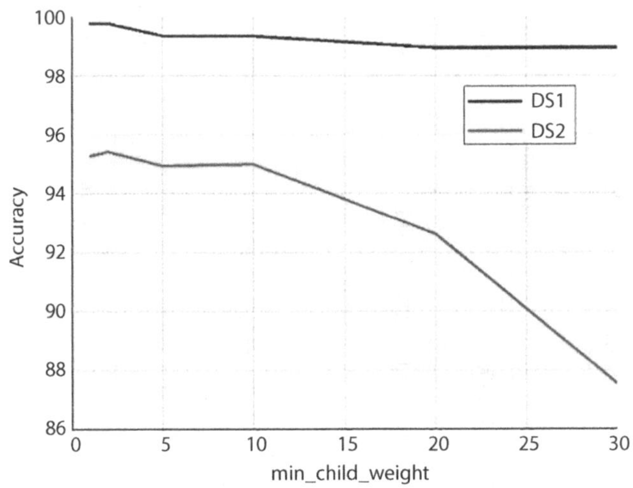

Figure 6.10 min_child_weight versus accuracy for XGBoost.

6.5.6 Stacking

Table 6.18 shows the important parameters of stacking and its results for DS1. DT, NB, KNN, SVM, and SGD are used as the base estimators, stacked together. Furthermore, one classifier is used as a final estimator, used to combine the base estimators. Logistic regression is the default classifier for the final estimator. It gives an accuracy of 99.58%. DT gives 99.37%, NB gives 99.17%, KNN gives 99.79%, SVM gives 99.58%, and SGD gives an accuracy of 99.79%. It is observed that when the final estimator is KNN or

Table 6.16 Subsample versus accuracy in XGBoost.

subsample	Accuracy (%)	Accuracy (%)
	D1	D2
0.01	98.96	84.11
0.1	98.96	94.16
0.2	99.58	94.72
0.3	99.79	94.94
0.4	99.79	95.08
0.5	99.79	95.15
0.6	99.58	95.22
0.7	99.58	95.50
0.8	99.79	95.22
0.9	99.79	**95.57**
1	**99.79**	95.29

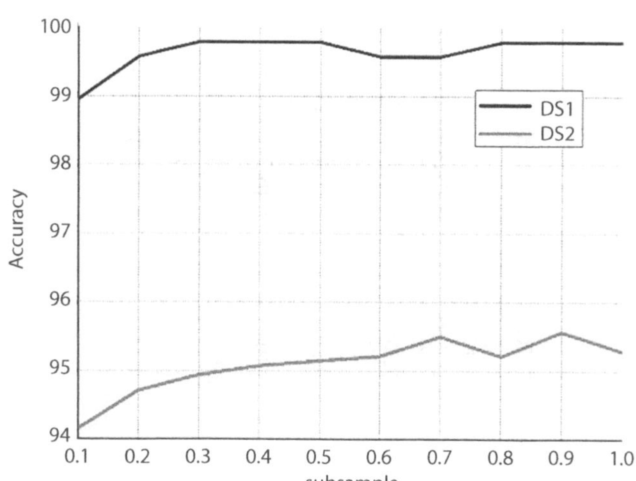

Figure 6.11 Subsample versus accuracy for XGBoost.

Table 6.17 XGBoost best parameters.

Extreme boost parameters	Parameter grid	Best parameter	Accuracy (%)	F1 score	Precision	Recall
Learning_rate	{0.1,0.2,0.3,0.4,0.5,0.6,0.7, 0.8,0.9,1.0}	0.3	99.79	1	1	1
min_child_weight	{1,2,5,10,20,30}	1	99.79	1	1	1
max_depth	{6,2,3,5,10,50}	6	99.79	1	1	1
Subsample	{1,0.01,0.1,0.2,0.3,0.4,0.5, 0.6,0.7,0.8,0.9}	1	99.79	1	1	1
colsample_bytree	(1,0.01,0.1,0.2,0.4,0.6,0.8}	1	99.79	1	1	1

Table 6.18 Stacking parameters for DS1.

Stacking parameters		Accuracy (%)	F1 score	Precision	Recall	Training time (sec)
Estimators	DT, NB, KNN, SVM, SGD					
Final_estimators	Logistic Regression	99.58	1	1	1	15.737
	DT	99.37	0.99	0.99	0.99	15.796
	NB	99.17	0.99	0.99	0.99	15.971
	KNN	99.79	1	1	1	15.725
	SVM	99.58	1	1	1	15.857
	SGD	99.79	1	1	1	15.939

SGD, then the accuracy is high. The table shows the training time also, but not much difference is observed.

6.5.7 Comparison with Single Classifiers

As we can see in Table 6.19, DT is giving accuracy of 99.17%, NB is giving 90.26%, SVM is giving 99.58 %, KNN is giving 95.65%, SGD is giving 98.55%, Bagging Meta Estimator (BME) is giving 99.79%, Random Forest (RF) is giving 99.79%, AdaBoost is giving 99.79%, Gradient Tree boosting (GTB) is giving 99.58%, XGBoost (XGB) is giving 99.79% and stacking is giving accuracy of 99.79% for the dataset DS1. It is observed that ensemble classifiers are giving better accuracy than individual classifiers, as expected. In the dataset DS2, DT is giving accuracy of 92.83% NB is giving 92.83%, SVM is giving 94.86%, KNN is giving 93.74%, SGD is giving 94.58%, BME is giving 95.22%, RF is giving 95.36%, Adaboost is giving 94.94%, GTB is giving 95.22%, XGB is giving 95.57% and stacking is giving accuracy of 95.63%. Here the results once again show that ensemble classifiers are giving better results. Another observation is that SVM is giving similar results to ensemble classifiers. Figure 6.12 represents the accuracies of different classifiers.

Table 6.19 Accuracy of Single classifier and ensemble classifier in DS1 and DS2.

Classifiers	DS1	DS2
DT	99.17	92.83
NB	90.26	92.83
SVM	99.58	94.86
KNN	95.65	93.74
SGD	98.55	94.58
Bagging Meta Estimator (BME)	99.79	95.22
Random Forest (RF)	99.79	95.36
AdaBoost	99.79	94.94
Gradient Tree boosting (GTB)	99.58	95.22
XGBoost (XGB)	99.79	95.57
Stacking	99.79	95.36

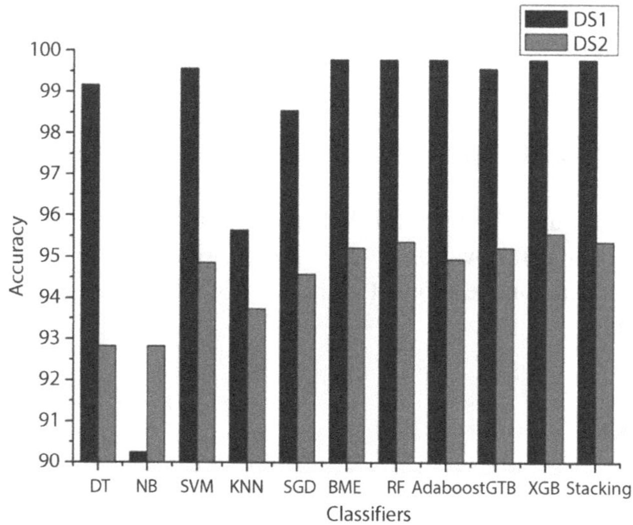

Figure 6.12 Accuracy of single classifiers and ensemble classifiers.

6.6 Conclusion

When used for web page classification, Ensemble classifiers such as Bagging Meta Estimator, Random Forest, AdaBoost, XGBoost, and stacking show good accuracy with 99.79% maximum accuracy for the two datasets. Gradient Tree boosting and SVM also shows good accuracy with 99.58%. In this study, five individual classifiers, namely DT, NB, SVM, KNN, and SGD, were trained and tested. Ensemble models, namely Bagging Meta Estimator, Random Forest, AdaBoost, Gradient Tree boosting and XGBoost, were trained and tested. A comparison has been made between individual classifiers and Ensemble classifiers. The performance of the classifiers has been evaluated using accuracy as the performance metric. Parameter tuning for Ensemble models was performed to find out the optimum values which give the best accuracy. This study was used to compare a manually collected dataset (DS1) and a public dataset (DS2) to increase the confidence in the prediction. In DS1, the Testing Accuracy of individual classifiers ranges from 90.26% to 99.58%, and the maximum overall accuracy is 99.79% obtained for the Bagging Meta Estimator, Random Forest, AdaBoost, XGBoost and stacking Ensemble. In DS2, the testing accuracy of individual classifiers ranges from 92.83% to 94.86%, and maximum accuracy of 95.57% is obtained for the XGBoost ensemble. In both the dataset, it is observed that XGBoost is faster than Gradient

Tree Boosting. The study concludes that Ensemble models are giving good accuracy compared to individual classifiers.

Acknowledgement

The authors would like to thank WIZnet Co. Ltd., South Korea, and the Centre for Digital Innovation at the CHRIST (Deemed to be University), Bangalore, India, for the support without which this work would not have been possible.

References

1. "Machine Learning for Web Page Classification: A Survey | lassri | International Journal of Information Science and Technology." https://innove.org/ijist/index.php/ijist/article/view/137 (accessed Mar. 23, 2020).
2. X. Acm Reference Format: Qi and B. D. Davison, "Web page classification: Features and algorithms," *ACM Comput. Surv*, vol. 41, p. 31, 2009, doi: 10.1145/1459352.1459357.
3. "Web search using automatic classification." http://www.ra.ethz.ch/CDstore/www6/Posters/725/Web_Search.html (accessed Aug. 17, 2020).
4. D. M. and R. R. Selvakuberan K, "Classification of web pages using machine learning techniques," in *Machine learning: Concepts, methodologies, tools and applications*, 2012, pp. 50–65.
5. P. Vinod, Pooja & Prajapati, "Comparative Study of Web Page Classification Approaches," *Int. J. Comput. Appl.*, vol. 179, no. 6–9, 2018.
6. O. W. Kwon and J. H. Lee, "Web page classification based on k-nearest neighbor approach," in *Proceedings of the 5th International Workshop on Information Retrieval with Asian Languages, IRAL 2000*, Nov. 2000, pp. 9–15, doi: 10.1145/355214.355216.
7. A. Sun, E.-P. Lim, and W.-K. Ng, "Web classification using support vector machine," in *Proceedings of the Fourth International Workshop on Web Information and Data Management - WIDM '02*, 2002, p. 96, doi: 10.1145/584931.584952.
8. Z. Mašetic, A. Subasi, and J. Azemovic, "Malicious Web Sites Detection using C4.5 Decision Tree," *Southeast Eur. J. Soft Comput.*, vol. 5, no. 1, May 2016, doi: 10.21533/scjournal.v5i1.109.
9. B. Altay, T. Dokeroglu, and A. Cosar, "Context-sensitive and keyword density-based supervised machine learning techniques for malicious web-page detection," *Soft Comput.*, vol. 23, no. 12, pp. 4177–4191, Jun. 2019, doi: 10.1007/s00500-018-3066-4.

10. P. Manchanda, S. Gupta, and K. Kumar Bhatia, "On The Automated Classification of Web Pages Using Artificial Neural Network," *OSR Journal of Computer Engineering (IOSRJCE)*, vol. 4, no. 1, pp. 20–25, Sep-Oct 2012.
11. P. D. P. Woogue, G. A. A. Pineda, and C. V. Maderazo, "Automatic Web Page Categorisation Using Machine Learning and Educational-Based Corpus," *Int. J. Comput. Theory Eng.*, vol. 9, no. 6, pp. 427–432, 2017, doi: 10.7763/ijcte.2017.v9.1180.
12. A. S. Patil and B. V. Pawar, "Automated Classification of Web Sites using Naive Bayesian Algorithm," *undefined*, 2012.
13. R. Polikar, "Ensemble based systems in decision making," *IEEE Circuits and Systems Magazine*, vol. 6, no. 3. pp. 21–44, 2006, doi: 10.1109/MCAS.2006.1688199.
14. Yan-Shi Dong and Ke-Song Han, "A comparison of several ensemble methods for text categorisation," Dec. 2004, pp. 419–422, doi: 10.1109/scc.2004.1358033.
15. Z. Han *et al.*, "An Ensemble Learning-based model for Classification of Insincere Question," 2019.
16. S. Wan and H. Yang, "Comparison among methods of ensemble learning," in *Proceedings - 2013 International Symposium on Biometrics and Security Technologies, ISBAST 2013*, 2013, pp. 286–290, doi: 10.1109/ISBAST.2013.50.
17. Y. W. Kim, J. Innila Rose, and A. V. N. Krishna, "Accuracy Enhancement of Portrait Segmentation by Ensembling Deep Learning Models," in *Proceedings - 2020 5th International Conference on Research in Computational Intelligence and Communication Networks, ICRCICN 2020*, Nov. 2020, pp. 59–64, doi: 10.1109/ICRCICN50933.2020.9296196.
18. Y.-W. Kim, Y.-C. Byun, and A. V. N. Krishna, "Portrait Segmentation Using Ensemble of Heterogeneous Deep-Learning Models," *entropy*, vol. 23, no. 2, pp. 1–20, Feb. 2021, doi: 10.3390/E23020197.
19. "Github link." https://github.com/anurag19997/Web-Page-Classification-using-LSTM.
20. Y. Wang, J. Hodges, and B. Tang, "Classification of Web Documents Using a Naive Bayes Method," in *Proceedings of the International Conference on Tools with Artificial Intelligence*, 2003, pp. 560–564, doi: 10.1109/tai.2003.1250241.
21. M. Tsukada, T. Washio, and H. Motoda, "Automatic Web-Page Classification by Using Machine Learning Methods," Springer, Berlin, Heidelberg, 2001, pp. 303–313.
22. A. Onan, "Classifier and feature set ensembles for web page classification," *J. Inf. Sci.*, vol. 42, no. 2, pp. 150–165, Apr. 2016, doi: 10.1177/0165551515591724.
23. R. Xia, C. Zong, and S. Li, "Ensemble of feature sets and classification algorithms for sentiment classification," *Inf. Sci. (Ny).*, vol. 181, no. 6, pp. 1138–1152, Mar. 2011, doi: 10.1016/j.ins.2010.11.023.
24. G. Wang, J. Sun, J. Ma, K. Xu, and J. Gu, "Sentiment classification: The contribution of ensemble learning," *Decis. Support Syst.*, vol. 57, no. 1, pp. 77–93, Jan. 2014, doi: 10.1016/j.dss.2013.08.002.

25. H. T. Elshoush and E. A. Dinar, "Using adaboost and stochastic gradient descent (SGD) algorithms with r and orange software for filtering e-mail spam," in *2019 11th Computer Science and Electronic Engineering Conference, CEEC 2019 - Proceedings*, Sep. 2019, pp. 41–46, doi: 10.1109/CEEC47804.2019.8974319.
26. W. A. AWAD, "Machine Learning Algorithms in Webpage Classification," *Int. J. Comput. Sci. Inf. Technol.*, vol. 4, no. 5, 2012.
27. S. K. Dwivedi and C. Arya, "News web page classification using url content and structure attributes," in *Proceedings on 2016 2nd International Conference on Next Generation Computing Technologies, NGCT 2016*, Mar. 2017, pp. 317–322, doi: 10.1109/NGCT.2016.7877434.
28. W. Petprasit and S. Jaiyen, "E-commerce web page classification based on automatic content extraction," in *Proceedings of the 2015 12th International Joint Conference on Computer Science and Software Engineering, JCSSE 2015*, Aug. 2015, pp. 74–77, doi: 10.1109/JCSSE.2015.7219773.
29. S. Qaiser and R. Ali, "Text Mining: Use of TF-IDF to Examine the Relevance of Words to Documents," *Int. J. Comput. Appl.*, vol. 181, no. 1, pp. 25–29, Jul. 2018, doi: 10.5120/ijca2018917395.
30. "What is TF-IDF?" https://monkeylearn.com/blog/what-is-tf-idf/ (accessed Aug. 17, 2020).
31. L. Breiman, "Bagging predictors," *Mach. Learn.*, vol. 24, no. 2, pp. 123–140, Aug. 1996, doi: 10.1023/A:1018054314350.
32. Y. Freund and R. E. Schapire, "A Decision-Theoretic Generalization of On-Line Learning and an Application to Boosting," *Journal of Computer and System Sciences* 55, 119-139, 1997.

7

Feature Engineering and Selection Approach Over Malicious Image

P.M. Kavitha[1]* and B. Muruganantham[2]

[1]Department of Computer Applications, SRM Institute of Science and Technology, Ramapuram, Chennai, India
[2]Department of Computer Science and Engineering, SRM Institute of Science and Technology, Kattankulathur, Chennai, India

Abstract
Raw data gets transformed into features representing the problem in a more improved manner. The problem gets represented in predictive models and in turn the accuracy is at a higher rate on the data. With feature engineering, the input data is turned into the machine learning algorithms, using the domain knowledge. Various feature extraction techniques like color, texture, shape, position and edge are worked upon a gray scale image to analyze the malware and its features. The malware image being a gray scale image, the featuring methods are performed to work on the malware. An irrelevant or partially relevant model feature may depreciate the model performance. When a model is first designed, feature selection and data cleaning are the foremost steps to be attended to. The feature selection models that are in practice are Univariate Selection, Feature Importance, Heatmap. These models can be utilized to improve the result on a model when applied.

***Keywords*:** Heatmap, malware variant, handling outliers, Binning, GAN

7.1 Introduction

For the construction of all the machine learning and deep learning models, the data is the most important and compromising aspect. The contribution to the dataset choose to the model planning will be of various sorts

**Corresponding author*: pmkavimca@gmail.com

Kukatlapalli Pradeep Kumar, Aynur Unal, Vinay Jha Pillai, Hari Murthy, and M. Niranjanamurthy (eds.) Data Engineering and Data Science: Concepts and Applications, (173–194) © 2023 Scrivener Publishing LLC

like content, pictures, numerals and so on. Feature is the property of data available in the dataset which helps out in prediction analysis. Features influence a lot in the result of predictive models. The main objectives of feature engineering task are to prepare proper dataset and to take care of performance improvement of the predictive model. Also, the predictive power of the machine learning algorithms will tend to increase with a right feature engineering process. It creates features with the raw data and works more predictively on a machine learning algorithm. A new feature may be created with the old features either by combining two new features or simply by creating one new feature by overseeing the existing features. In cases, some features that look unwanted may be removed too. Various steps involved in the feature engineering process include brainstorming features, creating features, the working progress of the model and finding the appropriate feature for the model. The main goal of feature engineering and selection lies in providing tools for representing the predictors to develop a good framework [7, 19]. Nowadays the internet is used for almost all activities, which helps attackers to formulate new malware variants with the help of automated software. It is said that there is a tremendous increase in malware day by day. With a rapid growth in malware, an enhanced malware analysis is much required to end the malicious activities that are expanding day by day. A more accurate and highly efficient model is needed that characterizes the malware variants. Feature extraction is a bit of the aspect non-acceptance measure, in which a necessary movement in information is segregated and diminished to more sensible social events. So when it is needed to be managed it will be not be flexible. The ultimate moto of these instructive assortments is that they have innumerable variables. These components require a huge load of figuring resources to handle them. So feature extraction helps in finding the appropriate data by choose and unite factors into features, thus, reasonably diminishing the proportion of data. These features are not hard to quantify, but simultaneously prepared to depict the authentic educational record with exactness and innovativeness. The technique for removing the component is useful when you have an enormous educational assortment and need to reduce the amount of resources without losing any huge or significant information. Feature extraction helps with minimizing the percentage of high volume of data from the educational assortment. In the end, the abatement of the data helps with building the model with less machine's undertakings and besides accelerate learning and theory steps in the data science cycle. Feature designing alludes to the way toward utilizing area information to choose and change the most important factors from crude information while making a unique model utilizing the entire information. The objective of feature engineering and determination is to work on the exhibition of AI (ML) calculations.

Picture document designs are the normalized method for getting sorted out and putting away advanced pictures. Figure 7.1 represents the phases that are involved in feature engineering task. Rasterization changes over the image data into a grid of pixels. Every pixel has various pieces to assign its tone. Rasterizing a picture record for a particular gadget considers the quantity of pieces per pixel (the shading profundity) that the gadget is intended to deal with. The size of raster picture documents is decidedly corresponded with the quantity of pixels in the picture and the shading profundity (bits per pixel). Pictures can be compacted differently. A pressure calculation stores either a precise portrayal or an estimation of the first picture in fewer bytes. The fundamental importance of Image handling insinuates getting ready of cutting edge picture, i.e., taking out the commotion and any kind of irregularities present in an image using the advanced technique. The irregularity might move into the image, everything considered during its turn of events or during change and so on. For mathematical examination, an image may be described as a 2D limit f(x, y) where x and y are spatial (plane) coordinates, and the abundance of any pair of sorts out (x, y) is known as the image's force or dim level at that place. We call the image a high-level picture when x, y, and the force potential gains off are all constrained, discrete sums inside and out. It is critical that a high-level picture be composed of a predetermined number of parts, each of which has a distinct region and value. Picture segments, picture parts, pels, and pixels are all names for these segments. Pixel is the most commonly used term to refer to the segments of an electronic image.

With Feature Engineering, the raw data transfigures into features which are used by a predictive model either in Machine learning or deep learning. The combinations of business knowledge, mathematics and statistics are put together in transforming the data as per the machine learning models. Then the hidden patterns are uncovered, thereby predicting and boosting the machine learning data. The main aspect focuses on the formulation of the data set that fits into an ML model. Besides selecting the right

Figure 7.1 Feature engineering phase in model creation.

features and using them in a significant model, the performance of the same is refined, along with the performance improvement in ML models. Additionally, the exact features are chosen, thereby reducing the size of the feature set results in feasibility of ML algorithms and data analytic algorithms. Data scientists have gained a great significance with this engineering as it accelerates the total time taken in extracting the variables. Also, better models are created with higher accuracy, which promotes the usage among the organizations and data scientists.

The main contribution of this chapter can be summarized as the feature selection and a few feature engineering techniques that have been implemented in the existing 9,700 samples. This enables us to enhance the malicious images [11, 12]. A good volume of image has been generated, which may be used as input dataset in the CNN algorithm in future. Choosing the best features for a model gives the most accurate result and Feature Engineering takes its place. The best-suited algorithm for the ML problem is chosen by the Feature Engineering. The contrast reduction, heat map, and noisy image identification are the ML techniques implemented in here. These techniques are chosen based on the Feature Engineering, which on the other hand improves the flexibility of the problem case. The focus is on multiplying the input data set, which is image here. Upon multiplying the images without restoration of the quality, the ultimate aim is to increase the quantity of the input data set. Thus, the machine can be trained on with numerous inputs and will be capable of identifying and working on many data. One important aspect of Feature Engineering is the reduction of the redundancy without compromising on the quality of the input, which is most foremost of all the other techniques.

7.2 Feature Engineering Techniques

Feature engineering consists of creation, transformation, extraction, and selection of features. A few of the processes that are involved in feature engineering are Creating the feature, Transformation, Feature Extraction and Selection. While getting statistical information over a few decades the image processing strategies have been imported [1, 3]. Most strategies are developed to improve images obtained from automated shuttles, space tests, and military observation flights. Image Processing frameworks are getting more importance as they are readily available and accessible as in PCs, enormous size memory gadgets, and designs.

7.2.1 Methodologies in Feature Engineering

The feature engineering task consists of various methodologies. A few of the methodologies are Imputation, Handling outliers, Binning, Log transform, One-Hot Encoding, Grouping Operations, Feature split, Scaling and Extraction date. In order to manage the task of missing value in data set a few rows and columns will be dropped in the dataset based on 70% ratio. The algorithms we use also accept the dataset. This type of technique is named as imputation. Imputation is categorized into numerical and categorical based on the values in the dataset. In order to manage with statistical data we will deal with outlier technique in feature engineering. Outlier detection also deals with percentile and dilemma also mentioned as Drop or cap. Another numerical technique to identify exceptions is to utilize percentiles. You can expect a specific percentage of the worth from the top or the base as an anomaly. The central issue here is to set the rate esteem indeed, and this relies upon the dispersion of your information as referenced before. Another choice for taking care of exceptions is to cover them as opposed to dropping. The fundamental inspiration of binning is to make the model more vigorous and forestall overfitting, in any case, it has an expense for the exhibition [2, 4]. The compromise among execution and overfitting is the central issue of the binning cycle. Logarithm transformation (or log change) is perhaps the most regularly utilized numerical changes in highlight designing. It assists with taking care of slanted information and after change, the conveyance turns out to be more rough to ordinary. One-hot encoding is quite possibly the most well-known encoding technique in AI. This technique spreads the qualities in a segment to various banner segments and doles out 0 or 1 to them. These paired qualities express the connection among assembled and encoded section [14, 16]. Mathematical sections are gathered utilizing total and mean capacities in a large portion of the cases. Both can be ideal as per the significance of the component. For instance, assuming you need to get proportion sections, you can utilize the normal of paired segments [6–8]. Splitting the features is a decent method to make them valuable as far as AI. More often than not the dataset contains string sections that disregards clean information standards. By extricating the utilizable pieces of a section into new highlights. The numerical features of the dataset do not have a specific reach and they vary from one another. All things considered, it is garbage to anticipate that age and income columns should have a similar reach. Scaling tackles this issue. The nonstop highlights become indistinguishable as far as the reach, after a scaling cycle. This interaction is not compulsory for some calculations; however, it very well may be as yet ideal to apply. The two common methods of scaling are normalization and standardization. Normalization is calculated using Eq. 7.1.

Table 7.1 Techniques available in feature selection process.

Feature selection techniques		Methods
Filter	A generic set of methods which will not incorporate a specific algorithm	Information Gain Chi-Square test Fisher Score Correlation Coefficient Variance threshold
Wrapper	The method which evaluates Specific ML Algorithm	Recursive Feature Elimination Sequential Feature Selection Algorithm Genetic Algorithm
Embedded	The method that embeds various features	L1(LASSO) Regularization Decision Tree

$$X_{norm} = {(X - X_{\min})}/{(X_{max} - X_{min})} \tag{7.1}$$

Where the range is between 0 and 1. By the same way standardization is also calculated using the mean function.

$$z = \frac{x - \mu}{\sigma} \tag{7.2}$$

Even though the date segments typically give significant data about the model objective, they are dismissed as an information or utilized counter-intuitively for the AI calculations. It very well may be the justification this, that dates can be available in various organizations, which make it difficult to comprehend by calculations. Table 7.1 explains the various feature selection techniques available.

7.2.2 Strides in Feature Engineering

The competence of feature engineering may differ based on the feature engineer who is working on the related task. Nevertheless, the steps involved in the performance of feature engineering are almost the same. Every data involved in a feature engineering task come across few common strategies and strides. At most Machine learning and deep learning

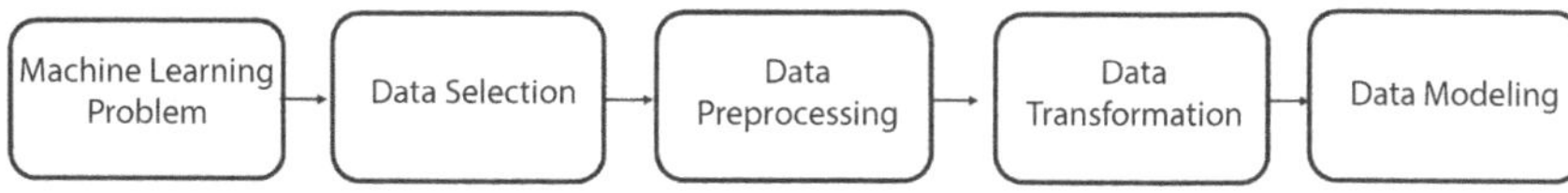

Figure 7.2 Strides involved in feature engineering task.

algorithms include the below-stated strides. Data Selection, Data Pre-Processing, Data transformation and Data modeling. Data selection is the process in which the model is associated with data integration, data denormalization of data in dataset. Data pre-processing involves formatting and cleaning the data from noise. Data transformation involves transforming data based on the requirement of feature engineering task. Data modeling deals with the creation of model and evaluating the model developed based on the requirement. Figure 7.2 demonstrates the various steps in the task performed by feature engineering. The figure represents the flow of data from one stage to another. Each stage is most prime factor that may impact in the task.

7.2.3 Feature Extraction

Feature extraction can be stated as the method of acquiring a new set of features from raw data available in dataset. For tabular type of information, this may incorporate projection strategies like image processing and Unsupervised clustering techniques. For picture information, this may incorporate line or edge location. Contingent upon the area, picture, video and sound perceptions loan themselves to enhance in the large numbers of similar variety of DSP strategies. Key to highlight extraction is that the strategies are programmed (despite the fact that may should be planned and built from more straightforward techniques) and tackle the issue of unmanageably high dimensional information, most regularly utilized for simple perceptions put away in computerized designs [5, 9]. On the off chance that the quantity of highlights becomes comparative (or significantly greater!) than the quantity of perceptions put away in a dataset then this can in all likelihood prompt a Machine Learning model experiencing overfitting. To avoid overfitting, it is important to apply either regularization or dimensionality decrease procedures (Feature Extraction). In Machine Learning, the dimensionality of a dataset is equivalent to the quantity of factors used to address it. Utilizing Regularization could surely assist with decreasing the danger of overfitting; however, utilizing rather Feature Extraction strategies can likewise prompt different kinds of benefits, like Improvement in accuracy, reduction risk in overfitting, training the data with high speed, improvement in data visualization and Increase in the explainable features of the model developed. Feature Extraction tends to reduce the number of respected features available in the dataset by organizing new features which are framed by existing features. These new decreased plans of features should then have the alternative to summarize by far most of the information contained in the principal course of action of features.

7.2.4 Feature Selection

Feature selection is the process normally utilized strategy to lessen the quantity of highlight in a dataset. The contrast between Feature Selection and Feature Extraction is that highlight choice points rather to rank the significance of the current highlights in the dataset and dispose of less significant ones. Feature Selection is the process of selecting the features either by physical or logical selection and observing the methodologies to be used. Having excess volume of features tends our extraction process to extract the information which reduces the originality of the features included in dataset available. The benefit in applying the feature selection strategies is it supports in reducing Overfitting, Improvement in accuracy and reduces the training time. Many feature selection techniques exist. A few of them are Univariant Selection, Feature importance, and Correlation matrix. LASSO and RIDGE are very popular methods in the feature selection process. Univariant selection can be mentioned as measurable tests can be used to pick those features that have the most grounded relationship with the yield variable. The scikit-learn library gives the Select K Best class that can be used with a set-up of different authentic tests to pick a specific number of highlights. By including significance we can get the part meaning of every component of your dataset by using the component importance property of the model. Highlight importance gives you a score for every segment of your data, the higher the score more huge or appropriate is the component towards your yield variable. Highlight importance is an inbuilt class that goes with Tree Based Classifiers; we will use Extra Tree Classifier for isolating the best 10 features for the dataset. Relationship states how the features are related to each other or the goal variable. Relationship can be positive (extension in one worth of feature fabricates the value of the goal variable) or negative (development in one worth of feature reduces the value of the goal variable). Figure 7.3 showcases the various selection engineering tasks performed by the model in order to give the most required output.

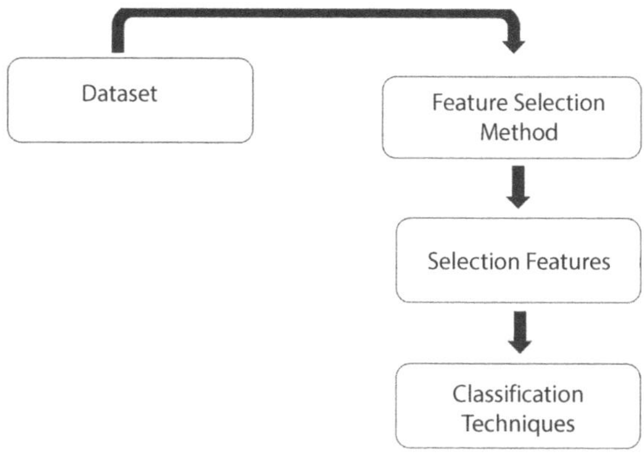

Figure 7.3 Feature selection technique applied in classification technique of machine learning.

7.2.5 Feature Engineering in Image Processing

Choosing the most significant highlights is a pivotal advance during the time spent grouping issues particularly in handwritten dataset. It is important to track down all conceivable component subsets that can be shaped from the underlying set which bring about tedious; each element is significant for essentially some of separations, and varieties inside intra-class and between class is not an excessive amount of high. Past a specific point, the consideration of extra highlights prompts a more regrettable as opposed to better execution. A decent list of features contains segregated data, which can recognize one item from different class. Features can be of two classifications. One is local feature and other is global feature. Local features are generally geometric and global features are topological.

7.2.6 Importance of Feature Engineering in Image Processing

Feature engineering is an imperative piece of image processing. Without this progression, the exactness of your AI calculation lessens altogether. A machine learning processing begins with information assortment and exploratory investigation. Information cleaning comes straightaway. This progression eliminates copy esteems and remedying mislabeled classes and features. Feature designing is the subsequent stage. The outcome include designing is taken care of to the prescient models, and the outcomes are cross-approved. A calculation that is taken care of the crude information is ignorant of the significance of the highlights. It is making expectations in obscurity. It may be considered feature designing as the directing light in this situation. The intricacy of the calculations diminishes. Regardless of whether features are utilized a calculation that is not ideal for the circumstance, the outcomes will in any case be exact. More straightforward models are frequently more clear, code, and keep up with. The most legitimate response to the inquiry is feature designing that it is a manual for calculations.

7.3 Malicious Feature Engineering

Malignant entertainers frequently need to get data of interest from designated PC conditions. To accomplish this objective, they typically choose to plant some sort of programming that will give that data constantly. Over the long haul, the cautious frameworks improved and moved to the next level at distinguishing such executable inserts. In this waiting game, the two sides attempt to work on their apparatuses and, as guarded instruments improve, malware entertainers attempt to discover better approaches for sneaking malignant programming into a framework. There are a few mainstream methods of doing this. Over the long haul, security arrangements are turning out to be progressively more mindful of such dangers. One of the cleverer techniques that got our attention is stowing away malware inside picture designs like PNG, BMP, GIF or JPEG. Malware, being a human developed program that steals the information and performs malicious actions should be identified on an image. Machine learning helps in the study of feature selection and feature extraction on an image. When the features of the image are identified, it helps in identifying the malware and its type, which is done through correlating the features of the image. By doing this, the malware can be classified properly.

Image designs are fascinating to malware creators since they are by and large thought to be undeniably less hurtful than executable records. Pictures can be utilized to send malware in blend in with a dropper, where the dropper goes probably as a charitable executable which parses dangerous substance concealed inside a picture. One district where this strategy can be used are web moves. Various destinations enable moving picture content, anyway improperly filter through executables and scripts. In such cases, harmful code can be squeezed into an image and moved to a web laborer containing a potential shortcoming which enables execution of its substance. Probably the most normal sort of such payloads are PHP web shells. Risky performers find and experience shortcomings in applications used to parse picture plans. To remain undetected and make an effort not to attract the thought of security gadgets, they usually endeavor to make records which hold quick to the image plan specific whenever possible. The most un-complex way to deal with introduce malignant substance into an image is to append it to the image end, or, as it is consistently insinuated, the overlay. Vindictive entertainers ordinarily take a kindhearted picture document and affix some substance. This makes it a notable strategy that is very simple to distinguish.

7.4 Image Processing Technique

Image Processing is used to control the pictures by the use of calculations. Advanced Image Processing is utilized to control the pictures by the utilization of calculations. One such programming utilized for handling the pictures is Photoshop. This innovation gives a stage wherein different activities like picture improving, handling the simple and computerized signals, picture signals. The pictures get adjusted in various organizations.

7.4.1 Steps Involved in Image Processing Technique

Image processing techniques comprises a few steps. Those steps include acquisition tools, image analyzation and report generation based on image analysis. The steps are as follows:

- Acquisition tools are used in importing the image.
- Image gets analyzed and manipulated.
- The output gets generated where an image may be altered and the report may be based on image analysis.

7.4.2 Image Processing Task

The tasks performed while an image is processed consist of image sharpening and restorations, medical applications, machine/robot vison, etc. Supposing we have the image processing task which could be represented in 2 dimension then the random fields will be f(x,y). Here we consider f(x,y) as existing image and g(x,y) as modified image.

7.4.2.1 Image Enhancement

Switching up of the digital images into a more versatile form to support multiple image analysis, is Image Enhancement process. This step includes the process like turning on/off the sound, sharpening and turning on the image which results in identifying the key features of the image in a more precise way.

7.4.2.2 Image Restoration

The unethical, noisy or any unused image is removed from the set, which is Image restoration. Exploitation of the image may result in the process of blurring, sound and camera focus. This helps in improving the quality of the image by reducing the noise and reclaiming the loss of decision.

7.4.2.3 Coloring Image Processing

Color vision Physiology and Physics of light is the concept under-wined here. Two principal factors motivated in the image processing are "Color, a powerful descriptor that helps in object identification" and "Extraction from a scene" [13, 15]. The two major areas include Full-color images are obtained with full-color sensor, and Pseudo-color processing - a shade of color is assigned to a monochrome color intensity.

7.4.2.4 Wavelets Processing and Multiple Solutions

One such important aspect of Image processing is Wavelet transformation, which allows the usage of multiple solutions on an image. Wavelets are similar to brushes, where different sizes of brushes that fits the topographies. These are useful in representing the scale of features of an image including their position. This gets applied to 1D images as well.

7.4.2.5 Image Compression

The digital image gets encoded with lesser bits than the unencoded representation, with certain encoding schemes. At lower resolution, the common compression methods, JPEG blurs the appearance of the lines [17, 10]. On the other hand, the lossless compression methods retain its original quality rather than compressing the image. TIFF, GIF and PNG are the best suited examples in this case.

7.4.2.6 Character Recognition

OCR, commonly used by Optical character binding uses a machine operated or an electronic replacement for the scanned images as to a computer formattable text. This feature greatly helps to access the new data source from the next minute it gets in [18, 20, 21]. The digitalized manuscripts are sorted electronically, then searched and stored in the machine. OCR is a plethora for the research intelligence.

7.4.2.7 Characteristics of Image Processing

- Certain software are free of cost and are readily available on the internet.
- Clear images are processed.
- Image Processing do image upgradation to recollect the information through images.

7.5 Image Processing Techniques for Analysis on Malicious Images

In image preprocessing, picture data recorded by virus total software is collected and stored in binary variables. These mix-ups are modified using legitimate mathematical models which are by similar symbolic clear or real models. Picture overhaul is the change of picture by changing the pixel magnificence regards to chip away at its enhanced visualization. Picture overhaul incorporates a variety of techniques that are used to chip away at the visual appearance of an image, or to change the image over to a construction which is more equipped for human or machine understanding. A few of the picture handling procedures are Contrast extending, Noise separating, histogram change and so on.

In this chapter the malicious images are taken from the dataset malimg_paper_dataset_imgs from Kaggle. The stated dataset consists of malicious images of around 25 different families of malwares. In total, the images available in the dataset are around 25,000 malware images. Each image is of different pixel and size. The image is of grey scale. In a few cases malicious images can also be worked with RGB images. The images are equally resized using feature engineering technique. The pictures in dataset must undergo the process of equalization. That is all the images of the dataset must be of same size. Only when the images are of equal size will the outcome from the feature extraction and engineering be appropriate. Figure 7.4 is the sample representation of malware images from 25 families. The malware families are Adialer.c, Agent.FYI, Allaple.A, Allapel.L, Autorun.K, C2LOP.p, Fakerean, Lolyda.AA1, Skrintrim.N, Obfuscator.AD, Malex.gen.

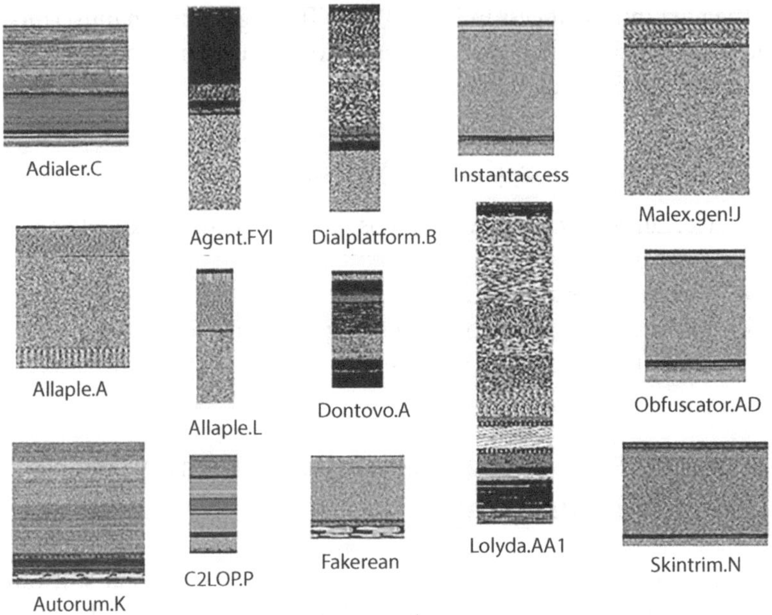

Figure 7.4 Malware images of malware families.

The contrast switching strategies are planned only for habitually experienced circumstances. Distinctive extending methods have been created to extend the limited reach to the entire of the accessible dynamic reach.

ICA or Independent Component Analysis isolates the multivariate sign into its hidden segment. The required segment gets removed from the mix of different parts or signals. ICA changes the signs so that the potential relationships are disposed of the difference of the segment gets equivalent to 1.

Noise Filtering is used to channel the unnecessary information from an image. It is furthermore used to wipe out various kinds of upheavals from the photos. Different channels like low pass, high pass, mean, middle and so on.

Histogram has a lot of importance in picture improvement. It reflects the characteristics of the picture. By changing the histogram, picture characteristics can be modified. One such model is Histogram Equalization. Histogram change is a nonlinear stretch that redistributes pixel regards so that there is around a comparable number of pixels with every value inside a compass.

Figure 7.5 is the histogram representation of original malicious image as collected from the dataset and histogram of the image whose contrast is been adjusted.

Figure 7.6 consist of the histogram representing comparison of original malicious image with image applied with heat map.

Figure 7.7 represents noisy malicious image along with the original image.

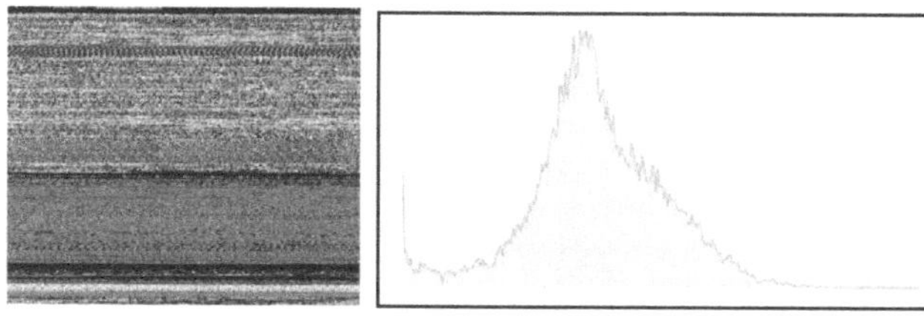

(a) Original Image with the histogram representation

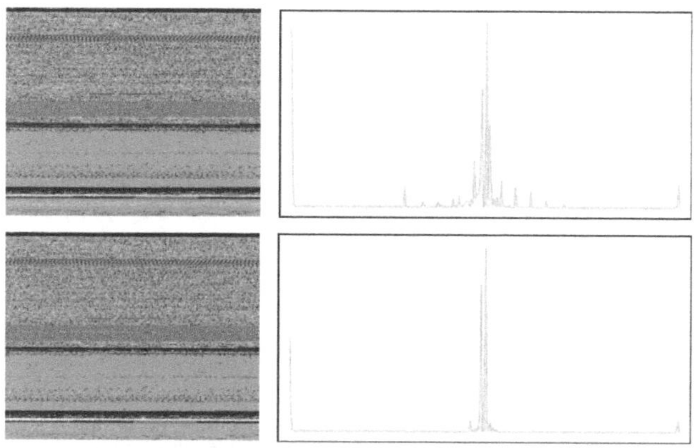

(b) Reduced Contrast Images with histogram representation

Figure 7.5 Representation of contrast streching with histogram.

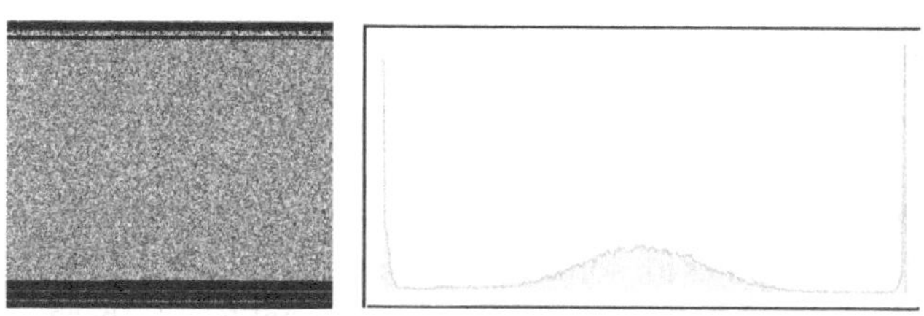

(a) Original Image

(b) Image implemented with heat map

Figure 7.6 Representation of malicious image over heat map application.

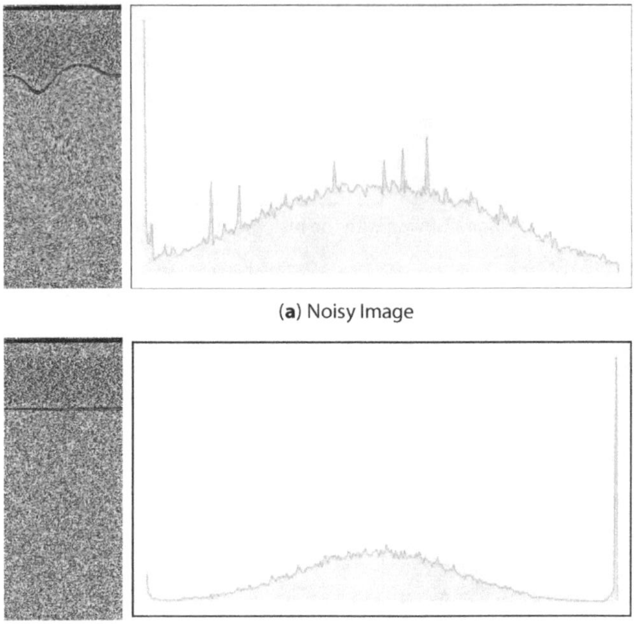

(**a**) Noisy Image

(**b**) Original Image

Figure 7.7 Representation of noisy malicious image over Agent.FYI family.

Figure 7.8 is the image that represents glitch in the image and Figure 7.9 is the representation of assorted image.

Figure 7.10 is the representation of malicious image applied with filter.

Layout coordinating is the strategy is utilized in looking and finding the format in a bigger picture. The format picture gets slide into the bigger picture like that of the convolution cycle.

A picture disintegrates for some reasons; for example any of the older images taken long back would have been lost in its original form. These images may be restored after a physical stress. A degradation model may be used to restore the image using a linear shift-variant and convolution. The image before degradation is known as "True Image" and after the degradation it is known as "Observed image". OpenCV method is used to work on the image restoration process, where the image is split up into Damaged image and a Masked image. A masked image and the noise has the same spatial dimensions.

Linear filtering is the output pixel's value is the combination of the neighboring input pixels. Thus convolution technique is used to perform this process. Convolution process may be defined as the addition of every

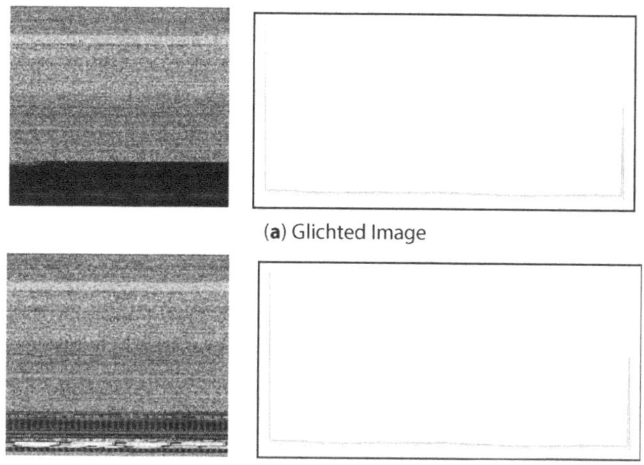

(a) Glichted Image

(b) Original Image

Figure 7.8 Representation of malicious glichted image.

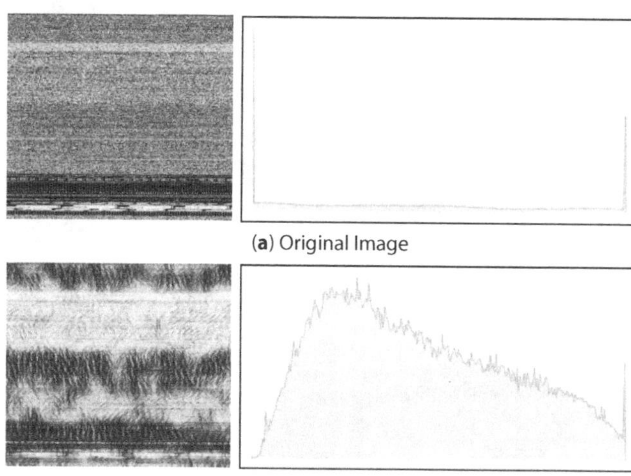

(a) Original Image

(b) Resorted Image

Figure 7.9 Representation of resorted malicious image.

element of the image into its local neighbors which is weighted by the kernel. A filter is present at the convolution layer which acts as a sliding window and convolves the image. Here, each pixel gets multiplies with their corresponding kernel and then its sum is calculated. The new sum is now the new pixel on the output image.

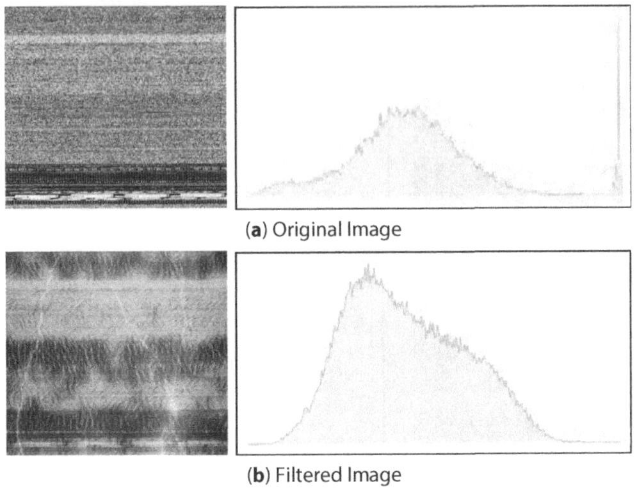

Figure 7.10 Representation of filtered malicious image.

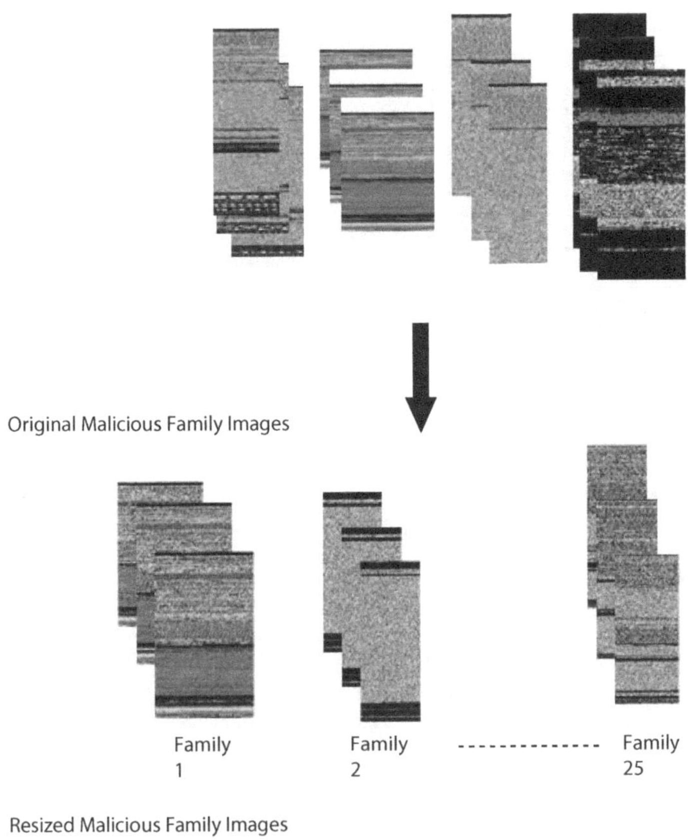

Figure 7.11 Representation of original malicious image with resized image.

Generative Adversarial Network is useful in training the learning model on any data image which in turn generates the similar type of image data. GAN, invented by Ian Goodfellow in 2014, described two distinct models – Generator and Discriminator. Generator helps in generates the fake image and discriminator classifies between the fake and the original image. Training is done when the generator outsmarts the discriminator and generates the fake images and the discriminator helps in improving and differentiating between the real and the fake image.

Pixelation can be represented as the images when they are resized, that is, when made bigger, the image gets pixelated. The image gets enlarged where the pixels in the image get stretched beyond their actual size. Figure 7.11 represents the resized structure of the malicious family images.

7.6 Conclusion

Feature engineering and selection assists in helping a large number of issues these days. The malware gets detected by a tedious process that chooses subset of discriminating features from the samples. The feature selection method in general detects the malware and efficiently finds the subset features. The accuracy wholly depends on the selection of the features which in turn enhances the performance of the process. When a feature is used without filtering the samples the performance rate is improved. The filter-based feature subset selection technique gives high computational rate yet they fail in handling multicollinearity issues affecting the accuracy. This can be studied in the future. Having a single feature may not be effective in producing high accuracy and thus a combination of the permissions should be handled to produce a high accuracy in malware detection methods. Another great feature work is to analyze the characteristics of the data and machine learning models that favor feature selection process.

References

1. G. Kumar and P. K. Bhatia, A Detailed Review of Feature Extraction in Image Processing Systems, *2014 Fourth International Conference on Advanced Computing & Communication Technologies*, pp. 5-12, doi: 10.1109/ACCT.2014.74, 2014.
2. B. Chitradevi and P. Srimathi, An overview on image processing techniques, *International Journal of Innovative Research in Computer*, vol. 2, no. 11, pp. 6466–6472, 2014.

3. S.Manikandan, Chhabi Nigam, J P Vardhani and A. Vengadarajan Gradient based Adaptive Median filter for removal of Speckle noise in Airborne Synthetic Aperture Radar Images, ICEEA, 2011.
4. K. Banjarey, S. Prakash Sahu and D. Kumar Dewangan, A Survey on Human Activity Recognition using Sensors and Deep Learning Methods, *5th International Conference on Computing Methodologies and Communication (ICCMC)*, 2021, pp. 1610-1617, doi: 10.1109/ICCMC51019.2021.9418255, *2021*.
5. Chirag Joshi, Vishal Bharti, Ranjeet Kumar Ranjan. Chapter 25, Analysis of Feature Selection Methods for P2P Botnet Detection, Springer Science and Business Media LLC, 2020.
6. ABM.Adnan Azmee, Pranto Protim, Md. Aosaful, Orko Dutta, Muhammad Iqbal. Performance Analysis of Machine Learning Classifiers for Detecting PE Malware, *International Journal of Advanced Computer Science and Applications*, 2020.
7. Oivind Due Trier, Anil K. Jain, Torfinn Taxt, Feature extraction Methods for character recognition-A Survey, *Journal of Pattern Recognition*, Vol. 29, No. 4, pp. 641-642, 1996.
8. Khaled Mohammed bin Abdl, Siti Zaiton Mohd, Azad Kamilah Muda, Feature Extraction and Selection for Handwriting Identification: A Review, pp. 375-381.
9. Gaurav Y. Tawde, Jayshree Kundargi, An Overview of features Extraction Techniques in OCR for Indian Scripts Focused on Offline Handwriting, *International Journal of Engineering Research and Applications*, Vol. 3, Issue 1, pp. 919-926, Jan-Feb 2013.
10. M. Wagner, F. Fischer, R. Luh, A. Haberson, A. Rind, D. A. Keim, *et al.*, A Survey of Visualization Systems for Malware Analysis, *Eurographics Conference on Visualization (EuroVis)*, 2015.
11. L. Nataraj, S. Karthikeyan, G. Jacob and B. S. Manjunath, Malware images: Visualization and automatic classification, *Proc. 8th Int. Symp. Visualization for Cyber Security VizSec*, pp. 4-7, 2011.
12. L. Nataraj, V. Yegneswaran, P. Porras and J Zhang, A comparative assessment of malware classification using binary texture analysis and dynamic analysis, *Proc. 4th ACM Workshop on Security and Artificial Intelligence AISec*, pp. 21-30, 2011.
13. Said Zainudeen Mohd Shaid and Mohd Aizaini Maarof, Malware Behavior Image for Malware Variant Identification, *IEEE International Symposium on Biometric and Security Technologies (ISBAST)*, 2014.
14. Aziz Makandar and Anita Patrot, Computation Pre-processing Techniques for Image Restoration, *International Journal of Computer Applications*, vol. 113, no. 4, pp. 11-17, March 2015.
15. Dolly Uppal, Vishakha Mehra and Vinod Verma, “Basic survey on Malware Analysis Tools and Techniques, *International Journal on Computational Sciences & Applications (IJCSA)*, vol. 4, no. 1, February 2014.

16. G. A. N. Mohamed and N. B. Ithnin, Survey on Representation Techniques for Malware Detection System, *Am. J. Appl. Sci.*, vol. 14, no. 11, pp. 1049-1069, 2017.
17. Zhaoqi Zhang, Panpan Qi, Wei Wang, Dynamic Malware Analysis with Feature Engineering and Feature Learning, *Proceedings of the AAAI Conference on Artificial Intelligence*, 10.1609/aaai.v34i01.5474, 2020.
18. Swarna Priya R.M., Praveen Kumar Reddy Maddikunta, Parimala M., Srinivas Koppu, Thippa Reddy Gadekallu, Chiranji Lal Chowdhary, Mamoun Alazab, An effective feature engineering for DNN using hybrid PCA-GWO for intrusion detection in IoMT architecture, *Computer Communications*, 10.1016/j.comcom.2020.05.048, 2020.
19. Mubarak Albarka Umar, Zhanfang Chen, Yan Liu, A Hybrid Intrusion Detection with Decision Tree for Feature Selection, *Information & Security: An International Journal*, 10.11610/isij.4901, 2021.
20. B. Sudhakar, V. B. Narsimha, G. Narsimaha , Detection of Intrusion using Hybrid Feature Selection and Flexible Rule Based Machine Learning, *International Journal of Engineering and Advanced Technology*, 10.35940/ijeat.f8783.088619, 2019.
21. Adel Hamdan Mohammad, Intrusion Detection Using a New Hybrid Feature Selection Model, *Intelligent Automation & Soft Computing*, 10.32604/iasc.2021.016140, 2021.

Blog

1. Omnisci.2021,https://www.omnisci.com/technical-glossary/feature-engineering
2. Machine learning Mastery, Aug 2020.https://machinelearningmastery.com/discover-feature-engineering-how-to-engineer-features-and-how-to-get-good-at-it/
3. Reversing Labs Blog, Karlo Zanki, March 2021. https://blog.reversinglabs.com/blog/author/karlo-zanki
4. Neptune Blog, Gopal Singh Panwar July 2021, https://neptune.ai/blog/image-processing-techniques-you-can-use-in-machine-learning
5. Analytics Vidhya, 2020 https://www.analyticsvidhya.com/
6. Sebastian Raschka blog, 2021 https:/ /sebastianraschka.com /faq/ docs/ feature_sele_ categories.html

8

Cubic-Regression and Likelihood Based Boosting GAM to Model Drug Sensitivity for Glioblastoma

Satyawant Kumar[1]*, Vinai George Biju[2], Ho-Kyoung Lee[1] and Blessy Baby Mathew[3]

[1]Dept. of CSE, Kyungpook National University, Daegu, South Korea
[2]Dept. of CSE, CHRIST, School of Engineering and Technology, Bangalore, India
[3]Dept. of Biotechnology, Dayananda Sagar College of Engineering, Bangalore, India

Abstract

Generalized additive models (GAM) being versatile and efficient do present some difficulties in model selection and inference based on context and fitting with linear smoothers. GAM is being extended to embed smooth functions of variables described by a parameter which governs curve smoothness or approximate forecast precision in a wide range of non-linear models. Boosted GAM offers smooth function estimates of covariate influence functions and estimated degrees of freedom. GAM boosting solves the problem of limiting explanatory elements to relatively few and choosing smoothing settings. The challenges of drug Sensitivity towards the Glioblastoma are modelled for tissue samples, using the proposed Cubic-Regression-based GAM, and further analyzed using boosted GAM approach. Various spline variations and smooth interaction functions have been used to examine the behavior of the GAM model. Similarly, different base-learners have been explored in boosted GAM, and a hyperparameter is tuned to analyze the model behavior.

***Keywords*:** Cubic regression GAM, glioblastoma, boosted GAM, selection of spline functions, drug sensitivity

**Corresponding author*: kumarsatyawant460@gmail.com

Kukatlapalli Pradeep Kumar, Aynur Unal, Vinay Jha Pillai, Hari Murthy, and M. Niranjanamurthy (eds.) Data Engineering and Data Science: Concepts and Applications, (195–224) © 2023 Scrivener Publishing LLC

8.1 Introduction

Glioblastoma Cancer stem cells are important for initiating, growing, and resisting cancer. Our knowledge regarding the processes and how they deal with the heterogeneity of glioblastoma is restricted. Understanding cellular and molecular variability in glioblastoma (GBM), the most frequent and fatal recurrent brain tumor, is essential for designing effective treatments. In addition to the interpatient variability, the involvement of multiple cell populations in tumors involves the creation of molecular signature modelling strategies that drive tumor evolution and therapeutic failure.

The main issue in the treatment of glioblastoma is the inevitability of recurrence after radio-chemotherapy. Intratumor heterogeneity in glioblastomas is well known, manifesting itself in a variety of cell types and molecular patterns. The failure of cytotoxic therapy to appropriately target glioma progenitor cells is thought to be the cause of glioblastoma recurrence. Glioblastoma recurrence is believed to be influenced by therapy-induced molecular evolution, according to new research. According to growing evidence, intratumor variability, longitudinal variations in molecular biomarkers, and the specific potential impacts of glioma stem cells must all be properly considered to increase the precision of molecular diagnostic methods which still rely on indicators from a single tumor sample.

Glioblastoma, the most frequently identified brain cancer in adults, exemplifies the issues. The estimated duration of relapse for patients with the disease is seven months following radiation treatment and temozolomide (TMZ). A greater component of a cell which forms the basis of our knowledge for both the disorder's molecular and genetic underpinnings and a smaller, less differentiated piece of stemming cells are the two main cell divisions. Drug resistance and its associated exceedingly poor effectiveness are driven by molecular and genomic heterogeneity and persistence following radiotherapy and chemotherapy. The research network of the Cancer Genome Atlas (TCGA) developed interpatient heterogeneity through genomic and transcriptomic analyses. Transcriptomic data from entire tumors, predominantly obtained from differentiated cells, were examined, and categorized into four basic subsets: proneural, neural, classical, and mesenchymal. The neural subgroup is no longer included in the current classification. To distinguish GBM-specific mRNAs from those found in non-malignant cells, investigators employed three distinct methods: comparing tissue samples to their corresponding cell lines, decoding RNA from solitary GBM cells, and comparing core vs. peripheral biopsy specimens. GBM and non-GBM cell transcriptomes were further segregated, revealing three subcategories with mRNAs associated with classical, proneural, and mesenchymal categories. However, none of the subcategories were enriched with mRNAs associated with the neural subgroup, showing that neural subgroup resulted from non-tumor tissue involvement with the initial collections.

Despite bulk tumor sequencing approaches having been useful in developing GBM subgroup classification frameworks, they only provide a minimal grasp of GBM cancer progression. Preliminary studies were always focused on the molecular complexity of cells mostly from the tumor core. Single-investigation approaches are currently unable to address concerns such as the nature of invading GBM cells or unknown diversity of other neural, immunological, glial and vascular cell lines found in and around GBMs. Each cell type's interaction within the tumor surroundings is thought to play a pivotal role in tumor progression and medication resistance.

GBMs are diffusely infiltrating and typically revert, even in remote segments of the head, contrasting brain metastases, which have a local control rate of up to 80% with radiation and surgery. Local therapies are inefficient because migratory cells outside of the tumor core are frequently unaffected by them and are often responsible for GBM recurrence in a patient due to the diffuse nature of GBMs. A deeper understanding of the genetic properties of tumors is crucial for an implementation of potential treatment, with a strong emphasis on efforts to obtain and distinguish the migratory cells liable behind recurrence.

8.1.1 Glioblastoma

Glioblastoma is an aggressive form of brain and spinal cord cancer. GBMs are incurable tumors with a median overall survival of only 12-18 months despite rigorous treatment such as surgical resection, chemotherapy, and radiotherapy. The cancer originates from the astrocyte cells which support nerve cells. Glioblastoma could occur at any age but tends to take place in elderly people more frequently. Additional symptoms such as chest pain, nausea, vomiting and seizures can intensify the progression of the disease. Glioblastoma can be extremely hard to deal with, and is often referred to as "glioblastoma multiforme". Certain therapies can reduce the rate of cancer progression and minimize symptoms, but to discover transcriptome features of recurrent glioblastoma in whole-tissue specimens of glioblastoma or glioblastoma stem cells, a multisampling method is selected which is categorized under longitudinal approach. In the study, tissue samples from such tumors which were diagnosed early, some in recurring stages and a few samples have been considered from secondary recurrent glioblastoma.

8.2 Literature Survey

Despite a significant variety among tumor cells, infiltrating GBM cells had a comparable gene profile across sufferers, indicating a similar infiltration pathway. In addition, transcriptionally different myeloid cell populations inhabiting the tumor body and neighboring peritumoral zone were discovered while examining the immune response to malignancies [1]. Glial progenitor-like cells are at the center of a well-known neural tri-lineage cancer hierarchy. The progenitor population which is indeed frequently the creator of additional cell lines via RNA velocity contains a high proportion of the cancer's cycling cells. Treatment targets for cancer stem cells, also known as progenitor cells, can be found using a hierarchical map. According to the findings, regular brain evolution explains glioblastoma growth, gives a logical basis for the glioblastoma hierarchy, and aids in the identification of cancer stem cell-specific targets [2]. A classification strategy was used based on GBM scRNA-Seq data, in which differentiated cell types (neoplastic and normal cells) were selected as classes using sparse logistic regression. The objective was to uncover gene features that differentiated the classes and those which were shared by distinct neoplastic cells. The latter was tried to deal by using the network-based twiner regularizer to find genomic fingerprints shared by cancer core neoplastic cells and infiltrating tumor peripheral neoplastic cells as potential disease biomarkers to target multiple neoplastic cells [3].

The current standard of care for GB is debulking surgery, followed by a combination of DNA alkylating medication temozolomide (TMZ) and ionizing radiation [4]. Unfortunately, conventional therapy's performance and efficiency is restricted due to low effectiveness of TMZ, which only helps in around half of GBs [5]. Tumor recurrence following first treatment is probably unavoidable, and it poses a major clinical challenge in the cure of GB. Recurrent GB (recGB), which appears to be a deadly illness, currently has no viable therapeutic alternatives. So yet, despite intensive attempts to refine cytotoxic monotherapy therapeutic dose treatments for recGB, no level one confirmation has indeed been achieved [6–8]. Recent findings imply that therapy-driven structural variation is indeed a major factor in GB reappearance. Additionally, there seems to be a growing consensus that conventional diagnostic approaches based on a single tumor biopsy may be ineffective for diagnostic techniques of GBs, which are renowned for their tremendous intra-tumoral spatial variability [9–13]. The core premise behind GB's treatment opposition is that conventional radio chemotherapy fails to effectively target a certain percentage of glioblastoma cells known as glioma stem cells (GSCs). GSCs, according to the cancer hierarchy, seem to be the most clinically significant compartment capable of continuing tumor development in the event of cytotoxic therapy [14–17].

Despite being recognized as extremely efficient metastatic progenitors, circulating tumor cells (CTCs) clusters remain a little-known element of the metastatic process due to their dilution factor in a patient's plasma and the difficulties of isolating them without disrupting cellular interactions. Additional aggravating factors, including the circulatory mechanics, initial tumor's site and capillaries bed retention, can all seem to have an influence on the total of CTC aggregates seen in a patient's peripheral bloodstream. Clusters of glioblastomas in the bloodstream can breach the brain-blood barrier and enter the peripheral circulation [18]. For cancer to thrive, a fraction of cancer cells that acts similarly to stem cells seen in normal tissues are required. Metabolomics, or the study of metabolites which represent cellular phenotype, could aid in revealing the critical function of cells in cancer formation and progression, including the creation of treatment interventions [19]. The division of GBM variability into two classes—one which grows quickly but is more able to respond with extremely long survival and another which grows slower while being less responsive with shortened survival—suggests that several sufferers who obtain regular treatments may benefit more from alternative treatment options [20]. Even though microRNA dysregulation is a prevalent feature of glioblastoma, each gene's neurobiological impact is often small and clinically insignificant. During neural differentiation, miR-124, miR-128, and miR-137 form a microRNA module that is co-expressed and deleted at the same time during glioma genesis [21].

Approaches presented in Wood (2000) can be used to solve the nonlinear multiple smoothing parameter estimation problem in GAM, but real implementation of such an approach within a GAM environment is still a long way off [22]. Use of a basis gives the chosen curve a lot more flexibility while elevating the residual sum of squares by a roughness penalty [23]. The approach to model building is based on regression divisions with multiple variable fraction polynomials [24]. On a piece-wise polynomial of degree at most k, the continuous derivatives of all other orders up to k - 1 were used [25]. Natural cubic splines mostly with varying numbers of internal knots and thin-plate regression curves were one of the spline techniques explored. The degree of propensity score overlaps between outcome prevalence, therapy prevalence, treatment groups, and authentic marginal relative risk were changed all throughout iterations. The researchers compared approaches across simulations based on real marginal relative risk coverage, their bias and root mean squared error (RMSE) [26]. To describe time-varying parameters and provide some alternative functional forms for various time intervals, the natural cubic spline is adopted. To put it another way, the model fits a curve which smoothly goes across specified sample points, resulting in a flexible fitted curve [27].

The R package "mboost" makes it simple to get started with boosting. In the same way that almost all scholars are used to, the module uses a model-based boosting strategy to build interpretable systematic additive models. Fitting methods have interfaces that are quite comparable to common implementations such as *lm*() and *glm*(), making them reasonably simple to utilize. When something comes to the effect of prospective predictors and the type of risk feature to be adjusted, the fitting procedures of the module provide a wider range of options. As a result, mboost provides far more modeling options than traditional fitting methods for many model classes. Furthermore, by adding new families, and also the base-learners, the user can quickly extend mboost [28]. GAMBoost is a generalized additive model fitting methodology which tackles the drawbacks of conventional fitting approaches. Smoothing parameter selection, which is typically a p-dimensional task, is shortened to only one parameter value, the number of iterations. The selection of variables to be really revised in every iteration automatically manages the magnitude of smoothing that differs across predictors. Simply following stage-wise fitting of feeble learners, generalized additive model boosting is being used. Fitting strategies are available for distributions like Poisson, exponential, binomial, and for normal target variables too. The approach combines variable selection with determining the right amount of smoothing. In order to fit generalized additive models (GAMs), the technique outperforms existing procedures. Most likely, the only approach which works well in high-dimensional situations, in particular [29].

A plethora of mathematical and computational tools are accessible in fundamental cancer research. They have demonstrated that they can help in gaining a better knowledge of numerous aspects of cancer, developing novel theories and forecasts, and directing scientists to the most impactful experiments [30]. Researchers employed a successful and commonly used membrane model-based technique to simulate deformable cancer cells and duplicate experimental data from individual cell lines. A data-driven, phenomenological methodology yields high agreement for huge deformations while also defining how to describe different cancer cell lines explicitly. The model is simple to incorporate into cell-resolved hemodynamic transport simulations, implying that it has a great deal of potential to supplement studies and bring new insights into cancer progression [31]. The fast rise of large data resources, fueled by advances in high-throughput technology, has created a fundamental change in cancer therapy research. Several significant clinical applications have resulted from the integration of computational approaches with genomics data. Precision cancer medicine's usefulness relies on the implementation of data science to better identify the interactions between the host immunity, tumor environment and the ecosystem. Because of the increasing availability of massive data resources and computational tools, big data approaches may play an integral role in the continued development of precision cancer care [32]. Data science and liquid biopsy based on circulating tumor DNA (ctDNA) have synergistic effects which aid therapeutic potential in advanced cancer detection. ctDNA is a potentially cancer biomarker which incorporates data science. Data science has played a significant role in contemporary techniques such as data rectification, data selection and data mining, to address technical hurdles in ctDNA detection [33]. Predictive analytics and big data, especially in data-rich domains like oncology, have a tremendous number of possibilities for improving risk classification [34]. Large datasets may now be analyzed at a low cost and in a short amount of time, thanks to technological improvements. Deep learning algorithms have proven to be an effective tool in medical imaging for disease monitoring, diagnosis, surgical procedure assistance, and disease management. Most oncology-related diagnoses rely on the application of AI in radiology for various modalities with enhanced performance, such as magnetic resonance imaging (MRI), computed tomography (CT/CAT), positron emission tomography (PET), X-rays, digital pathology, and ultrasounds [35]. In the cancer realm, combining multidimensional diverse data with various feature extraction and classification models can likely result in useful forecasting tools. Machine learning tools' potential to identify important aspects in huge databases clearly illustrates its value. Support Vector Machines (SVMs), Artificial Neural Networks (ANNs), Decision Trees (DTs) and Bayesian Networks (BNs), are just a few of the techniques that have already been widely used in cancer research to build prediction models that result in effective and precise strategic planning [36].

8.3 Materials and Methods

Basically, R programming language has been used, along with some python libraries to perform exploratory data analysis on the datasets which have been used. Different packages or libraries which are available in R and Python have been explored.

Data pre-processing has been performed using python libraries. Initially the datasets contained 155 rows and 12 columns. Outliers, extraneous attributes, and irrelevant columns or features have been removed from the dataset. After pre-processing, the datasets are left with 120 rows and 8 columns, containing 7 different independent variables and 1 response variable at the last column. The explanation about all the variables have been illustrated thoroughly in the datasets description section. To illustrate the link between the response and the independent factors, an exploratory data analysis is performed. Then, "mgcv" and "mboost" packages in R have been used to carry out the research study.

8.3.1 Methodology

GAM methodology has been extended to boosting GAM for the work.

8.3.1.1 Generalized Additive Models (GAMs)

GAM methodologies extend linear models to combine smooth utilities from explanatory factors with either a parameter that directly controls curve smoothness or predictive accuracy estimates. Smoothing terms, which are spline functions of unit explanatory factors, are the focus. Non-linear mixtures of spline basis terms can be used to create the functions. A non-linear combination of the spline base's terms is chosen to provide the best curve for the results. A foundation is laid that allows for a high degree of fitting in the chosen curve while also increasing the residual sum of squares with roughness penalties which are a multiple of the integral part of the squared derivative. Steps include:

1) Evaluate the smoothing splines and placing a knot.
2) Penalize the Spline to ensure that knots spread across the data.
3) Utilize cubic regression-based GAM approach for spline smoothing.
4) Use of Generalized Cross-Validation as a metric of fitting.

The univariate outcome y_i is derived from the exponential distribution where $\mu_i = E(y_i)$ is resulted in the descriptive variables – m_{1i}, m_{2i}. Substituting the strict parametric GAM model yields the following:

$$g(\mu_i) = \beta_0 + \beta_1 m_{1i} + \beta_2 m_{2i} + \ldots \tag{8.1}$$

The parameter-less computations in GAM structure evolves to be:

$$g(\mu_i) = \beta_0 + s_1(m_{1i}) + s_2(m_{2i}) + \ldots \tag{8.2}$$

The method being versatile and effective, does present some difficulties in selecting and deducting models to back fit the system based on linear smoothers. The construction of GAMs with penalized splines implies an approximation between splines used for regression. Depending on the processes, approximations are often significantly lower than those used in the GAM modelling. The parametric structure provides the range of potential models with a great deal of versatility but adds a lot to the model selection problem and reduces the degree of flexibility. If the parameters are calculated with linear smoothness and back fit, the effective degrees of smooth freedom must be selected and would be difficult to interpret. The reason being, it offers a way to test all the smooth models, using only algorithms appropriate for the individual evaluation of single smooth conditions to make the retrofitting so appealing. The issue is that the parameters must be applied to the entire model, not just the smooth terms, to select the successful degrees of model freedom. It has been observed that calculation of the smoothness of the model is judged whether the data points need to be included as part of the model. Need to respond because the automated parameter smoothing occurs separately. Method of selection can't reduce to zero for the degrees of term independence. Once the simple polynomial of higher order is penalized, the wiggliness of the term is zero, so that further improvements in the smoothing parameter are not feasible for simpler models beyond the present stage.

The curve of a spline is a partially polynomial curve, which connects the polynomial covariates. The positions connecting the joints are also recognized as knots. Two different types of splines are widely used: Natural cubic splines at the border knots having the derivatives of second order being zero. Hence, the curve resembles a straight line. At the boundary knots, B-spines are unconstrained. As a linear combination of the base functions, spline curves can be formed of any given degree. The Splines use function – *bs*() produces B-spline formulations and *ns*() creates natural spline formulations, which can be used for generating base terms. Many distinct choices of basis functions are available in both the cases.

The function $g(x)$ is a subjective function made up of p cubic curves that merge smoothly, with the knots at the boundary having null second derivatives. The basis function is $\tau_1(x), \tau_2(x), \ldots \tau_p(x)$, so it follows:

$$g(x) = l_0 + l_1\tau_1(x) + l_2\tau_2(x) + \cdots + l_p\tau_p(x) \tag{8.3}$$

Consider the representation of a single-variable function $m(n)$. Let $k_x(n)$: $j = x \ldots z$ be a set of functions that are selected to have appropriate characteristics, along with owing no unknown parameters. The expression for $m(n)$ is as:

$$m(n) = \sum_{x=1}^{z} a_x b_x(n) \tag{8.4}$$

α_x are z indefinite factors. The $m(n)$ is derived from a mixture of the basis functions $k_x(n)$ approximating g corresponding to α_x.

The model's smoothing can be controlled by adjusting the number of basic functions utilized, which can be tricky too. In particular, the model is likely to overfit the data that contain some noise when the total available basis functions are sufficient to evaluate the underlying real function. In comparison, when the total available basis functions are selected to be low enough to prevent the overfitting, possibly may be too narrow for approximating the underlying reality. The use of a relatively high number of basis functions will mitigate the problem but prevent overfitting by implementing a penalty throughout the model being designed to keep the smoothness intact.

The smoothness function is represented by the variable g, which is estimated by minimizing:

$$\sum_{i=1}^{n} (g(p_i) - q_i)^2 + \delta \int (g''(p))^2 \, dp \tag{8.5}$$

To design a smooth model, δ is a smoothing constraint which regulates the wiggliness between the fitted data. Selecting a basis for g permits a penalty matrix H and a design matrix X to be computed. The fitting problem could be mentioned as:

$$||X\alpha - q||^2 + \delta\alpha^T H\alpha \tag{8.6}$$

8.3.1.1.1 Splines

Functions consisting of simpler functions are splines. Simpler functions are basic functions and base functions are the foundation. Splines are formed from the combination of basic functions.

Polynomials are chained together over the range of the data to characterize the shape to fit 'piece-wise' polynomials. Splines are piece-wise polynomials called after the curve-drawing tools used by draftsmen. Cubic Physical splines were flexible strips that could be bent to shape and were weighted in place.

The splines can be as smooth or as wiggly as you want and can also be regulated either by changing the number of knots (k) or by using the smoothing penalty option. Splines can be more 'wiggly' if the number of knots is raised and may likely be closer to the data and have low error, starts to 'over-fit' the relationship and fit the noise in the data. The model's complexity is punished when the smoothing penalty is combined, which helps to reduce overfitting. As the number of knots increases dramatically and the smoothness penalty lowers, the smoothness or wiggliness of the curve changes.

The basis complexity, or "size," is set to k. The number of tiny functions that make up a curve can be thought of as maximum wiggliness. Following smoothing, curves will have fewer effective degrees of freedom (EDF):

$$EDF < k \tag{8.7}$$

k must be large enough, and the penalty λ will take care of the rest. If the space of functions representable by the basis is huge enough, the genuine function or a good approximation to the true function will be included. The larger the k, the higher the computing cost. The default k values in mgcv are arbitrary – after selecting the model terms, the value of k is the most important user choice.

8.3.1.2 Model-Based Boosting – Boosted GAM

The "mboost" module is used to fit a large selection of generalized linear and generalized additive models on incredibly high dimensional datasets exploiting component-wise (penalized) least squares estimations as base-learners. Time-to-event analysis, classification, regression, and even a variety of many other standard statistical issues including potentially high-dimensional data also can be addressed using the "mboost." The library can be used in the same approach as traditional R methods for statistical approaches mostly because of its user-friendly mathematical interface. The library also offers several functionalities for displaying results, picking hyperparameters and parallelizing computations.

GAM is made up of two types of assumptions: structural and distributional. The conditional distribution of the outcome is specified by the distributional assumption. The structural assumption dictates which effects will be currently employed in the model and, as a result, the system's deterministic structure. In general, it shows how effectively the predictors are associated to the conditional mean of the end outcome. The "mboost" can also be used to express effects for hazard rates, conditional quantile and conditional expectile allowing us to deal with a range of models in one place. The cost function that is expected to be minimized as the family is described by the distributional assumption. Employing the base-learners, the main structural assumption is expressed as a mathematical formula.

The cost function stated by the family is unaffected by the base learner's assessment. The cost function is primarily used to compute the negative slope for each boosting phase in the component-wise boosting statistical methodology. Regardless of the cost function, the predictors are then coupled with the data utilizing penalized ordinary least squares estimate. Therefore, users have complete authority over how structural and distributional assumptions are combined to answer novel estimation problems.

gamboost() has the following interface:

$$gamboost(formula, data = list(), \ldots) \tag{8.8}$$

To specify the model, the function requires a formula. In addition, as with linear models, a data collection must be supplied. The model is specified using *response* ~ *predictor*1 + *predictor*2 as a formula, and the data set is supplied as a data.frame via the data argument. Additional parameters for the mboost *fit*() functions include family, weights, and control. Weights can also be applied for the weighted regression estimation if needed. The control option, which is unique to boosting, permits us to configure the boosting algorithm's hyperparameter. To accomplish the task, the *boost_control*() function is utilized. For instance, one may specify:

$$boost\_control(mstop = 200, nu = 0.05, trace = TRUE) \tag{8.9}$$

Lastly, the distributional assumption, which is hidden in the '...' expression, can be provided through a family argument. The basic type of family is Gaussian.

8.3.1.2.1 Base-Learners

Base-learners can be used to establish the model's structural constraints or the forms of effects to really be employed. Each base-learner has a distinct impact associated with it somehow.

The base-learners must always be configured in such a way that a single base-learner's degrees of freedom are limited enough just to avoid overshooting. Typically, at least, 4 degrees of freedom or less are employed (standard for several base-learners). Considering the modest initial degrees of freedom, the final estimate obtained from the base-learner may likely incorporate high order degrees of freedom due to the cyclical aspects of its operations. If the base-learner is picked several times, it will finally be modified to a suitable amount of flexibility and smoothness.

The *bbs*() base-learner, which is founded on B-splines with a distinction penalty, allows for the creation of smooth effects (i.e., covariate functions which must be sufficiently smooth) (i.e., P-splines). P-splines are a powerful and versatile modeling tool which can be used to generate smooth effects, also can be used to express a range of effects. Shown below are just few examples:

- Effects which are Smooth
- Smooth bivariate effects (e.g., spatial effects)
- Coefficient terms which vary considerably
- Cyclic effects (= recurrent effects) plus several other options

The P-spline base-learner looks like:

$$bbs(..., by = NULL, knots = 20, boundary.knots = NULL, degree = 3, differences = 2, df = 4, lamda = NULL, center = FALSE, cyclic = FALSE) \tag{8.10}$$

The '...' parameter can be used to supply variables related to the base-learner. Just one variable is normally supplied for smooth effects, but two variables are allowed at most for *bbs*().

8.3.2 Datasets Description

Gene expression data has been used to carry out the study [37]. Gene expression is the process through which genetic information is used to make a functioning gene product that permits it to produce end products such as peptide or non-coding RNA and so alter phenotype. Each row in the dataset was a record of an individual sample. Record of eight feature variables include unmapped, Multimap, NoFeature, ambiguous, mapped_to_features, mapped_to_hgnc, uniquely_mapped and number_of_mismatches. Mapped reads are those segments from the sample that have been sequenced that could align straight to a single set of loci on the reference genome. On the other hand, unmapped reads are those that do not find a match anywhere on the reference genome. Sequence alignment algorithms characteristically discard the whole set of unmapped reads into a discrete file for simple segregation and analysis. Data of these kinds are often unnoticed and rejected without additional investigation. Ambiguous data lacks the required information and taking those makes the decision-making procedures more difficult. Thus, these data sets must be discarded and at least if it is applied anywhere then must make use of the classifier required, which makes the search of getting an appropriate solution much easier. Certain data sets still can't be classified and thus it is put under a different label for improved classification. All data points must be checked for ambiguity and must be re-classified if necessary. This assembly is inferred to extents without ambiguous data resulting in untainted data performance in the input space. When many reads are generated then all need to be aligned to the reference genome allowing errors. Aligners can efficiently map reads allowing a small number of mismatches, but sometimes they are not well suited for handling large numbers of mismatches. The number of mismatches potentially affects the effectiveness of anticancer drug temozolomide (TMZ) towards glioblastoma, since there are multiple distinct cell clusters surrounding the tumor and each cluster or tissue reacts differently when it encounters the drug. For model building, "number_of_mismatches" feature as a response variable has been considered and the rest 7 as an independent or predictor variable.

Figure 8.1 represents the correlation matrix describing the correlation of each feature variable with all other feature variables in graphics form. The terms on both the axis in heatmap are the same variables as in the datasets. When one out of the two variables changes, correlation shows how far both move in lockstep. A positive correlation is defined as when two variables move in the same direction. When they move up and down, there exists negative correlation. The response variable, "number_of_mismatches" is highly correlated with the "unmapped" independent variable having positive correlation value of 0.52, and least correlated with the "NoFeature" independent variable having negative correlation value of -0.046. Basically, if the value of the "unmapped" feature variable increases

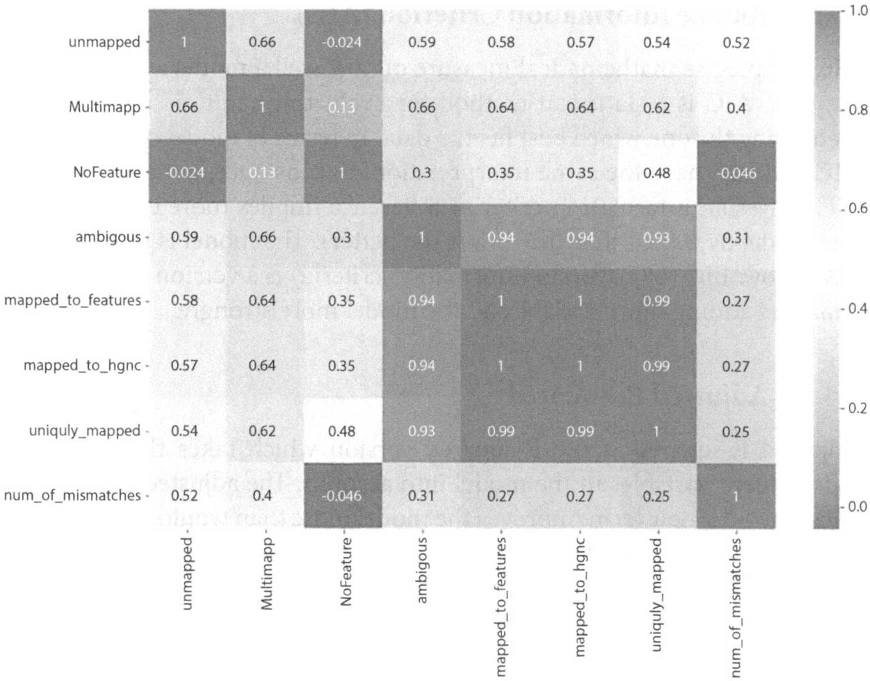

Figure 8.1 Correlation heatmap between variables.

or decreases then the value of "number_of_mismatches" response variable will also be impacted in the same fashion and the variation in "number_of_mismatches" will be much more unlike the variation, if the value of any other feature variable is increased or decreased. Instead, if the value of "NoFeature" feature variable increases or decreases then the value of "number_of_mismatches" response variable will be impacted in completely opposite fashion. The order of correlation of the response variable with the other independent variables follow:

Unmapped > Multimap > ambiguous > mapped_to_features > mapped_to_hgnc > uniquly_mapped > NoFeature

8.4 Evaluations, Results and Discussions

The research's outcomes will be listed and analyzed in depth here, with relevant reasoning (result analysis) leading to analytical conclusion.

8.4.1 Akaike Information Criterion (AIC)

AIC is a precise mathematical measure of how well a model appears to fit the data. AIC is a statistical method for evaluating multiple models and identifying the one which best fits the data. In terms of model comparison, AIC and BIC have the same interpretation. For instance, a bigger AIC or BIC (Bayesian information criteria) difference implies more evidence for one model over the other (the lower the better). The model is better if the AIC is low. BIC (or Bayesian information criteria) is a version of AIC that penalizes adding more variables to the model more strongly.

8.4.2 Adjusted R-Squared

Adjusted R-squared is the R-squared version which takes the count of independent variables in the model into account. The adjusted R-squared rises when the new term improves the model more than would be expected by chance. The adjusted R-squared drops when a predictor enhances the model by less than projected.

When compared to a model with fewer input parameters, a substantially lower adjusted R-squared implies that perhaps the additional input parameters are not offering significance to the model. When compared to a model with fewer input parameters, a higher adjusted R-squared indicates that the additional input parameters are offering significance to the model.

8.4.3 Discussion

Initially a linear model using the *gam*() function has been fitted, without any smoothing term or parameter. Standard linear model can be seen as a special case of GAM. Then an actual generalized additive model using cubic regression spline as the basis has been fitted. Here, again the gam function for basic model fitting has been used, but now a function "s" within the formula to denote the smooth terms has been used. Within the function, the type of smooth is also specified, though a default is available. *bs* = *'cr'* has been chosen, denoting cubic regression splines. After inserting a smoothness parameter, a low p-value (k-index < 1) for an independent variable "multimap" has been found. Effective degree of freedom (edf) for the variable was also very close to "k", which is the maximum wiggliness parameter. Value of k-index less than 1 may indicate that "k" is too low, especially if edf is close to "k". So, the value of "k" to 16 for the "multimap" feature has been increased manually to improve the model performance. The model summary and diagnostics result for the cubic regression-based model after increasing wiggliness has been tested and found that the k-index for each feature variable is 1 or more than 1, and the "k" value is not very close to the effective degree of freedom (edf) for any of the feature variables. Figure 8.2 shows the residual plots of the same GAM model. Figure 8.2(a) illustrates the QQ plot on Quantiles analysis for the model. Figure 8.2(b) indicates the predictor vs. residuals analysis for the fitting based on GAM. Figure 8.2(c) indicates the histogram of residuals and similarly Figure 8.2(d) indicates response vs. fitted values.

Normally, the value for the method argument in the gam function can be automatically selected by the gam, and in most of the cases "GCV" is the default. But the argument value can be assigned manually too. So, how behavior of the model changes by selecting method arguments (*method = 'REML'*) manually other than GCV is tested. Figure 8.3 shows the residual plots when 'REML' family argument is opted. Figure 8.3(a) illustrates the QQ plot on Quantiles analysis for the model. Figure 8.3(b) indicates the predictor vs. residuals analysis for the fitting based on CR-GAM.

Double penalty approach has also been used to penalize the null space (basis functions which are smooth or constant or linear). Double penalty approach works best but applies to every smooth and doubles the number

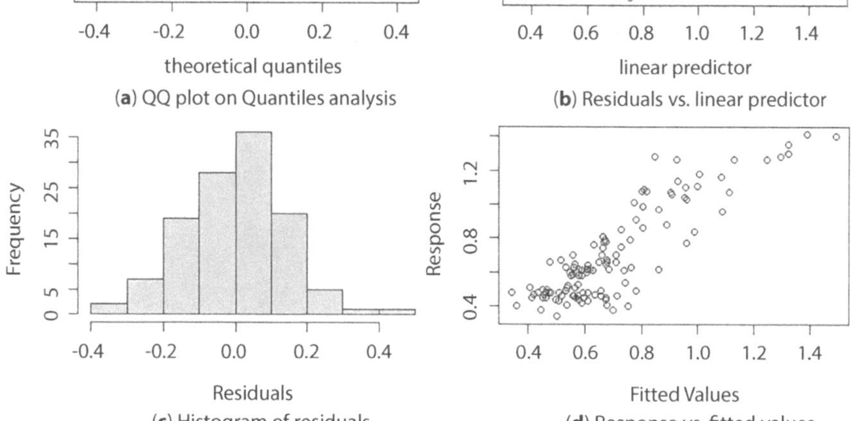

Figure 8.2 Residual plots of cubic regression-based model after increasing wiggliness.

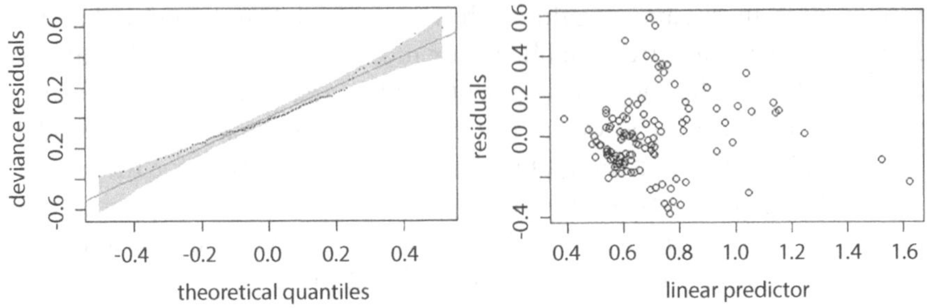

(**a**) QQ plot on Quantiles analysis (**b**) Residuals vs. linear predictor

Figure 8.3 Residual plots for REML-based model.

of smoothness parameters to estimate. Applying an extra penalty didn't give such a great result when compared to other models. Figure 8.4 shows the residual plots when the double penalty approach is used via "*select* = *True*" argument. Figure 8.4(a) illustrates the QQ plot on Quantiles analysis for the model. Figure 8.4(b) indicates the predictor vs. residuals analysis for the fitting based on CR-GAM.

Investigation about the behaviors of the model based upon the two most correlated independent variables (unmapped and multimap) with the response variable via tensor product smooth has also been done. Figure 8.5 shows the residual plots when tensor product smooth-based model has been built. Figure 8.5(a) illustrates the QQ plot on Quantile's analysis for the model. Figure 8.5(b) indicates the predictor vs. residuals analysis for the fitting based on CR-GAM.

The question of representing a response variable to interaction of several inputs with multiple units is addressed by tensor product smooths. Cost of arbitrary variable relative scaling decisions are eliminated, because they are

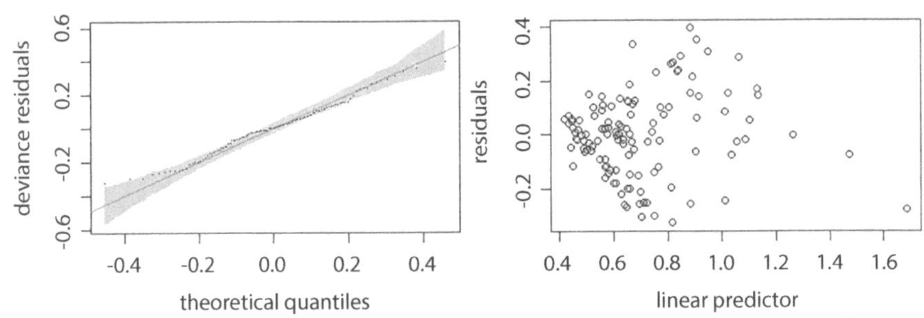

(**a**) QQ plot on Quantiles analysis (**b**) Residuals vs. linear predictor

Figure 8.4 Residual plots for double penalty-based model.

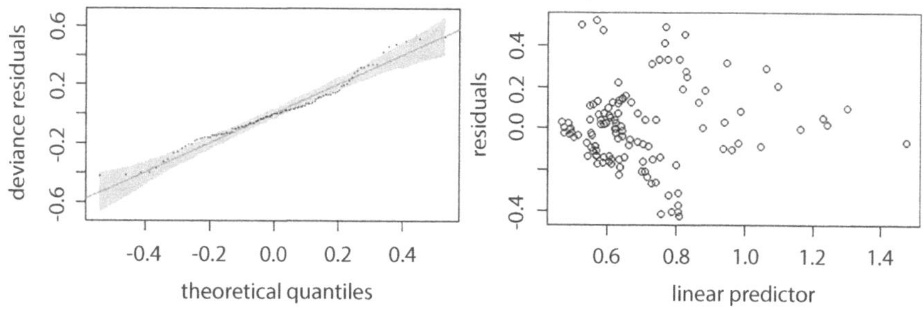

(**a**) QQ plot on Quantiles analysis

(**b**) Residuals vs. linear predictor

Figure 8.5 Residual plots for tensor product smooth-based model.

independent of the units in which the attributes are recorded. Figure 8.6 represents the summed effects of an interaction of the selected two feature variables, rather than partial effects of the tensor product smooth-based model. The three-dimensional plot describes how the number_of_mismatches (response feature in this study) differ even if the generated reads are getting aligned to the reference genome by an aligners, due to the variation in the number of unmapped and multimap reads from the sample.

Every time the family argument of the gam function is being selected automatically, here in the above cases "Gaussian" was always and the link function was also "identity" in every scenario.

Tables 8.1 and 8.2 represent different measures of performance and results of statistical tests using ANOVA function respectively for all the different GAM-based model approaches. The function conducts hypothesis testing on one or more fitted gam units.

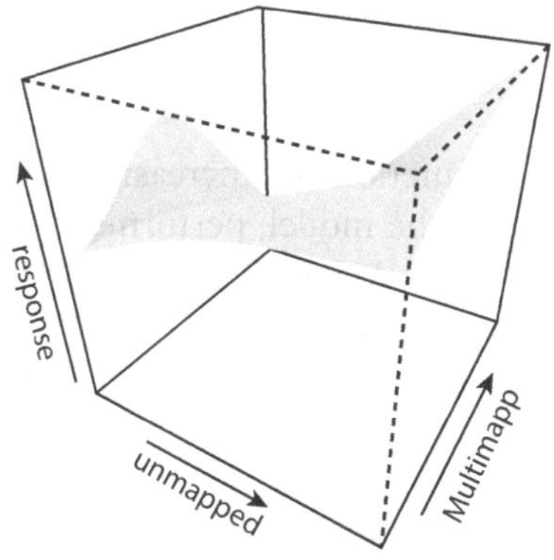

Figure 8.6 Summed effects of GAM.

Table 8.1 Different measures of performance for GAM.

S. no.	Model description	AIC	GCV	REML	Adjusted R-squared
1	Linear Model	27.354	0.072	-	0.096
2	Cubic Regression Spline	-52.007	0.04	-	0.602
3	Cubic Regression Spline, Increased wiggliness	**-58.113**	0.035	-	**0.617**
4	REML, instead of default	-34.504	-	-15.407	0.5
5	Double Penalty Approach	-56.653	0.037	-	0.607
6	Tensor Product Smooth	-30.557	0.044	-	0.456

Table 8.2 Statistical test via anova function for GAM.

S. no.	Resid.Df	Resid.Dev	Df	Deviance
1	113.000	7.794	-	-
2	86.573	2.671	26.426	5.123
3	86.925	2.621	-0.351	0.05
4	102.432	3.99	-15.507	-1.369
5	89.5	2.774	12.931	1.216
6	106.883	4.46	-17.383	-1.686

Using the smoothness function and increasing the wiggliness parameter (k) of a feature variable in the model, performed the best out of all fitted GAM models. Least value for AIC (-58.113) and higher value for adjusted R-squared (61.7%) for the third model in Table 8.1 shows that the model is fitted well as compared to others. Similarly, in Table 8.2 even if the value of resid.Df is very much close to 86 for both the second and third models, but least value for resid.Dev (2.621) for the third model also resembles better than others. The acceptable model is the one which fits well without any noise. Here, in the current scenario the model having 61.7% of adjusted R-squared seems mapping most of the predictor features quite well with the response variable "number_of _mismatches" which generally gets affected due to the genomic factors underlying tumor cells, and influence medication sensitivity.

After getting an insight about the GAM-based model behavior, boosted gam mechanism has been adopted, to compare as well as to improve the model performance. To integrate smooth effects to all the predictor variables, *gamboost*() with the P-spline base-learner *bbs*() is used at first. Figure 8.7 shows the partial effects of different base-learners when P-spline as a base-learner *bbs*() has been used via the boosted GAM approach. The model included 4 out of 7 predictor variables – unmapped, multimap, NoFeature and uniquely_mapped. The "unmapped" and "Nofeature" show linear variance, but there is a non-linear pattern at the middle of the predictor-grid (where most of the data are). In Figure 8.7(a), the linear pattern resembles that the variation in "unmapped" reads, which are basically the number of those reads which do not find a match anywhere on the reference genome, will affect the "number_of_mismatches" response variable linearly, and hence will affect the sensitivity of temozolomide (TMZ) drug towards the cancer cells in similar manner. In the case of multimap and

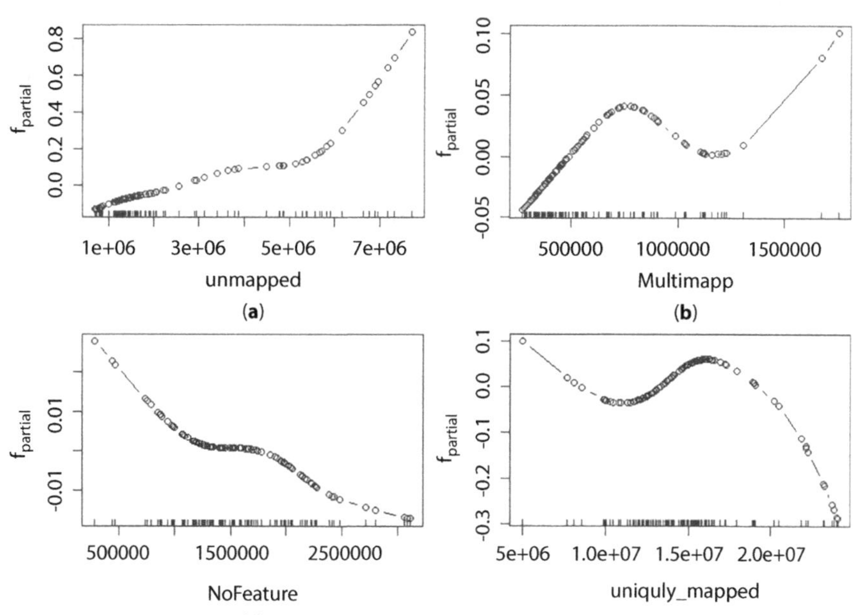

Figure 8.7 Partial effects of four measurements on the "number_of_mismatches" when "P-spline" as a base-learner is used.

uniquely_mapped, there exists a completely non-linear pattern. In Figures 8.7(b) and 8.7(d), the nonlinearity trend resembles if "multimap" and "uniquely_mapped" features differ then the "number_of_mismatches" feature may get influenced randomly without following any special fashion and indirectly the sensitivity of the drug too. Multimap and Uniquely_mapped features generally consider those reads which have multiple matches and different unique matches on the reference genome respectively.

Default settings for *boost_control*() is "*mstop* = 100", for all the boosted models. However, to achieve the best forecast for the response variable, an attempt has been made to optimize the tuning parameter in terms of predictive accuracy. Because variable selection is done simultaneously while tuning "mstop", models with just the most informative predictors are created. Therefore, first a model has been fitted with all available feature variables, then 25-fold bootstrapping has been used to optimize "mstop". For every term in the formula where no base-learner is explicitly specified using the base-learner option in *gamboost*(), the default base-learner has been specified, which is *bbs*(). The optimal number of boosting rounds is calculated, and a graph for "Cross-validated predictive risk with 25-fold bootstrapping" is created. Plot in Figure 8.8 shows the predicted risk for *mstop* = 1, …., 100 on the 25 bootstrap samples. The ideal iteration is the one which minimizes the mean hazard across all 25 samples. Optimal "mstop" is extracted using the function *mstop*(), and sets the model's "mstop" to the ideal value. Without optimizing, the model included all the same 4 predictor variables with the same pattern as described in Figure 8.7.

To make sure that nothing got lost, again the mstop has been increased to 1000 and then to 2000. Selected base-learners have also been checked at an initial number of iterations, i.e., 100, then at 1000 and again at 2000

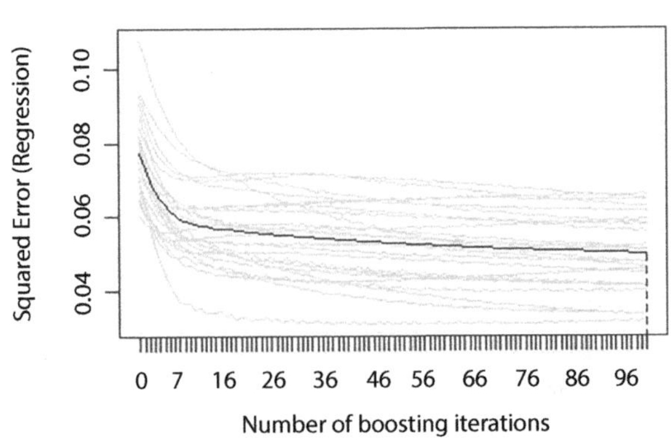

Figure 8.8 Cross-validated predictive risk with 25-fold bootstrapping.

using the function names (*coef(z)*). Here, "z" is the object returned by the *gamboost()* function. The model's complexity is controlled by the stopping iteration "mstop", which is the major tuning element for boosting. Larger mstop values resulted in larger and more complicated models, whereas smaller values resulted in a significant reduction in complexity. The model used all available variables as predictors for the "number_of_mismatches" (response variable) at iteration 1000, but only four predictors at the optimal iteration 100. Tuning the stopping iteration usually leads to the selection of the most influential predictors. Considering all the seven predictor variables by the model resembles that, all the features are taken into consideration for predicting the target variable "number_of _mismatches". To better understand the influence of "number_of_mismatches" on the sensitivity of temozolomide (TMZ) on individual cancer cell line clusters near the tumor which may be contributing to the progression of the tumor, all the potential entities must be considered by the model.

The model's behavior may also be seen by changing the base-learners to *bols()* for each predictor. While using *bols()* base-learner, the model became more complex without even optimizing the stopping iteration. The base-learner *bols()* included 6 out of all the 7 predictor variables. "Mapped_to_hgnc" predictor feature is not considered. While in the previous cases, there were only 4 out of 7 predictor variables without optimizing the stopping iteration. Since the *bols()* base-learner considered more predictor variables than the *bbs()* base-learner, the modelling with *bols()* as the base-learner suited better, and the extra cost of optimizing the model to include all the potentially important features as in the previous case is also eliminated. The partial effects of all the six base-learners on the response variable showed a complete linear trend without any small sort of non-linearity. Partial effects of the base-learners including feature variables "unmapped", "Multimap" and "ambiguous" resembled linearly increasing while the other base-learners including "Nofeature", "mapped_to_features" and "uniquly_mapped" feature variables resembled linearly decreasing. Hence, the model gives an insight that increase in the "unmapped" and "multimap" reads may increase the "number_of_mismatches" value significantly which will lead to the significant variation of the drug's efficacy differently on each tumor contributing cells also.

Different base-learners - *bbs()*, *bols()* and *bspatial()* are included in a single model. In the *bspatial()* the two most correlated predictors with the target variable are included. Figure 8.9 shows the output results which have been obtained from combining different base-learners in a single model. Figure 8.9(b) resembles an interaction surface between "unmapped" and "multimap" feature variables because of *bspatial()* base-learner.

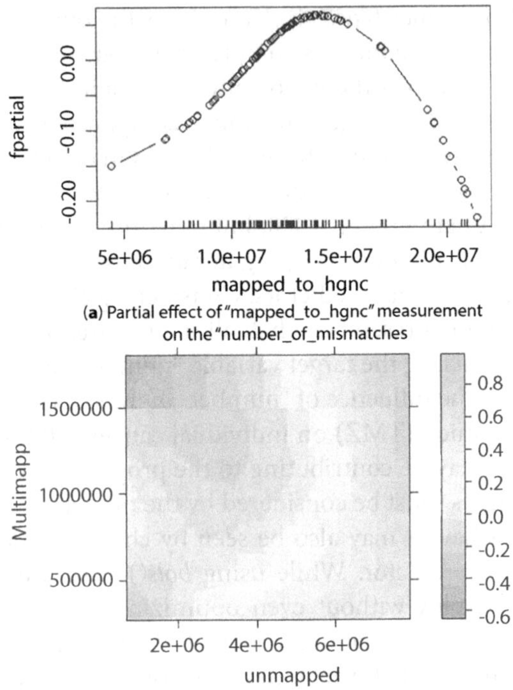

(**a**) Partial effect of "mapped_to_hgnc" measurement on the "number_of_mismatches

(**b**) Level plot interaction surface fitted using *bspatial*()

Figure 8.9 Results of combining different base-learners.

By default, the family for every boosted model is "*Gaussian*()". But there exists the flexibility to initialise other values for the family parameter. So, *family* = *QuantReg* (*tau* = 0.5) and *boos_control*(*mstop* = 500) is tested. The model included all the 7 predictor variables and that also without performing any optimization, which is quite good.

Figure 8.10 shows the partial effects of different base-learners which are obtained using base-learner *bbs*() & "*family* = *QuantReg*" for model building. Quantile regression is a set of statistical approaches for estimating and inferring conditional quantile functions which is still evolving. Some of the features like "unmapped", "ambiguous" and "mapped_to_hgnc" show a somewhat linear pattern, whereas others depict non-linear trend. Linear trend resembles that the number_of_mismatches value will be affected in linear fashion. But variations in the variables showing non-linearity trend will not impact the value of "number_of_mismatches" value linearly.

In the analysis of boosted gam-based models, predictor variables are considered differently by each model.

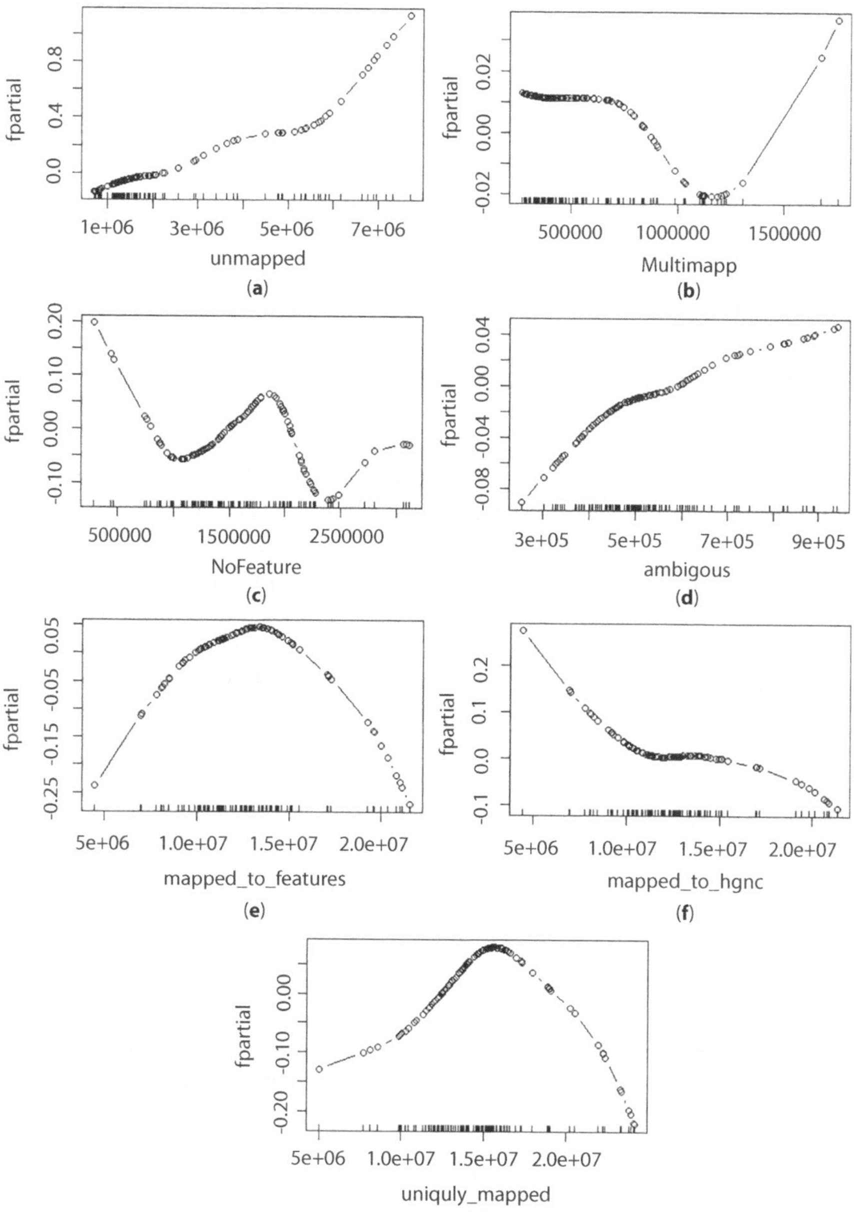

Figure 8.10 Partial effects of seven measurements on the "number_of_mismatches" when "*bols*()" base-learner & "*QuantReg*" family is used.

Conclusion

The challenges of Drug Sensitivity exemplified for glioblastoma is modelled using generalized additive models (GAMs) as well as likelihood based boosting GAM. The GAM model is extended using smoothness terms, cubic spline as the basis, double penalty approach and different smooth interaction functions and method arguments in the GAM function. The cubic regression or cubic spline setting in GAM could also model the non-linear expression levels of the samples. Boosted GAM is implemented using different base learners and family arguments and optimizing the stopping iteration – the tuning parameter for the boosting which controls the complexity of the model. The outcomes of the boosted GAM include how the partial effects of an individual base-learner affects the target feature in different scenarios. The workflow could potentially identify non-linear patterns in the samples that could be missed during conventional forms of regression frameworks. Boosting approach is computationally more efficient than the normal GAM-based methodology for modeling nonlinear patterns. Therapeutic efficacy of the drug is analyzed due to the variation in number of mismatches and the study resembles that each feature map contributes differently to the generation of mismatches which affect the sensitivity of the drug towards tumor cells.

References

1. Darmanis S, Sloan SA, Croote D, Mignardi M, Chernikova S, Samghababi P, Zhang Y, Neff N, Kowarsky M, Caneda C, Li G. Single-cell RNA-seq analysis of infiltrating neoplastic cells at the migrating front of human glioblastoma. *Cell reports*. 2017 Oct 31;21(5):1399-410.
2. Couturier CP, Ayyadhury S, Le PU, Nadaf J, Monlong J, Riva G, Allache R, Baig S, Yan X, Bourgey M, Lee C. Single-cell RNA-seq reveals that glioblastoma recapitulates a normal neurodevelopmental hierarchy. *Nature Communications*. 2020 Jul 8;11(1):1-9.
3. Lopes MB, Vinga S. Tracking intratumoral heterogeneity in glioblastoma via regularized classification of single-cell RNA-Seq data. *BMC Bioinformatics*. 2020 Dec;21(1):1-2.
4. Stupp R, Mason WP, Van Den Bent MJ, Weller M, Fisher B, Taphoorn MJ, Belanger K, Brandes AA, Marosi C, Bogdahn U, Curschmann J. Radiotherapy plus concomitant and adjuvant temozolomide for glioblastoma. *New England Journal of Medicine*. 2005 Mar 10;352(10):987-96.
5. Hegi ME, Diserens AC, Gorlia T, Hamou MF, De Tribolet N, Weller M, Kros JM, Hainfellner JA, Mason W, Mariani L, Bromberg JE. MGMT gene

silencing and benefit from temozolomide in glioblastoma. *New England Journal of Medicine*. 2005 Mar 10;352(10):997-1003.

6. Campos B, Olsen LR, Urup T, Poulsen HS. A comprehensive profile of recurrent glioblastoma. *Oncogene*. 2016 Nov;35(45):5819-25.
7. Gallego O. Nonsurgical treatment of recurrent glioblastoma. *Current Oncology*. 2015 Aug;22(4):273-81.
8. Lim M, Xia Y, Bettegowda C, Weller M. Current state of immunotherapy for glioblastoma. *Nature Reviews Clinical Oncology*. 2018 Jul;15(7):422-42.
9. Morrissy AS, Cavalli FM, Remke M, Ramaswamy V, Shih DJ, Holgado BL, Farooq H, Donovan LK, Garzia L, Agnihotri S, Kiehna EN. Spatial heterogeneity in medulloblastoma. *Nature Genetics*. 2017 May;49(5):780-8.
10. Szerlip NJ, Pedraza A, Chakravarty D, Azim M, McGuire J, Fang Y, Ozawa T, Holland EC, Huse JT, Jhanwar S, Leversha MA. Intratumoral heterogeneity of receptor tyrosine kinases EGFR and PDGFRA amplification in glioblastoma defines subpopulations with distinct growth factor response. *Proceedings of the National Academy of Sciences*. 2012 Feb 21;109(8):3041-6.
11. Sottoriva A, Spiteri I, Piccirillo SG, Touloumis A, Collins VP, Marioni JC, Curtis C, Watts C, Tavaré S. Intratumor heterogeneity in human glioblastoma reflects cancer evolutionary dynamics. *Proceedings of the National Academy of Sciences*. 2013 Mar 5;110(10):4009-14.
12. Aubry M, de Tayrac M, Etcheverry A, Clavreul A, Saikali S, Menei P, Mosser J. Correction: From the core to beyond the margin: a genomic picture of glioblastoma intratumor heterogeneity. *Oncotarget*. 2016 Oct 11;7(41):67685.
13. Kim J, Lee IH, Cho HJ, Park CK, Jung YS, Kim Y, Nam SH, Kim BS, Johnson MD, Kong DS, Seol HJ. Spatiotemporal evolution of the primary glioblastoma genome. *Cancer Cell*. 2015 Sep 14;28(3):318-28.
14. Aderetti DA, Hira VV, Molenaar RJ, van Noorden CJ. The hypoxic periarteriolar glioma stem cell niche, an integrated concept of five types of niches in human glioblastoma. *Biochimica et Biophysica Acta (BBA)-Reviews on Cancer*. 2018 Apr 1;1869(2):346-54.
15. Lathia JD, Mack SC, Mulkearns-Hubert EE, Valentim CL, Rich JN. Cancer stem cells in glioblastoma. *Genes & Development*. 2015 Jun 15;29(12):1203-17.
16. Liebelt BD, Shingu T, Zhou X, Ren J, Shin SA, Hu J. Glioma stem cells: signaling, microenvironment, and therapy. *Stem Cells International*. 2016 Oct;2016.
17. Fidoamore A, Cristiano L, Antonosante A, d'Angelo M, Di Giacomo E, Astarita C, Giordano A, Ippoliti R, Benedetti E, Cimini A. Glioblastoma stem cells microenvironment: the paracrine roles of the niche in drug and radioresistance. *Stem Cells International*. 2016 Jan 6;2016.
18. Krol I, Castro-Giner F, Maurer M, Gkountela S, Szczerba BM, Scherrer R, Coleman N, Carreira S, Bachmann F, Anderson S, Engelhardt M. Detection of circulating tumour cell clusters in human glioblastoma. *British Journal of Cancer*. 2018 Aug;119(4):487-91.

19. Grande S, Palma A, Ricci-Vitiani L, Luciani AM, Buccarelli M, Biffoni M, Molinari A, Calcabrini A, D'Amore E, Guidoni L, Pallini R. Metabolic heterogeneity evidenced by MRS among patient-derived glioblastoma multiforme stem-like cells accounts for cell clustering and different responses to drugs. *Stem Cells International.* 2018 Feb 18;2018.
20. Rayfield CA, Grady F, De Leon G, Rockne R, Carrasco E, Jackson P, Vora M, Johnston SK, Hawkins-Daarud A, Clark-Swanson KR, Whitmire S. Distinct phenotypic clusters of glioblastoma growth and response kinetics predict survival. *JCO Clinical Cancer Informatics.* 2018 Feb;2:1-4.
21. Bhaskaran V, Nowicki MO, Idriss M, Jimenez MA, Lugli G, Hayes JL, Mahmoud AB, Zane RE, Passaro C, Ligon KL, Haas-Kogan D. The functional synergism of microRNA clustering provides therapeutically relevant epigenetic interference in glioblastoma. *Nature Communications.* 2019 Jan 25;10(1):1-3.
22. Wood SN, Augustin NH. GAMs with integrated model selection using penalized regression splines and applications to environmental modelling. *Ecological Modelling.* 2002 Nov 30;157(2-3):157-77.
23. Maindonald J. Smoothing terms in GAM models. Smoothing Terms in GAM Models. 2010 Aug 29.
24. Royston P, Sauerbrei W. Multivariable modeling with cubic regression splines: a principled approach. *The Stata Journal.* 2007 Feb;7(1):45-70.
25. Gülüm M, Yesilyurt MK, Bilgin A. The performance assessment of cubic spline interpolation and response surface methodology in the mathematical modeling to optimize biodiesel production from waste cooking oil. *Fuel.* 2019 Nov 1;255:115778.
26. Tian Y, Baro E, Zhang R. Performance evaluation of regression splines for propensity score adjustment in post-market safety analysis with multiple treatments. *Journal of Biopharmaceutical Statistics.* 2019 Sep 3;29(5):810-21.
27. Yao J. Modelling Stillbirths with Natural Cubic Splines (Doctoral dissertation, University of Pittsburgh).
28. Hofner B, Mayr A, Robinzonov N, Schmid M. Model-based boosting in R: a hands-on tutorial using the R package mboost. *Computational Statistics.* 2014 Feb;29(1):3-5.
29. Tutz G, Binder H. Generalized additive modeling with implicit variable selection by likelihood-based boosting. *Biometrics.* 2006 Dec;62(4):961-71.
30. Bekisz S, Geris L. Cancer modeling: From mechanistic to data-driven approaches, and from fundamental insights to clinical applications. *Journal of Computational Science.* 2020 Oct 1;46:101198.
31. Balogh P, Gounley J, Roychowdhury S, Randles A. A data-driven approach to modeling cancer cell mechanics during microcirculatory transport. *Scientific Reports.* 2021 Jul 27;11(1):1-8.
32. Jiang P, Sellers WR, Liu XS. Big data approaches for modeling response and resistance to cancer drugs. *Annual Review of Biomedical Data Science.* 2018 Jul 20;1:1-27.

33. Li M, Xie S, Lu C, Zhu L, Zhu L. Application of data science in circulating tumor DNA detection: a promising avenue towards liquid biopsy. *Frontiers in Oncology*. 2021;11.
34. Parikh RB, Gdowski A, Patt DA, Hertler A, Mermel C, Bekelman JE. Using big data and predictive analytics to determine patient risk in oncology. *American Society of Clinical Oncology Educational Book*. 2019 May 17;39:e53-8.
35. Dlamini Z, Francies FZ, Hull R, Marima R. Artificial intelligence (AI) and big data in cancer and precision oncology. *Computational and Structural Biotechnology Journal*. 2020 Aug 28.
36. Kourou K, Exarchos TP, Exarchos KP, Karamouzis MV, Fotiadis DI. Machine learning applications in cancer prognosis and prediction. *Computational and Structural Biotechnology Journal*. 2015 Jan 1;13:8-17.
37. Kim EL, Sorokin M, Kantelhardt SR, Kalasauskas D, Sprang B, Fauss J, Ringel F, Garazha A, Albert E, Gaifullin N, Hartmann C. Intratumoral heterogeneity and longitudinal changes in gene expression predict differential drug sensitivity in newly diagnosed and recurrent glioblastoma. *Cancers*. 2020 Feb;12(2):520.

9

Unobtrusive Engagement Detection through Semantic Pose Estimation and Lightweight ResNet for an Online Class Environment

Michael Moses Thiruthuvanathan[1]*, Balachandran Krishnan[2] and Madhavi Rangaswamy[2]

[1]Department of CSE, School of Engineering and Technology, CHRIST (Deemed to be University), Bengaluru, Karnataka, India
[2]Department of Psychology, School of Social Sciences, CHRIST (Deemed to be University), Bangalore, Karnataka, India

Abstract

Analysing student engagement in a class through unobtrusive methods enhances the learning and teaching experience. During these pandemic times, where the classes are conducted online, it is imperative to efficiently estimate the engagement levels of individual students. Helping teachers to annotate and understand the significant learning rate of the students is critical and vital. To facilitate the analysis of estimating the engagement levels among students, this paper proposes a dual channel model to precisely detect the attention level of individual students in a classroom. Considering the possible inaccuracy of emotion recognition, a dual channel is configured with a Lightweight ResNet model for macro-level attention estimation and a 3d pose estimation using Euler angles for Pitch, yaw and roll that is trained, validated and tested on the Daisee database. The Emotional detection extracts the context of Engaged, frustrated, confused and disgust as higher levels of classroom attention cognition while the facial pose coordinates provide the real-time movement of the faces in the video to provide a series of engaged and disengaged coordinates. The Lightweight ResNet Model achieves a 95.5% accuracy and the Pose estimation test is able to distinguish the test videos at 92% as

**Corresponding author*: michael.moses@christuniversity.in

Kukatlapalli Pradeep Kumar, Aynur Unal, Vinay Jha Pillai, Hari Murthy, and M. Niranjanamurthy (eds.)
Data Engineering and Data Science: Concepts and Applications, (225–254) © 2023 Scrivener Publishing LLC

Engaged and Bored on the Daisee Dataset. The Overall Accuracies using the Dual channel was curated to 87%.

***Keywords*:** Student engagement detection, residual networks, convolutional neural network, emotion detection, facial pose estimation, facial recognition

9.1 Introduction

A classroom is a nurturing ground for students to anchor into the future through learning and experience. Commonly a class contains variety of students from diverse backgrounds and diverse cognition and emotions. The students can be primarily divided into the following categories of gender, learning capability, understanding capability, attentiveness, distractedness, etc. Generally, a classroom consists of more than 30 students and one teacher delivering a lecture or teaching. During such a process on a day-to-day basis, it is quite hard for a teacher to read and understand all the students' activity, engagement, mood, behaviour and understanding together. The feedback on how well the classes is received can be measured by student feedback forms, exams and interaction with all the students individually. The feedback can be made available only after a certain number of weeks into teaching. Daily reports on the class's understanding are not predominantly available and will create a monotonous reply by the students. Students are generally attentive if the classes are interesting and if they are able to understand the topic really well. Associating an automatic analysis through Visual mapping of students' gestures in a class for a period of time and instantly providing the analysis report to the students will help the teacher to know the status of learning in the classroom. Recent advancements in the visual receptors and computer vision methods enables tthe creation of an automated analysis system based on behaviour and facial expressions. Over the last few years, research into student learning status has garnered a lot of attention.

The attention status of the students can be analysed from a video captured through a surveillance camera, web camera, or dedicated monitoring cameras. The video may contain one student or many students. The students captured in the frames need to be tracked for a complete understanding of their emotional state. Identifying, recognizing and tracking many students in a video is always a non-trivial task. However, tracking the facial information of the students is important for the analysis and storyboarding. To create a report on how a student has received the teaching from the instruments created for teaching and learning helps in assessing the learning process. Integration of variables into the instruments across multiple verticals enhances the analysis. Conventional systems that are used to retrieve the feedback from the students are forms, questionnaires, short feedback reports, etc. The drawback of these methods is that the majority of the data received can be fabricated by the students; the questionnaires can be answered in favour of the teacher or against the teacher based on past experiences. The questionnaires have to be subject-specific and need to be created by the teachers, which gives rise to specific bias. Also, the questionnaires are generally answered after the class is completed, and a majority of the questionnaires have to be answered for a significant outcome. Using computer vision, one can also receive posed or acted features from the students since they are aware of a monitoring system that is analysing their activity. Capturing the level of understanding by the students through the vision techniques is non-trivial. A computer-vision-based system uses facial recognition, emotional analysis, behavioural analysis, gesture analysis, etc. Classroom videos can be analysed watching them individually; however, the analysis is subjective to the content of the search, and also, identifying the state of mind individually is very time consuming. Generally, video summarizations are carried out on surveillance videos, sports videos to reduce the number of frames.

Deep learning is a type of machine learning that uses artificial neural networks exclusively. Because neural networks are designed to imitate the human brain, deep learning is likewise a human brain mimic. We do not have to explicitly program anything in deep learning. Deep learning is not a new concept. It has been around for quite some time. It is popular now because we did not have as much processing power and data back then. As processing power has increased tremendously over the last 20 years, deep learning and machine learning have entered the scene. Deep Layers enables the network to learn high dimensional features and enables the network to produce high classification and prediction accuracy. Though Deep learning is a subset of Machine learning, there are major differences that deep learning functions on. Deep learning uses large datasets and high computation requirement, and it takes less time to learn and a shorter time for testing. These networks use a huge amount of data for learning and they produce state-of-the-art results. There are numerous applications like automatic text generation, healthcare, image recognition, autonomous vehicles, e-commerce, personal assistant, etc. Deep learning is chosen due to their best-in-class performance on problems, reduced Feature engineering requirement, need for labelling the data is minimal and their efficacy in generating quality results.

Prediction and classification models utilising machine learning have been the focus of recent study. Some of the renowned networks are Convolutional Neural Network (CNN), Recurrent Neural Networks (RNN), Autoencoders (AE), Deep Belief Networks (DBN), Long Short-Term Memory (LSTM); these are the most prominent deep learning approaches that have seen widespread adoption. However, no condensed list of deep learning algorithms has been given. As a result, there is a research gap in introducing deep learning methods and summarizing the methodology and applications in a concise, yet informative publication [1]. Image recognition and image classification techniques make use of models that are inspired from CNN. Some of the networks that are predominantly used are AlexNet [2], Visual Geometry Group (VGG) [3], DenseNet [4], ResNet [5], and MobileNetV2.

AlexNet, which was released in 2012, was a breakthrough improvement; it improved on standard Convolutional Neural Networks (CNNs) and became one of the optimum image categorization models. VGG is a cutting-edge object-recognition model that can accommodate up to 19 layers. VGG, which was designed as a deep CNN, outperforms baselines on a wide range of tasks and datasets outside of ImageNet. VGG is still one of the most widely used image-recognition models today. A DenseNet is a form of convolutional neural network that employs dense connections between layers through Dense Blocks, which connect all layers (with matching feature-map sizes) directly with one another. To maintain the feed-forward nature, each layer receives extra inputs from all preceding levels and transmits its own feature-maps to all following levels. Deep Residual Network was undoubtedly the most significant achievement in the computer vision/deep learning community in the previous several years, following AlexNet's acclaimed triumph at the ImageNet classification challenge. ResNet allows you to train hundreds or even thousands of layers while still achieving impressive results. The benefit of ResNet is its great representational capabilities, which has improved the performance of numerous computer vision applications other than picture classification, such as object identification and face recognition. Andrew G. Howard *et al.* of Google Research suggested the MobileNetV2 as a deep learning model in their paper [6]. This model was presented as part of a series of mobile-first computer vision models for TensorFlow, with the goal of maximizing accuracy while keeping limited resources in mind. MobileNets are tiny, low-latency, low-power models that may be customized to match the resource restrictions of various use cases. They may be used for classification, detection, embedding, and segmentation in the same way as other popular large-scale models are. MobileNetV2 is a convolutional neural network design that aims to be mobile-friendly. It is built on an inverted residual structure, with residual connections between bottleneck levels. As a source of non-linearity, the intermediate extension layer employs lightweight depth wise convolutions to filter features. Overall, MobileNetV2's architecture includes a fully convolutional layer with 32 filters, followed by 19 residual bottleneck layers. The model offers to employ this lightweight architecture to enable deep learning with minimal computing requirements, despite the fact that the network is designed to work well on mobile devices.

A holistic approach is required in order to create analysis that can touch various aspects of teaching and learning among students. A system needs to be integrated to conduct these studies on emotions that can be conclusive and decisive on how the learning and teaching systems can be developed to suit the needs of the students. This work uses a ResNet that is tailor made to improve accuracy by significantly decreasing the number of operations and the computation required. This network uses Inverted Residuals and linear bottlenecks that enables training by expanding to a high dimensional representation with a deep convolutional network.

As shown in Figure 9.1, the Current Research in many fields of interest use ResNets and mobileNetV2. Though AlexNet, DenseNet and VGG proved to be superior in their inception, as data and features needed for analysis increased, the deeper networks started to fail in their performance [7]. The inception of newer networks paved the way in improvising the training and validation accuracy in their respective domains. ResNets are currently being utilized widely and a few variants of custom-made architectures are being utilized to improve the time and computational complexity.

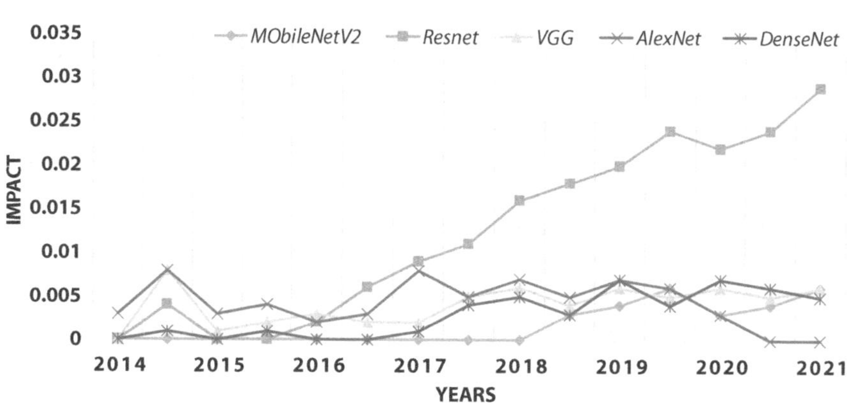

Figure 9.1 Deep learning model usage over time.

In the current scenario, Variants of ResNets and mobileNetV2 are critically acclaimed and used for their performance and accuracy. This work also utilizes ResNets for detection and benchmarking. The proposed work focuses on a holistic analysis story boarding based on the following:

- Integration of emotion recognition, deep learning, high-performance GPU computing, and feedback systems.
- Utilize 3D Pose Estimation to detect engaged Coordinates for a face in a video.
- Quantifying emotions using the Multimodal methods to estimate engaged and disengaged status.

The rest of this chapter is organised as follows. Section 2 describes the related work and section 3 explains the proposed methodology. The details of experimentation are provided in section 4. The detailed results are in section 5 and finally, section 6 concludes the chapter.

9.2 Related Work

9.2.1 Analysis for a Classroom Environment

The advent of emotional analysis for student fraternity has benchmarked higher interest and belief to improve understanding and perseverance of students' learning curve from a classroom. Cognition is the process of acquiring knowledge through thinking, experiencing and recognizing. Learning is the process of acquiring knowledge through study, teaching and analysis. Cognitive learning is a process of perceiving attention to the subject being taught and express interest or dislike based on the content being delivered. Emotions are traditional methods of articulating a perceived notion through the face to allow others to understand the state of mind. Therefore, emotions serve as indicators to understand students that can help teachers better organize their teaching and improve students' performance in learning [2]. There are two methods to capture and measure students' emotions in learning scenarios. Firstly, by self-evaluation by conducting individual reports based on the perceived data. Secondly, by automating a system that can efficiently create reports on the entire class and its interest level. However, data collected by self-evaluation methods can be biased and unreliable as the data can be cooked up, based on the requirement. Also, the data reported through Self-Appraisal need not be complete and detailed as the reports are generally obtained after the session. Auto evaluation methods are proposed to solve this problem. In this paper, we apply computer vision techniques to analyse students' emotions in a classroom based on videos captured from a smart classroom. Visualization can facilitate emotion analysis in learning scenarios. A plethora of experiments have been performed to evaluate and interpret the emotion data collected, analysing the evolution of both the individual and the group of people's emotions in classroom videos. Zhao *et al.* [8] proposed a system named PEARL, to consider the different emotion types generated from social media through the use of text analysis. Hernandez *et al.* [9] and Srivastava *et al.* [10] applied Computer vision methods used to derive sentiments from images or videos. Nevertheless, they simply provided basic visual symbols for representing emotion knowledge without further study. But this work focused primarily on studying the development of human feelings while not contemplating for a group of people. H. Zeng *et al.* focus on analysing the emotion of both an individual and a group of people in videos; however, a need for self-evaluation is required to validate the entire process [11]. Kempter *et al.* designed a gadget described Emotion Watch that graphically shows how a group of people's emotions evolve [12]. This technique also prevented users from monitoring and evaluating the development of human emotions. Jeffrey D. Wammes *et al.* measured the usage of media and related technologies during lectures through studies 1 and 2. The media multitasking rates were fairly high and increased with the amount of time spent in lectures, compared with prior research [13], whereas in Study 2 energy rates stayed relatively constant. Aitor *et al.* developed an emotional intelligence questionnaire for a group of students in one class, showing convergence, stability and internal consistency [14].

9.2.2 Pose Estimation

In recent years, much research has been conducted on the assessment of head postures. It employs three dimensions to determine the angle of the head's Euler angle and incorporates spatial information to modify it [15].The three vectors in a rotation matrix are proposed as the representation of head position estimation in the study proposed by Zhiwen *et al.* based on a novel neural network [16]. For Euler angle discontinuities, a novel annotation approach based on three vectors to describe head locations and a new Mean Absolute Vice Error (MAEV) measurement has been proposed. Using the AFLW dataset, Kumar *et al.* have proposed a better architecture for Google LeNet [17] and using the AFLW dataset to learn posture judgments in rough form while working together [18]. Ranjan *et al.* [19] created a multi-task convolution neural network in order to concurrently recognise face, gender, key points and head position. AlexNet is enhanced using the R-CNN algorithm. Ruiz *et al.* [20] have developed the RGB method, combined with regression limitations to the prediction of Euler angles, to improve pose estimation. Fast, efficient, easy-to-use R-CNN model is based on all picture sides with 6DoF regress and no face identification prior to training. During the training, the distance between the input photo and the arbitrary cropping created changes and the model's evaluation is translated and preserved as a constant. Box training labels for detection can be replaced with face postures [21]. The techniques above are not intended to be used to a low-resolution face. As the model learns from different resolutions, it will be able to overcome the limitations of low resolution and become more robust. Currently, many techniques can only calculate attention in one direction. Model stability will be substantially affected by variations in head position in complex settings. Since the Euler angles must be measured from three different directions, our technique is more accurate, while also providing spatial information to offset changes in Head of the Euler Angle caused by uncontrolled variables. It is possible to produce initial alignment and pose estimations by merging two standard neural network model systems for face detection and medium form learning [22]. The concept consists of a two-target optimization approach that refines estimates repeatedly.

9.2.3 Face Alignment and Landmark Estimation

These landmarks record the rigid and non-rigid deformation of the face owing to head movement and facial emotions. Thus, they play a key role in the numerous facial analysis tasks that need their use. A number of algorithms have been developed over the years to recognise these landmarks automatically, and a classifier called Gentle Boost is utilised. As a result, it is uncertain which feature representation and classifier to employ in the local appearance model. SIFT, HOG, and SVM classifiers have been taught using deep learning methods, and the learned features are equivalent to HOG and SIFT [23]. Unified model for face occlusion is presented, which clearly predicts landmark positions and landmark occlusion [24]. For simultaneous recognition of face action units and facial landmarks, a combined cascaded regression architecture is presented in [25]. Facial action unit activation probability and landmark positions are updated in a cascaded iterative way based on the joint connections between face action units and facial forms. As a result, Xiong *et al.* have suggested an iterative approach for landmark recognition in which the pose-dependent model is automatically determined based on estimations from previous iterations [26]. Using a "coarse to fine" search technique, the original face shape is continually modified based on the estimation from the previous step. A closer initialization will be created, and finer facial landmark detection results will be simpler to get. In addition, the cascaded regression technique does not provide a mechanism to assess the accuracy of landmark prediction or to modify the number of cascaded steps for various pictures [27]. Following the cascading regression paradigm, CNNs are coupled with deformable face shape models to estimate the head position and facial landmarks on pictures with large head poses [28]. As regression models for face alignment, an ensemble of regression trees is utilised. By changing the goal function, the system may use pictures with partly labelled facial landmark positions as training images. Compared to previous approaches, the cascaded regression method is quicker [29]. CLM techniques, regression-based methods for facial landmark detection algorithms have been studied in depth by Yue *et al.* Under various facial expressions, head postures and occlusion, it compares the performance of these theories on controlled and in-the-wild benchmark data sets [30]. These techniques enable the models to detect landmarks on the face and estimate the alignment to the camera.

9.2.4 Deep Networks for Emotional Analysis

ResNets have lately been used to explore the emotional analysis of faces. According to Bin *et al.*, the CNN algorithm coupled with ResNet-50 will produce more accurate results for multi-classification problems. The adoption of this approach offers accuracy and a systematic recognition effect with average cognitive [31]. "Neutral", "Anger", "Disgust", "Fear", "Happiness", "Sadness" and "Surprise" were classified using a ResNet-18 architecture pre-trained on the FER+ dataset by Singh *et al.* As part of the second workshop and competition on emotional behaviour analysis in the wild (ABAW2) [32]. Three deep neural networks were thoroughly explored for their configurations and performances: ResNets, LSTM and hybrid ResNets-LSTM [33]. On three Indian datasets that mostly work on classroom engagement models, Michael *et al.* developed a methodology for constructing convolutional neural networks (CNN) and equipping them with residual connections that can boost the learning rate of the network and improve classification. Due to Residual Learning, which incorporates extra learning from the previous batch of layers, and the OHP (Optimized Hyper Parametric) configuration [34], this suggested network works well. Many authors use modified CNN in order to achieve greater results by adding multiple deep layers that enhance the performance of the system [36–38]. By adding additional deep layers that increase the performance of the system, several writers modify CNN to get better results [36–38]. Gradient vanishing is one of the most common problems with CNNs, which can lead to a loss of accuracy due to a growing memory need for training. It is necessary to enhance the scope of detection rate and the precision of detection percentage for specific emotions in all significant research projects listed in this section. In order to enhance detection in the data, it is necessary to have a system that can minimise the mistake rates when training. During the process of classification, engagement detection is used to draw conclusions on both two-level (not engaged and engaged) and a three-level (not engaged, typically engaged, and extremely engaged) scale. On the Daisee Dataset, they used Kernel Principal Component Analysis (KPCA) to capture nonlinear correlations among the retrieved features and Local Directional Pattern (LDP) to extract person-independent edge features for the various face expressions [35]. A multimodal framework using video summarization and micro emotional classification to reduce the computational requirements for video processing was proposed by Michael *et al.* [39]. Deep Networks vary in feature extraction process, depth of the network, categorization of the network and many other aspects. However, each network can be built and trained for the requirement.

Dataset: This is the first multi-label video classification dataset consisting of 9068 video clips taken by 112. For each of the four emotional states, the dataset contains four degrees of labelling - very low, low, high, and very high [40].

9.3 Proposed Methodology

In this section we present the data acquiring methods, data labelling, analysis of the attention levels observation, nonverbal template validation and the overall behavioural analysis.

9.3.1 Data Description

The aim was to record the student behaviour in a class and estimate the level of attention the student has expressed. The experimental classes were between 10-15 minutes of class of different teachers and students' scenario. This analysis was initially conducted for a single student and later sessions were conducted for a group of students not exceeding 15 in a class. If the number increased, then main issues of occlusion become more prominent and the results were not appropriate. The optimum number of students were found to be 15 as the size of each face needs to be at least 96 x 96. Face images play an important role in the study. Face detection, face positioning are the important tasks that need to be carried out in each frame. A camera that can capture videos with 1920 x1080 was used fixed at 30 frames per second.

9.3.2 Facial Detection and Recognition

The vital need in this study is the face, as the face delivers the required information for the necessary attentiveness detection in a class. A single face detection and tracking the same face is a trivial task. However, detecting, recognizing, tracking and detecting the important features for a long video is a non-trivial task.

9.3.2.1 *Face Detection*

Face detection for pose estimation is carried out using the dLib library. As the name implies, the dLib CNN face detection method is a mix of Convolutional Neural Network (CNN) and dLib. The CNN algorithm is a deep learning system that analyses images as an input. dLib is an open-source library that contains a number of machine learning methods for solving difficult tasks. To detect faces, the system employs the dLib toolbox in addition to CNN characteristics. The CNN characteristics offer the algorithm an advantage in detecting methods. dLib CNN employs MMOD, or Maximum-Margin Object Detector, in addition to CNN-based characteristics. In contrast to the previous techniques, this algorithm automates the process of picking the filter to extract features from images [35].

The first step in the face detection is to load the pre-trained. As previously stated, this approach automates the process of picking filters to apply to an image in order to extract features. The face detection model is included in a model weights file, which needs to be obtained in order to run the method. The second step is to initialize, the weights file. The weights file that was downloaded in the first stage is used in the second phase to initialise. The weights file must be provided as an input before starting the dLib CNN face detector [16].

After the face detector has been configured, we must apply it to the test pictures, which will result in the detection of faces. The binary classifier format of CNN's output is extremely specific in this case. The dLib CNN performs well with both frontal and non-frontal faces, as well as unusual angles, while HOG-based detectors do not. The dLib CNN, in a nutshell, identifies faces from virtually any angle.

The dLib CNN is extremely simple to use. As previously said, the stages in this algorithm are quite simple, making it very simple to train. It works much better with the GPU than with the CPU. With a variety of facial occlusions, the algorithm performs admirably.

9.3.2.2 *Facial Landmark Detection*

Important face characteristics such as the eyes, eyebrows, nose, mouth, and jawline are identified and represented by facial landmarks. In the context of shape prediction, face landmark identification is a subset of this problem.

Given an input image, a shape predictor attempts to identify important areas of interest along a form (and typically a ROI that identifies the item of interest). Face-landmarking algorithms are used to discover important facial characteristics on a face.

This method starts with an image from a set of annotated face landmarks in a training set. The (x,y)-coordinates of areas surrounding each facial structure are specified in these photographs. The probability depending on the distance between pairs of input pixels are known as priors. Using this training data, an ensemble of regression trees is developed to predict face landmark locations directly from pixel intensities. The position of 68 (x, y)-coordinates that correspond to facial structures is estimated using the dLib library's pre-trained facial landmark detector. These annotations are from the 68-point iBUG 300-W dataset, which was used to train the dLib face landmark predictor. It is worth mentioning that many types of facial landmark detectors exist, including the 194-point model that can be trained with the HELEN dataset [41]. Regardless of which dataset is utilised, the same dLib framework may be used to train a shape predictor on the input training data.

9.3.3 Emotion Quantification

Emotion recognition systems focus on computational approaches for understanding and recognizing expressive behaviours of emotion. In this paper, we assume that the perceived emotion label of human annotators describes the emotional cues in a given dataset, and use this label as a ground truth emotion of the dataset. The dimensions for an emotional recognition system are represented using continuous values, described in the dimension of emotions. The Emotions considered for study are categorised as Angry, Happy, Sad, Frustrated, Neutral, Surprise and Fear proposed by Eckman. Dimensions involving Valence and arousal determines the attentiveness of a student. The primary challenge in quantifying these emotions is due to subjectivity. A Deep learning model is well trained to capture emotions and average them based on perceived labelled emotional annotations depicted by multiple students. This helps in reducing subjectivity challenge and also improves the accuracy of the system for emotional analysis. The emotion quantification is carried out using a Lightweight ResNet model. This model utilizes the strength of Residual connections in the network that enhances the learning. CNN's are prone to vanishing gradients and the learning probability decreases as the depth of the layers increases. The Lightweight network utilizes a three-skip network model in a 74-layer deep network. The intent of this model is to train a network that can classify four behavioural emotions namely Engagement, Confused, Frustrated and Bored.

The model uses the residuals that is built on convolutions layers for a nonlinear transformation. The residual shortcuts are directly connected to the Convolutional ReLu blocks. The kernel size is 3 x 3 throughout the convolutional layers.

Residual networks can have several blocks and inter connections. In our model we construct a three-block model with 3 skip network connections to classify 4 behavioural emotions. The trade-off hyper parameters (H_p) that are tuned in the network are input size (i_s), Filter Size (F_s), Block size (B_s) and the number of output classes. As CNN are prone to vanishing/exploding gradient issues as the network grows deep, introduction to the skip networks to the CNN have enabled the system to skip the block of network that causes degradation of the performance in the network. The regularization coefficients used in the network maintenance the correlation from the output of the previous block into the input of the next block

The network is enabled with a growing shrinking filter (16,32,64,128 and 256). This enables in depth feature extraction from each image. The Initial Learn Rate (Lr) for the network was set at 1.00E-04 and l_2. Regularization Function is used to regularize irregularities. The Mini Batch Size (Bs) size used was 32 and the maximum epochs for training is 40. The network also used cross validation process and after every 500 iterations of training validation was carried out. Padding is introduced to the images to utilize every pixel of data for convolutions and the stride (1,1) was used.

$$y = F(x, \{W_i\}) + x \tag{3.1}$$

x and *y* are the input and output vectors of each layers $F(x)$, $\{w_i\}$ is the functional block for residual mappings that needs to be learnt. Adding a linear projection W_s with the skip connections helps to map the dimensions. Eq 3.1 can be rewritten as

$$y = F(x, \{W_i\}) + W_s\, x \tag{3.2}$$

Adding the weights from the previous block makes the equation more adaptable. Few considerations taken to create the network are, adding the identity connections will not increase the parameters, hence the computation complexity remains unaltered. The dimensions of x and F are to be same hence 1 x 1 dimensions are used to match the dimensions. The Function represents multiple Convolutional layers. The element wise addition is carried out on two feature maps on individual channels.

9.3.4 Pose Estimation

Pose Estimation of the face is a complementing task carried out to supplement the process to achieve accurate Engagement detection. The parallel process is carried out on the faces to estimate the intent of engagement by an individual student during online learning. Facial pose estimation is carried out. Pose Estimation of the face provides a greater understanding of the attentiveness of the face. While the face is constantly looking at the screen, it provides an information that the student is concentrating on the lecture. However, the student can stoop down to write notes, Turn Left, right or up to be distracted. Considering these poses a model is proposed to detect facial key points and estimate the pitch, yaw and roll movements of the face.

Pose Estimation in the computer vision arena refers to the relative orientation of the object with respect to the camera. Pose estimation often refers to a Perspective n-Point Problem (PnP). Due to the extreme complex steps necessary to solve it, head position estimation is a difficult problem in computer vision. The face must first be placed in the frame, followed by the different facial markers. In today's world, identifying a person's face appears to be a simple process, and this is especially true when the person is facing the camera. When the face is at an angle, the issue emerges. The steps involved in estimating pose are Face Detection, Facial Landmark Detection, Pose Estimation as listed in the section 9.3.4.1.

9.3.4.1 Facial Pose Estimation

To determine the pose of an item, a camera must be calibrated, and the locations of n 3D points on the object must be known, together with their corresponding 2D projections in the picture. With regard to a camera, a 3D rigid object has just two kinds of movements. Moving the camera from its current 3D location (X, Y and Z) into another 3D location (X`Y`Z`) is called translation. Transposition has three degrees of freedom: it can move X, Y, or Z. The vector *t* represents translation in equation (3.3).

$$t = (X' - X, Y' - Y, Z' - Z) \tag{3.3}$$

Rotation of the camera about a location X, Y and Z Axes. As a result, a rotation has three degrees of freedom. There are several methods to depict rotation. This work uses Euler's angle to represent the Yaw, Pitch and Roll of the face.

An upward facing face with a positive Euler X angle is said to be Euler X positive. Positive Euler-Y angle means the face is looking to the right of the camera, while negative Euler-Y angles mean it's gazing to the left of the camera. When a face is turned counter-clockwise, it has a positive Euler Z angle. The head has six degrees of freedom: three rotation angles and three linear shifts. In general, a head rotation is defined by three Euler angles: roll, pitch, and yaw (Figure 9.2). 3D rotation matrices of the head *R*, is attained through (R_x, α), (R_y, β), (R_z, γ) as pitch, yaw and roll around its respective axis.

$$R_x, \alpha = \begin{bmatrix} 1 & 0 & 0 \\ 0 & cos\alpha & -sin\alpha \\ 0 & sin\alpha & cos\alpha \end{bmatrix} \tag{3.4}$$

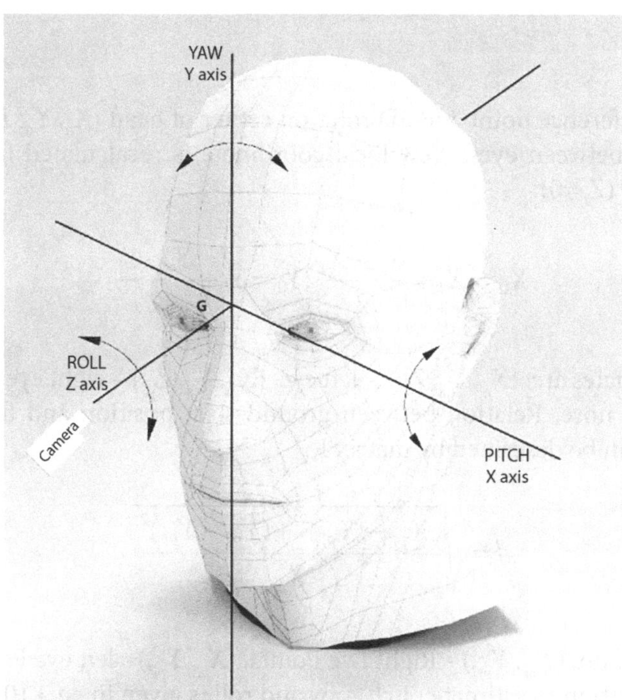

Figure 9.2 The definition of three head rotation angles.

$$R_x, \beta = \begin{bmatrix} cos\beta & 0 & sin\beta \\ 0 & 1 & 0 \\ -sin\beta & 0 & cos\beta \end{bmatrix} \tag{3.5}$$

$$R_x, \gamma = \begin{bmatrix} cos\gamma & -sin\gamma & 0 \\ sin\gamma & cos\gamma & 0 \\ 0 & 0 & 1 \end{bmatrix} \tag{3.6}$$

Thus, head 3D rotation matrixes R can be expressed:

$$R = \begin{bmatrix} cos\gamma.cos\beta & cos\gamma.sin\beta.sin\alpha - sin\gamma.cos\alpha & cos\gamma.sin\beta.sin\alpha + sin\gamma.cos\alpha \\ sin\gamma.cos\beta & sin\ \gamma.sin\beta.sin\alpha + cos\gamma.cos\alpha & sin\gamma.sin\beta.cos\alpha - cos\gamma.sin\alpha \\ -sin\gamma & cos\beta.sin\alpha & cos\beta.cos\alpha \end{bmatrix} \tag{3.7}$$

Local reference point and 3D rotation center of head $\{X_0, Y_0, Z_0\}$ centre is located between eyes. New local coordinate is recalculated from eyes coordinate (Z_0=0):

$$X_0 = X_l + \frac{X_r - X_l}{2}; Y_0 = Y_l + \frac{Y_r - Y_l}{2} \tag{3.8}$$

Coordinates are: $\{X_{pl}, Y_{pl}, Zp\}$ – left eye, $\{X_{pr}, Y_{pr}, Z_{pr}\}$ – right eye and $\{X_{pn}, Y_{pn}, Z_{pn}\}$ – nose. Relation between ground-start position and final head position can be described by matrix R:

$$d = \sqrt{\frac{(X_{pr} - X_{pl})^2 + (Y_{pr} - Y_{pl})^2}{X_E}} \tag{3.9}$$

X_E - 70 mm, (X_{pr}, Y_{pr}) - Right Eye Points, (X_{pl}, Y_{pl}) - left eye Points.
The equation to estimate pitch, yaw and roll is given in eq 3.10

$$R.\begin{bmatrix} X_{px}\ .d \\ Y_{px}\ .d \\ Z_{px}\ .d \end{bmatrix} = \begin{bmatrix} X_{fx}\ .\ d \\ Y_{fx}\ .\ d \\ Z_{fx}\ .\ d \end{bmatrix} \tag{3.10}$$

From eq 3.10, we deduce $W_p = (X_{px}.d, Y_{px}.d, Z_{px}.d)^T$, $W_f = (X_{fx}.d, Y_{fx}.d, Z_{fx}.d)^T$. From the equation we can rewrite eq (3.10) as

$$\partial_1 = \sum_{j=1}^{3} R_{1,j}.w_{pj}^{(i)} - w_{f1}^{(i)} \tag{3.11}$$

$$\partial_2 = \sum_{j=1}^{3} R_{1,j}.w_{pj}^{(i)} - w_{f2}^{(i)} \tag{3.12}$$

Where $i=1..3$

System of six equations can be solved using minimization least squares method:

$$D = \sum \partial_1^2 + \sum \partial_2^2 \tag{3.13}$$

Three angles α, β, γ, and were calculated using the least squares minimization approach which is characterized as head 3D rotation.

9.4 Experimentation

Images from the testing data is drawn at random and fed into the network; these images are tested for True Positive, false positive, True negative or false negative. Based on the confusion matrix calculations, for each dataset the following parameters, Accuracy, precision, sensitivity, Specificity and F1-score are calculated for each class. The network is trained on Daisee dataset. The dataset is created to detect engagement, confusion, frustration and boring classes of individual students in an environment where the students watch videos and excerpt of the emotions on various timelines are extracted as short videos to illustrate the various emotions. In this work the frames were extracted from each video belonging to a particular class and the faces were extracted and used for training, validation and testing purposes.

All the experiments were carried out using Intel Xeon E3 based workstations, NVIDIA Gforce GTX graphics card on a 32 GB RAM and Matlab2019b was used as the platform to train and validate the network. Images from the testing data is drawn at random and fed into the network, these images are tested for True Positive, false positive, True negative or false. The images are resized and fed into the Residual network with the sizes of 128 x 128 x 3 dimensions. The network was created from the scratch and the network was used to improve the efficiency of emotional understanding. The network hyper parameters (were determined using the Bayesian Optimization Model. This model builds the network based on probability of ascertaining the Objective Function (O_f) and uses it to select the hyper parameter to evaluate the true objective function. Since there are unlimited resources, computing each and every point of the O_f is not possible, hence a surrogate model (S_m) was used to represent the objective function that is essentially trained on the hyper parameter and true objective function pair,

$$S_m = P(H_p|O_f) \tag{4.1}$$

The Surrogate model uses the algorithm to determine the hyper parameters based on sequential modelling-based optimization. Complementing the Surrogate model is the acquisition function which is used to find the next hyper parameter. In order to select the next point, the Acquisition function is maximised. If the fitness function (X -> R) is expensive to evaluate, Algorithm 1 is used. On calculates an approximation of T (as a substitute for M). This model M is easier to compute than the previous model. the numerical optimisation of this surrogate (or any alteration thereof) is the inner loop of SMBO (line 3 in below code). Surrogate maximising point x* becomes a recommendation for where to assess the genuine loss function.

The model used is a Gaussian process for Expected Improvement (*EI*) which is modelled in step 6 of the algorithm as $P(y|x,D)$ the predictive distribution *(D)*. Based on the algorithm the in Table 9.1, specific hyper parameters for the Lightweight ResNet was determined. Table 9.2 details the various parameters and the values used for the network. The networks hyper parameter was estimated based on the Algorithm 9.1. Each parameter was chosen based on the Optimization provided by the B_o which was used to detect the True Objective Function based on the Surrogate

Table 9.1 The surrogate model algorithm.

Bo(T, M_0, I, S)
    $H \leftarrow \emptyset$,
    for $t \leftarrow 1$ ***to*** I,
        $x^* \leftarrow \text{argmin}_x S(x, M_{t-1})$,
        Evaluate $T(x^*)$,
        $H \leftarrow H \cup (x^*, f(x^*))$,
        Fit a New Model M_t *to* H,
    return H

Bo- Basis Approximation
H- Observation History (hyper parameter, O_f *score) pair*
$\emptyset$ *- Empty Set*
I- Max number of Iterations
T- True Objective Function
M- Surrogate function
S- Acquisition function
x^* *- The next chosen parameter to evaluate.*

Function. Pose Estimation was carried out in parallel on the video to detect if the student is not watching the camera or being attentive. Pose estimation complements the study by helping creating an analysis of the face over a period of time.

Based on various observations, distracted details are curated over the full video where the different angles of poses estimate the overall distracted and engagement phases of the student. Figure 9.3 elicits the Methodology of computation and estimation modules that were used to build the Engagement detection system. In the training and Computational module, the videos are fed into the ResNet model and pose estimation model in parallel. ResNet uses the deep feature assimilation to extract features and train the network to understand the four classes.

Similarly, the pose estimation carries out the analysis to determine the angle deviations of each face in each class of video. The end result being the trained model for behaviour computation and degree of freedom that can be used for Engaged/Disengaged classes. The trained network and pose angles are used in the testing and estimation module where the classroom videos collated during online classes are utilized to estimate the Engaged/Disengaged or the 4 classes (Engaged, Confused, Frustrated and Bored).

Estimating angles of the face, where the student is engaged can vary between angles of -180 degree to +180 degrees in all the three axes (X<Y<Z). However, the proposed approach estimates the angle of

Table 9.2 Parameter setting detail.

Parameter's name	Value
Networks Layers	74
Initial LearnRate (*Lr*)	1.00E-03
Regularization Function	l_2Regularization
Gradient Threshold	'l2norm'
Max Epochs	40
Mini BatchSize (B_s)	32
Verbose Frequency	50
Validation Frequency (V_f)	500
Shuffle	'Every-epoch'
Padding Direction	'right' (1,1)
Filter Size	(3 x 3)
Stride	(1,1)
Number of Filters (layer)	8,16,32,64,128
Optimiser	SGDM
Learning Rate Scheduler	Piecewise
Image size	128 x 128 x 3
Learning Drop rate	60 iterations

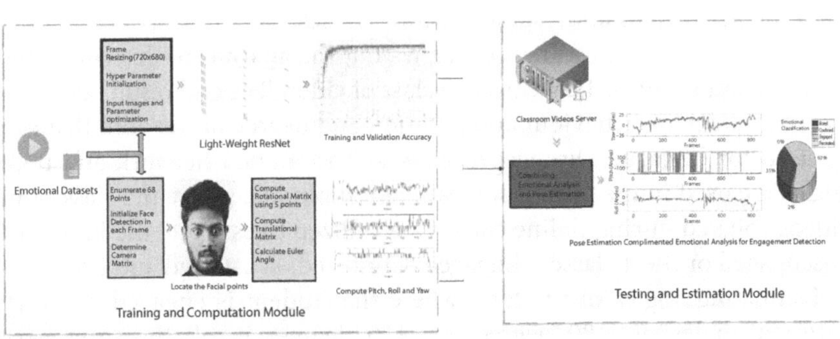

Figure 9.3 The overall engagement estimation methodology.

engagement through the videos of Daisee dataset. The Engaged videos are used to detect the position of the student in an online platform. There are 25 videos taken from two classes, Engaged and Boredom. Engaged videos are used for estimating the angles at which a student is present in front of the camera and sincerely engaging in learning. The videos from boredom are used to detect disorientation. Bored videos exhibit either stoop or droop in facial coordinate that eventually helped in discriminating the face posture into engaged and disengaged. Pose estimation was carried out on each video that has 300 frames.

Each video was individually fed into the pose estimation module to analyse the change in orientation of poses in X, Y and Z axes. The rotational and transformation matrix computed on the 5 points over the face assisted in computing the pitch, roll and yaw changes of the sphere's face. Degree of freedom (Dof) over the axes is estimated. Transverse direction envelopes that rotate on the Z axis. Y-axis tilts up and down. Left and right turns on the X-axis. (Yaw). The manual calculations to estimate the engaged coordinates were between -178° to -170° & 163° to 178° to Pitch, -43° to 43° to Yaw and -8° to 8° Roll. The manual notations were based on maximum extrinsic regions of face moving away from and moving closer to the camera on the respective axis. The coordinate extraction modelling was carried out on the dataset videos and it was noted that the positive Engagement of individual students face coordinates needs to lie between -36° to 36° on the X axis, 168° to 178° on the Y-Axis and -7° to 7° on the Z axis.

The rotational degree mentioned in the respective axis has been the summative approximation from the pose estimation procedure. This method is robust and processes at close to 15 frames per second which enables the estimation of engagement detection. Combining the models of Lightweight ResNet and pose estimation, decisive results are generated on test videos that exhibit two classes engaged and disengaged. However, a Macro estimation of emotions helps in approximating the 4 classes of Engaged, Boredom, Frustrated and Confused that helps in a 2 Channel feedback system for improved outcome.

9.5 Results and Discussions

The results are deduced based on the training and computation module and testing and estimation module. The Lightweight ResNet determines the emotional context of the students and the pose estimation to determine the portions of the video where the face is not engaged or not. The trained networks based on Daisee dataset is used in the videos collated in online classes conducted. The Lightweight ResNet was trained using the dataset Daisee. The training data had one student in each video. The network shuffled the input sequence after each epoch to overcome overfitting issues. The Validation frequency (Vf) is a crucial parameter since it evaluates the training effectiveness of the model at regular intervals. For this model, the value for Vf is set once per 500 iterations.

Table 9.3 derives the outcome of the calculation of Accuracy, Precision Sensitivity, Specificity and the F1-score parameters that are essential to grade a classifier. Each of the values mentioned are calculated based on True Positive (TP), True Negative (TN), False Positive (FP) and False Negatives (FN) acquired from the Confusion Matrix. To increase learning pace and decrease Validation Loss, a weighted average of the scores independently generated from individual layers using the posterior class probabilities is computed. After each iteration, the Training Accuracy (T) and Validation Accuracy (V) are plotted and presented for the model's efficiency index. The following metrics will provide insight on this categorization model. As a consequence, Lightweight ResNet's F1-score, accuracy, precision, recall, and specificity are determined.

The network was trained over a period of 40 epochs. The network is consistent, and there is a strong relationship between validation and training accuracy. On the Daisee dataset, the model obtained 89.93% validation accuracy; the accuracy was obtained from the 10-fold Cross validation. This proves that the Lightweight network establishes a strong learning procedure from the features extracted. The network used to test its performance on the trained network was tested on the test images and the overall accuracy was recorded at 96.2%. The Confusion Matrix to the test data is provided in Figure 9.4. The True positives and True Negatives derived by the classifier are based on the test images that were not included during

Table 9.3 Metrics using daisee dataset.

Data	Class	Acc	Pre	Sen	Spe	F1
Daisee	Boredom	97.69	1	0.91	1	0.95
	Confusion	96.95	0.89	0.96	0.97	0.93
	Engaged	97.69	0.94	0.97	0.98	0.96
	Frustrated	100	1	1	1	1

Acc= Accuracy, Pre=Precision, Sen= Sensitivity, Spe= Specificity, F1= F1-score

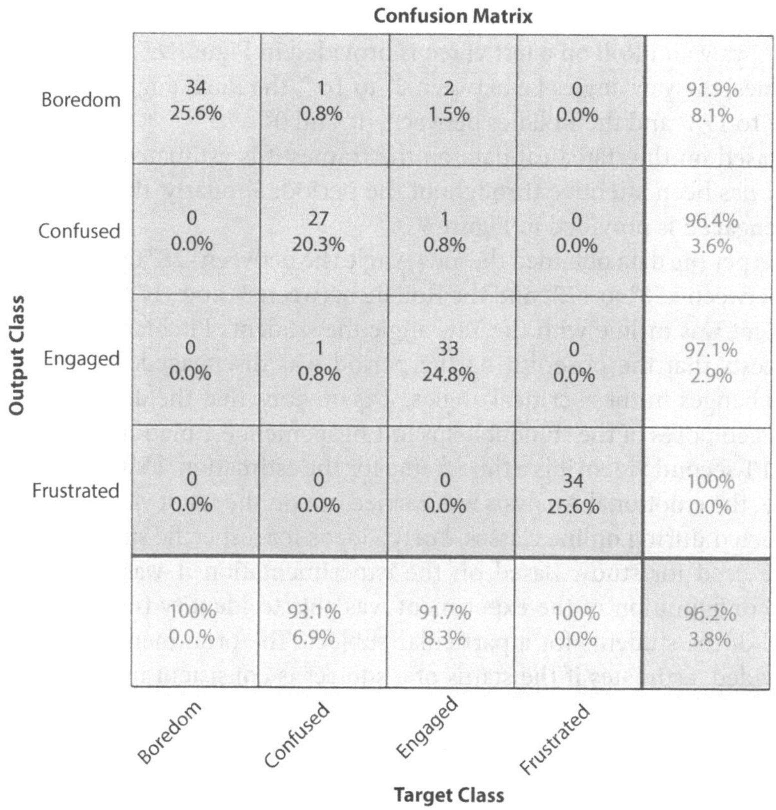

Figure 9.4 The confusion matrix using lightweight ResNet.

training the network. In addition to the C_M, the accuracy, precision, recall, sensitivity, and F1 score of each class are computed and listed in Table 9.3.

The results show that the residual connections added to the CNNs had a significant influence on the outcomes for detecting and identifying emotions. The matrix also shows the proportion of False Positives and False Negatives for each class, which offers a thorough insight into the data's performance for each class. Pose Estimation calculation was derived from the videos that Daisee dataset provided. The pose coordinates of 30 videos under engaged and Bored were used to precisely locate the angles of coverage for engaged and disengaged respectively. Based on the videos all the faces in the engaged videos lie within the range of -36° to 36° on the X axis, 168° to 178° on the Y-Axis and -7° to 7° on the Z axis. Generally, bored students tend to droop down or look away from the camera. These gestures provide the lineage to understand the disengaged gesture of the student. The estimation of coordinates for the engagement class using the Pitch, yaw and Roll on a test video is provided in Figure 9.5. There are 300 Frames, the yaw angles lie between 5° to 16 °, the Pitch angles lie between 171° to 178° and the Roll lies between -1° and 0°.

Based on the statistical data on the frames it is estimated that the student has been attentive throughout the period. Similarly, the estimate for Disengaged is provided in Figure 9.3.

As per the data obtained the Yaw Angles lie between -28° to -8°, the Pitch lie between 158° to 172° and the Roll lie between 8° and -172°. Though the student was in-line with the Yaw angle the students Pitch and Roll angles suggests that the pose during the period was disengaged. By observing the changes in these critical angles, it is obvious that the disengaged and engaged poses of the student helps in Engagement estimation. 300 Frames is a 10-second video; it is a time frame for the estimation. Using the trained Data, the emotional Analysis was carried out on the set of videos that were collected during online classes. Forty videos for a specific subject of study were used for study. Based on the experimentation it was noticed that the configuration of the experiment was able to identify the engagement level of the students for a particular subject. The prominence in the data provided, estimates if the status of a subject is consistent in attentiveness during a class. Though there are changes in peaks of data, the estimate is

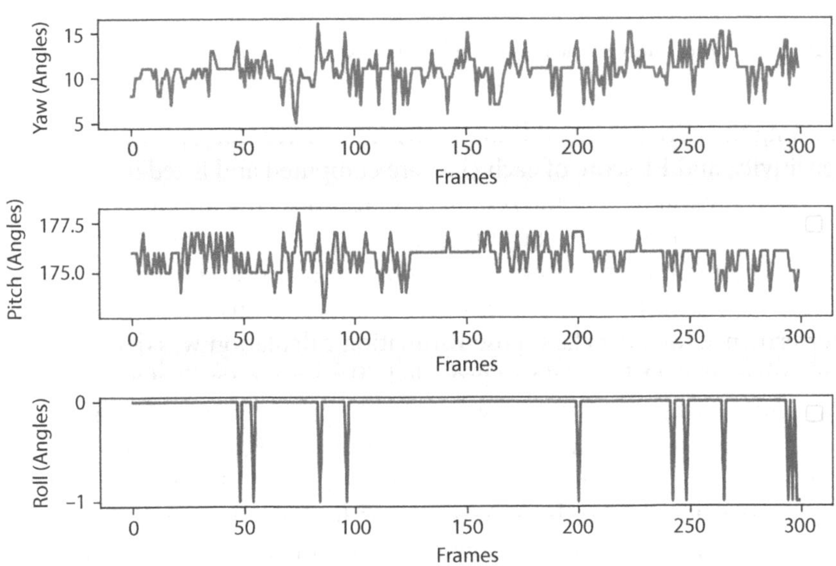

Figure 9.5 The pose estimation on an engaged student.

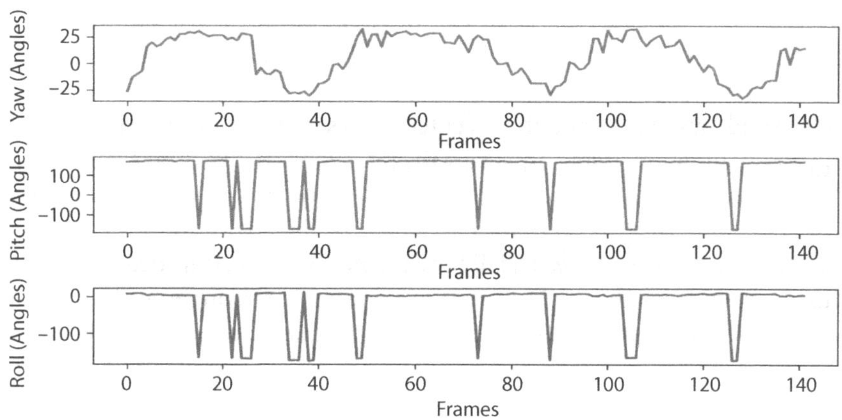

Figure 9.6 The pose estimation on a disengaged student.

based 3D calculation it is bound to change over time. The data suggests that there are variations in the engaged sequence and disengaged sequence by a 10-second frame analysis. Refer the above Figure 9.6.

The experiment was carried out in a live classroom where the emotional analysis and the pose estimation was carried out. During the testing the pose estimation was in sync with the emotional analysis with an 85% score with manual visualization and tracking. The details of the score are listed in Figure 9.4. The analysis on the graphs clearly informs that the pitch of the face is moving away from the angle of freedom, the face is disengaged in the frames shown in Figure 9.7. Though the emotional classification that provides 63% of engaged state and 35% of bored emotions as well.

Statistically, the pose estimation also supports the face posture of being disengaged for close to 35% of frames. By utilizing both these studies it is

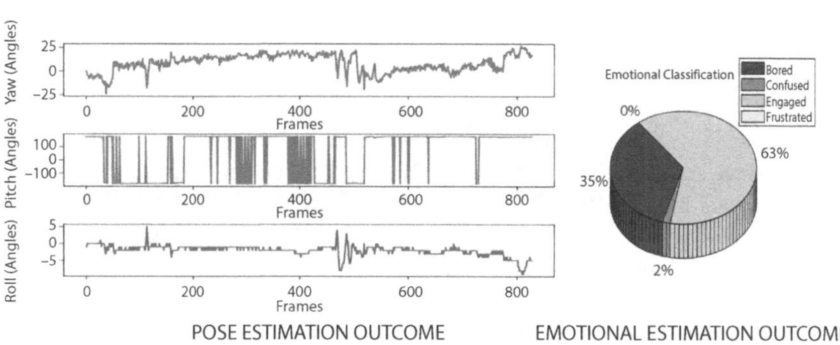

Figure 9.7 The two channel analysis on a test video.

convenient to conclude that in this segment of frames the student was attentive for 65% and disengaged for the rest of the period. Figure 9.6, is an estimate of a student who has been disengaged over a period of time. The values of Roll and Yaw are beyond the accepted values to be in engaged status. Also, the pose estimation graph, especially the pitch, does not vary on every frame; this suggests that the student is continuously moving from the line of the camera. When there is constant movement of the angles from the positive axis to the negative axis, especially the pitch, it gives a clear idea of the student's engagement status. As shown in Figure 9.7, the angles of Pitch, Roll and Yaw lie well within the defined boundary conditions of Engaged.

Conclusion

This paper concentrates on improving the inaccuracies of Emotion detection for engagement detection for a classroom environment. A dual channel model that uses Light Weight Resnet and Pose Estimation techniques are trained with the help of Daisee Datasets. The Attention estimation through emotional classifications resulted in 95.5% accuracy and the degrees of freedom where the face coordinates are in engaged state are also extracted by analysing each engaged video. The combined face coordinates analysis and facial emotion analysis enhances the accuracy of the result. The test videos were collated from the online platform where classes were conducted via platforms that enable teaching and learning through non-physical contact mode. Face detection through dLib methods and using important 5 point Coordinates enabled the system to be accurate in estimating the pose of the face in a video. This study was carried out for individual students in a classroom. As a future study, the analysis for a group of students' engagement status through valence and arousal estimate will be undertaken.

References

1. Mosavi, A., Ardabili, S. & Varkonyi-Koczy, A. R. List of deep learning layers. *Preprints* 1–28, 2019.
2. Krizhevsky, B. A., Sutskever, I. & Hinton, G. E. ImageNet Classification with Deep Convolutional Neural Networks. *Commun. ACM*, 60, 84–90, 2012.
3. Simonyan, K. & Zisserman, A. Very deep convolutional networks for large-scale image recognition. *3rd Int. Conf. Learn. Represent. ICLR 2015 - Conf. Track Proc.* 1–14, 2015.

4. Huang, G., Liu, Z., Van Der Maaten, L. & Weinberger, K. Q. Densely connected convolutional networks. *Proc. - 30th IEEE Conf. Comput. Vis. Pattern Recognition, CVPR 2017*, 2261–2269, 2017.
5. He, K., Zhang, X., Ren, S. & Sun, J. Deep residual learning for image recognition. *Proc. IEEE Comput. Soc. Conf. Comput. Vis. Pattern Recognit.*, 770–778, 2016.
6. Sandler, M., Howard, A., Zhu, M., Zhmoginov, A. & Chen, L. C. MobileNetV2: Inverted Residuals and Linear Bottlenecks. *Proc. IEEE Comput. Soc. Conf. Comput. Vis. Pattern Recognit.* 4510–4520, 2018.
7. From AlexNet to NASNet: A Brief History and Introduction of Convolutional Neural Networks, https://towardsdatascience.com/from-alexnet-to-nasnet-a-brief-history-and-introduction-of-convolutional-neural-networks-cf63bf3320e1, 2008
8. Zhao, J., Gou, L., Wang, F. & Zhou, M. PEARL: An interactive visual analytic tool for understanding personal emotion style derived from social media. *IEEE Conf. Vis. Anal. Sci. Technol. VAST 2014 - Proc.* 203–212, 2015.
9. Hernandez, J., Hoque, M. (Ehsan), Drevo, W. & Picard, R. W. Mood meter. 301, 2012.
10. Srivastava, R., Yan, S., Sim, T. & Roy, S. Recognizing emotions of characters in movies. *ICASSP, IEEE Int. Conf. Acoust. Speech Signal Process. - Proc.* 993–996, 2012.
11. Zeng, H. *et al.* EmotionCues: Emotion-Oriented Visual Summarization of Classroom Videos. *IEEE Trans. Vis. Comput. Graph.* 27, 3168–3181, 2021.
12. Kempter, R., Sintsova, V., Musat, C. & Pu, P. EmotionWatch: Visualizing fine-grained emotions in event-related tweets. *Proc. 8th Int. Conf. Weblogs Soc. Media, ICWSM 2014* 236–245, 2014.
13. Wammes, J. D. *et al.* Disengagement during lectures: Media multitasking and mind wandering in university classrooms. *Comput. Educ.* 132, 76–89, 2019.
14. Aritzeta, A., Balluerka, N., Gorostiaga, A., Alonso-Arbiol, Haranburu, M., Gartzia, L., "Classroom emotional intelligence and its relationship with school performance", *European Journal of Education and Psychology*, 9, 1-8, 2016.
15. Xu X., Teng X. "Classroom Attention Analysis Based on Multiple Euler Angles Constraint and Head Pose Estimation", *Multimedia Modelling. Lecture Notes in Computer Science*, 11961, 1-27, 2020.
16. Cao, Z., Chu, Z., Liu, D. & Chen, Y. A Vector-based Representation to Enhance Head Pose Estimation. *2021 IEEE Winter Conference on Applications of Computer Vision (WACV)*, 1187–1196, 2021.
17. Szegedy, C. *et al.* Going deeper with convolutions. *Proc. IEEE Comput. Soc. Conf. Comput. Vis. Pattern Recognit.*, 07, 1–9, 2015.
18. Kumar, A., Alavi, A. & Chellappa, R. KEPLER: Keypoint and Pose Estimation of Unconstrained Faces by Learning Efficient H-CNN Regressors. *Proc. - 12th IEEE Int. Conf. Autom. Face Gesture Recognition*, 258-265, 2017.

19. Ranjan, R., Patel, V. M. & Chellappa, R. HyperFace: A Deep Multi-Task Learning Framework for Face Detection, Landmark Localization, Pose Estimation, and Gender Recognition. *IEEE Trans. Pattern Anal. Mach. Intell.* 41, 121–135, 2019.
20. Ruiz, N., Chong, E. & Rehg, J. M. Fine-grained head pose estimation without keypoints. *IEEE Comput. Soc. Conf. Comput. Vis. Pattern Recognit. Work.* 2018-June, 2155–2164, 2018.
21. Albiero, V., Chen, X., Yin, X., Pang, G. & Hassner, T. img2pose: Face Alignment and Detection via 6DoF, Face Pose Estimation, CVPR 2021, 7617–7627, 2021.
22. Yang, X., Jia, X., Yuan, M. & Yan, D. M. Real-time facial pose estimation and tracking by coarse-to-fine iterative optimization. *Tsinghua Sci. Technol.* 25, 690–700, 2020.
23. Wu, Y. & Ji, Q. Discriminative Deep Face Shape Model for Facial Point Detection. *Int. J. Comput. Vis.* 113, 37–53, 2015.
24. Wu, Y. & Ji, Q. Robust facial landmark detection under significant head poses and occlusion. *Proc. IEEE Int. Conf. Comput. Vis.*, 3658–3666, 2015.
25. Wu, Y. & Ji, Q. Constrained Joint Cascade Regression Framework for Simultaneous Facial Action Unit Recognition and Facial Landmark Detection. *Proc. IEEE Comput. Soc. Conf. Comput. Vis. Pattern Recognit.*, 3400–3408, 2016.
26. Xiong, X. & De La Torre, F. Global supervised descent method. *Proc. IEEE Comput. Soc. Conf. Comput. Vis. Pattern Recognit.*, 2664–2673 (2015)
27. Wan, J. *et al.* Face alignment by coarse-to-fine shape estimation. *Chinese J. Electron.* 27, 1183–1191, 2018.
28. Zhu, X., Lei, Z., Liu, X., Shi, H., Li, S. Z., Face alignment across large poses: A 3D solution. *Proc. IEEE Comput. Soc. Conf. Comput. Vis. Pattern Recognit.*, 146–155 (2016).
29. Kazemi, V., Sullivan, J., One millisecond face alignment with an ensemble of regression trees. *Proc. IEEE Comput. Soc. Conf. Comput. Vis. Pattern Recognit.* 1867–1874, 2014.
30. Wu, Y. & Ji, Q., Facial Landmark Detection: A Literature Survey. *Int. J. Comput. Vis.* **127**, 115–142, 2019.
31. Li, B., Lima, D., Facial expression recognition via ResNet-50. *International Journal of Cognitive Computing in Engineering*, 2, 57-64, 2021.
32. Singh, S. & Schicker, D. Seven Basic Expression Recognition Using ResNet-18. 1–3, 2021.
33. Yu, D. & Sun, S. A systematic exploration of deep neural networks for EDA-based emotion recognition. *Inf.* **11**, 2020.
34. Thiruthuvanathan, M. M., Krishnan, B. & Rangaswamy, M. Engagement Detection through Facial Emotional Recognition Using a Shallow Residual Convolutional Neural Networks. *Int. J. Intell. Eng. Syst.* **14**, 236–247, 2021.
35. Dewan, M. A. A., Lin, F., Wen, D., Murshed, M. & Uddin, Z. A deep learning approach to detecting engagement of online learners. *Proc. - 2018*

IEEE SmartWorld, Ubiquitous Intell. Comput. Adv. Trust. Comput. Scalable Comput. Commun. Cloud Big Data Comput. Internet People Smart City Innov. SmartWorld/UIC/ATC/ScalCom/CBDCom/IoP/SCI 2018 1895–1902, 2018.
36. T. S, A. & Guddeti, R. M. R. Automatic detection of students' affective states in classroom environment using hybrid convolutional neural networks. *Educ. Inf. Technol.* **25**, 1387–1415, 2020.
37. Mohamad Nezami, O. *et al.* Automatic Recognition of Student Engagement Using Deep Learning and Facial Expression. Lect. Notes Comput. Sci. (including Subser. Lect. Notes Artif. Intell. Lect. Notes Bioinformatics) 11908 LNAI, 273–289 2020.
38. Huang, T., Mei, Y., Zhang, H., Liu, S. & Yang, H. Fine-grained engagement recognition in online learning environment. *ICEIEC 2019 - Proc. 2019 IEEE 9th Int. Conf. Electron. Inf. Emerg. Commun.* 338–341, 2019.
39. Thiruthuvanathan, M. M. & Krishnan, B. Multimodal emotional analysis through hierarchical video summarization and face tracking. *Multimed. Tools Appl.*, 2021.
40. Gupta, A., D'Cunha, A., Awasthi, K. & Balasubramanian, V. DAiSEE: Towards User Engagement Recognition in the Wild. **14**, 1–12, 2016.
41. Le, V., Brandt, J., Lin, Z., Bourdev, L. & Huang, T. S. Interactive facial feature localization. *Lect. Notes Comput. Sci. (including Subser. Lect. Notes Artif. Intell. Lect. Notes Bioinformatics)* **7574 LNCS**, 679–692, 2012.

10

Building Rule Base for Decision Making – A Fuzzy-Rough Approach

Sabu M. K.[1]*, Neeraj Krishna M. S.[2] and Reshmi R.[3]

[1]Department of Computer Applications, Cochin University of Science and Technology, Cochin, Kerala, India
[2]C3m India Private Limited, Ayyappankavu, Kochi, Kerala, India
[3]Department of Computer Science and Engineering, Federal Institute of Science and Technology, Mookkannoor, Kerala, India

Abstract

Models based on Fuzzy sets and Rough Sets are efficient in handling vagueness present in information systems. In Fuzzy Set Theory, to handle vagueness, a fuzzy membership function is applied, whereas in Rough Set Theory the equivalence classes of an equivalence relation are considered. Usually, Rough Set methods were inefficient to process real-valued features effectively. To address this drawback, in the proposed system, a data discretization using Fuzzy Set Theory is performed to assign linguistic labels to the feature values. This is typically implemented with the help of a standard fuzzification technique provided by the Fuzzy Set Theory. This paper conducts a novel fuzzy rough approach for constructing rule bases for decision making. In this approach, after performing the fuzzy discretization, decision rules are generated with the help of a popular rough set-based rule mining algorithm. To explain the process of rule induction, an experiment is conducted to predict the yield of coconut using the data collected from farmers cultivating coconuts. The generated rules are evaluated with the help of rules obtained from the same dataset using the same algorithm, but by applying a similar crisp discretization and then by generating rules by using Apriori algorithm. The outcomes of evaluation show that the proposed approach is effective and efficient in generating significant decision rules.

***Keywords*:** Rule induction, fuzzy set theory, rough set theory, discretization, decision making

**Corresponding author*: sabumk@cusat.ac.in

Kukatlapalli Pradeep Kumar, Aynur Unal, Vinay Jha Pillai, Hari Murthy, and M. Niranjanamurthy (eds.) Data Engineering and Data Science: Concepts and Applications, (255–278) © 2023 Scrivener Publishing LLC

10.1 Introduction

Fuzzy Set Theory (FST) is basically a many-valued logic that allows to consider any value from the interval [0, 1]. It is found that most of the real-world data are complex, and this complexity arises from uncertainty in the form of vagueness. For example, data associated with young, tall, very fast, intelligent, etc., defined in association with many of the real-world objects do not have properly defined sharp boundaries. These concepts are termed as fuzzy or vague concepts. To handle these vague concepts using computers, mathematical formulations equivalent to how the human brain processes these concepts are needed [2]. Since the conventional set theory fails to represent such vague concepts, we need a natural method by relaxing the strict notion of belongs to or not belongs to. This kind of an approach is followed in FST. To achieve this, FST provides a mechanism to generalize these vague concepts by attaching more values between 0 and 1 [2] for the objects belonging to the domain. In FST, the degree of belongingness of an object x in a certain fuzzy set A is determined with the help of a function $A(x)$ which assigns to each element x, a real number in the range [0, 1]. $A(x)$ is known as the membership function associated with the fuzzy set A. The method of generating the degree of membership of a fuzzy variable using a membership function is termed as fuzzification [2]. After this fuzzification, it is possible to attach a unique fuzzy label to each data value of the fuzzy variable by considering its degree of belongingness in the neighbouring fuzzy labels. This is the concept of fuzzy discretization adopted in this work.

From the discrete data, rules are generated using Rough Set Theory (RST). RST is a popular soft computing tool which constitutes a sound basis for Data Mining. One advantage of RST compared to other similar theories is that RST can perform computations directly on data contained in datasets without requiring any additional information about data such as grade of membership in FST. In RST, equivalence classes generated from the dataset with the help of an equivalence relation, known as the indiscernibility relation, is the only information needed to process the data [5, 6]. Acceptable mathematical models are available in Rough Set Theory for generating valid decision rules even from datasets which are inconsistent. Because of this reason, a popular rule mining algorithm LEM2 (Learning from Examples Module Version 2) [9] is used in this work to generate various decision rules.

The research conducts a novel fuzzy-rough rule base construction approach. The proposed fuzzy-rough system has two major steps: (1)

Discretization of the collected data. (2) Decision rule generation to build the rule base. To fuzzify the collected data for discretization, in this work, a triangular membership function is used. Fuzzy logic provides 11 built-in membership function types [2]. Triangular membership function is the simplest of all these. For rule generation, the LEM2 algorithm introduced by Grazymala-Busse [9] is implemented in MATLAB and used in this work. The rules thus obtained become the rule base for the prediction. The proposed algorithm is then experimented on a dataset prepared by collecting information from farmers cultivating coconuts [7]. Several predictor variables (conditional attributes) are considered while collecting the data. But only three significant predictors affecting the yield, namely air temperature, amount of rainfall, and the volume of fertilizers applied for trees are selected for the purpose of decision rule generation. The number of coconuts produced, the yield becomes the response (decision) variable. Using triangular membership function, the collected values of both predictor and the response variables are fuzzified by considering three different linguistic labels. Then data discretization is performed by considering these fuzzy values. From the discretized data, the decision rules are generated using LEM2 algorithm. These decision rules will function as a rule base when a fully automated prediction system is developed. After employing a crisp discretization in the collected data with the help of some non-overlapping intervals, the performance of the propounded method is compared with the rules generated using the same LEM2 algorithm. The same linguistic labels defined for the fuzzy discretization performed in the proposed approach are used for this purpose. For another comparative analysis, rules are generated from the discretized dataset using Apriori, a conventional rule mining algorithm. The number of 100% confidence (certain) rules obtained from this algorithm are compared with the number of certain rules obtained from the proposed approach. The comparison results show that the proposed approach is acceptable for generating significant rules for decision making.

Following are the remaining sections of the study. In sections 10.2 and 10.3, the review of the related works is discussed and the process of discretization using FST is explained. The details of the dataset used for the experiment is given in section 10.4. Section 10.5 deals with the proposed methodology adopted for the rule base construction. Then, the experiments conducted are presented in section 10.6. Evaluation of the results is done in section 10.7 and discussion of the findings in section 10.8. Finally, the conclusion of the work is stated in section 10.9.

10.2 Literature Review

Decision rule mining is used to identify the association among predictors and response variable of a given data set. A decision rule is represented as $\alpha \rightarrow \beta$, where α represents the conjunction of conditions involving predictors and β, the decision involved in the rule. For decision rule generation, several algorithms are available [8]. Apriori algorithm introduced by R. Agrawal and R. Srikant is the most popular and widely used one in this area. To classify the web users for sales improvement, Mahsa *et al.* designed a recommender system by integrating decision trees with association rule mining [12]. In [13], after clustering the course data to find similar students and their course, association rules are generated to recommend elective courses. Rabiatual *et al.* used association rule mining to explain the patterns exist in climate data and they predict the climate change in Malaysia using these patterns [14]. Apriori association rule mining algorithm is also applied to study the impact of the therapeutic actions while combining the two groups of drugs in Ayurveda Panchkula and Shadushna [15]. In an Irish sign language dataset, association rule mining has been carried out to explore the pattern of articulation among ISL verbs [16]. By combining the constraints originating from pair-wise positive and negative product associations generated from weekly sales data, Demiriz *et al.* proposed a cluster-based framework for apparent retailing firms [17]. In [18], Apriori algorithm is applied to analyse the associations between products and customers in Small and Medium Enterprise in Colombia. Integration of neuro computing with the symbolic structures and association rule-based methods is used to extract the relationship between disease and symptoms in [19]. In [20], Hemant conducted a study of incremental data mining using Apriori algorithm and genetic algorithm to generate frequent patterns for the addition, deletion, and modification of data in an efficient way. M. Shawkat *et al.*, in their study, demonstrated the use of the Apriori rule mining algorithm to trace and recognize COVID-19 cases leading to the efficient diagnosis of the disease [21]. With the help of association rule mining M. Tandan *et al.* identified the frequent patterns of symptoms helpful for detecting COVID-19 efficiently [22]. Conventional rule mining algorithms like Apriori, generate many rules and it is very difficult to classify interesting and important rules [11].

Z. Pawlak introduced RST which provides an objective methodology to process data having uncertainty and vagueness. The aim of RST is to characterize various concepts of the decision variable with the equivalence classes (granules) obtained from the indiscernibility relation defined over the values of the predictors by considering the similarity of these values.

The equivalence classes generated by this indiscernibility relation depict the lack of knowledge about the considered data domain [4]. A decision concept *X*, which is represented as a union of the granules generated by the indiscernibility relation is considered as a crisp set; otherwise, the decision concept is rough or vague [1]. Each rough set possesses certain border line objects which cannot be classified as members of the target concept or its complement with certainty. Consequently, vague concepts cannot be described in terms of the available information provided by the indiscernibility relation [1, 23]. However, it is attainable to approximate these concepts as a consequence of two crisp sets, called the lower and upper approximations of the vague concept *X*. While approximating a target concept X using this approach, the lower approximation implies a lower boundary by selecting all those objects that surely belong to *X* and the upper approximation, an upper boundary, consisting of all those objects which possibly belong to *X*. The difference among the lower and upper approximation indicates the boundary region consisting of the border line elements of *X*.

In this study, a popular RST-based rule induction approach LEM2 is used for rule base construction. The advantage of using RST in rule mining is its ability to remove redundant features during the process of rule generation itself. Hence, as an outcome of the rule induction process only a minimum number of significant rules are generated. To generate rules using LEM2, the basic idea used is that of a local covering [9]. Local covering represents a rule set describing a concept *X*. The notion of minimal complex, which is equivalent to a single rule, is used to define the local covering. Mathematically, to define a minimal complex, first consider the selected concept *X*. Let T represents the set of all predictor-value pairs t = (a, v) of the decision table. The set of all objects of the decision table for which a = v is called a block of t, denoted as [t]. If *X* satisfies the condition $\varphi \neq \cap\{[t] \mid t \in T\} \subseteq X$, then the considered concept is said to depend on T. If T is the least possible set of predictor-value pairs satisfying the condition of dependency, T is called a minimal complex of *X* [9]. A non-empty set of all such minimal complex satisfying $\cup\{[T] \mid T \in L\} = X$ is called a local covering L of *X* [9].

In the case of inconsistent datasets most of the concepts are rough sets. In such cases rule induction is possible only through the crisp sets (i.e., lower and upper approximations) used for approximating the selected rough concept, say *X*. Since lower approximation includes objects that are members of *X* with certainty, rules generated from lower approximation are certainly valid and hence all those rules are treated as certain rules. Similarly, decision rules derived from upper approximations are possibly reasonable and accordingly they are called possible rules. In the implemented LEM2 algorithm, if lower approximation of X ($\underline{X}$) is passed as an argument, the output is a local covering representing certain rules of X. When upper approximation of X ($\bar{X}$) is passed, the local covering given by the algorithm corresponds to the possible rules of X. In a similar manner, certain and possible rules are generated from all other concepts to complete the rule generation process.

10.3 Discretization of the Dataset Using Fuzzy Set Theory

To model a predictor variable (feature) with vagueness using FST, some linguistic labels acceptable to represent that feature are assigned first. Then membership functions appropriate to represent these linguistic terms are selected and inputs of the fuzzy variable are fuzzified based on the selected membership function. For example, if the variable "Fertilizers" representing the volume of fertilizers applied for the coconut tree during a certain period is considered, the linguistic terms assigned to this variable are {low, normal, high}. Fuzzy set is a mathematical framework for representing these linguistic terms [2, 4]. To fuzzify the data values pertaining to the selected fuzzy set, in this work, triangular membership function is used. If a certain data value belongs to more than one fuzzy label with two different fuzzy memberships, the label having maximum membership is selected and the data value is mapped on to that label [3]. In this way all data values including data values of the considered response variable are discretized. After this discretization, the original dataset is transformed into a new decision table suitable for rule induction [10].

10.4 Description of the Dataset

Data for the experiment was composed from 80 farmers cultivating coconuts. These farmers belong to a small geographical region located in Kannur district of Kerala State [7]. Stratified random sampling is the strategy forwarded for data collection by considering panchayath as the strata. To construct the decision table, in this work, Temperature, Rain and Fertilizers are considered as the predictors and Yield becomes the response variable. A subset of the actual data collected from the farmers is presented in Table 10.1 [7].

Table 10.1 Data collected as a part of the study.

Temperature [Celsius]	Rain [mm]	Fertilizers [Kg/Acre]	Yield [Nuts Per Acre]
32	114	120	5000
28	125	70	4000
28	117	130	5000
28	117	130	5000
28	125	70	4000
32	102	70	3000
38	112	110	6000
38	112	110	6000

10.5 Process Involved in Proposed Work

The process of the proposed fuzzy-rough rule base construction approach includes the following eight major steps:

(1) Input the decision table (dataset).
(2) Identify the linguistic variables to be assigned for the considered predictors and the response variable.
(3) To qualify the linguistic variables selected in step 2, assign various linguistic terms suitable to map the values of all the variables.
(4) Select the appropriate membership function and quantify the linguistic terms assigned to selected variables based on the actual values contained in the decision table.
(5) If a data value of a variable belongs to more than one linguistic label with two different fuzzy memberships, the label having maximum membership is selected and the data value is mapped on to that label.
(6) Apply steps (4) and (5) to build a new decision table consisting of the appropriate linguistic terms for the values of selected predictors and the response variable.
(7) Generate lower and upper approximations of various concepts of the response variable of the new decision table formulated in step 6.

(8) By passing these lower and upper approximations generated in step 7 as parameters to LEM2 algorithm produce various decision rules.

Steps 1 to 5 represent the discretization process where the collected dataset is first fuzzified. For fuzzification, in this work, all the three predictors are first assigned various linguistic terms. To quantify each linguistic term the triangular membership function is used. The response variable "Yield" is also fuzzified and quantified using the same procedure. Based on the assigned linguistic terms and with the help of the selected membership function, a new decision table consisting of the selected fuzzy values is constructed in step 6. From this newly constructed decision table, in step 7, lower and upper approximations of various concepts of the response variable are recognized. For generating the decision rules, these lower and upper approximations serve as inputs to the LEM2 algorithm.

10.6 Experiment

As a pre-processing, the collected values of the variables are first fuzzified as explained in section 10.5. The overlapping intervals assigned to define various linguistic terms are given in Table 10.2.

After pre-processing, a modified dataset in the form of a decision table suitable for the rule induction algorithm is obtained. The resulting decision table is given in Table 10.3. Rows of the decision table indicate several cases and columns indicate the considered predictors and the response variable.

In RST, the local covering generated from the lower (or upper) approximation of the concepts of the response variable {Yield} obtained from Table 10.3 leads to the certain (or possible) decision rules of {Yield} [11]. To generate local covering, in this work, LEM2 algorithm implemented in MATLAB is used. When implementing a complete fuzzy inference system, the rules generated through the proposed approach are added to the rule base by considering the corresponding confidence as weights.

In Table 10.3, elementary sets {1, 6, 16}, {2, 5, 7, 8, 9, 11, 12, 13, 17} and {3, 4, 10, 14, 15} represent the concepts given by the response variable {Yield}. These sets respectively represent the classification of farmers with Yield 'poor', 'average', and 'good'. The equivalence class structure induced by the indiscernibility relation IND ({Temperature, Rain, Fertilizers}) is {{1, 16}, {2, 8}, {3}, {4}, {5}, {6}, {7}, {9}, {10, 17}, {11}, {12}, {13}, {14}, {15}}. From Table 10.3, the elementary set {10, 17} is not a subset of any concept means the cases 10 and 17 are conflicting. This leads to the conclusion

Table 10.2 Linguistic labels assigned for the variables.

Temperature [Celsius]			**Rain [mm]**			**Fertilizers [Kg/Acre]**			**Yield [in 1000/Acre]**		
23-33	28-38	33-43	82.8-113.2	101.8-132.2	120.8-151.2	0-80	50-130	100-180	-800-4800	2700-8300	6200-11800
cold	warm	hot	light	moderate	heavy	low	normal	high	poor	average	good

Table 10.3 Decision table.

Cases	Temperature	Rain	Fertilizers	Yield
1	warm	moderate	normal	poor
2	warm	light	low	average
3	warm	heavy	high	good
4	cold	moderate	normal	good
5	hot	heavy	high	average
6	cold	heavy	low	poor
7	hot	moderate	normal	average
8	warm	light	low	average
9	cold	light	high	average
10	warm	moderate	high	good
11	warm	light	high	average
12	cold	heavy	high	average
13	cold	moderate	low	average
14	warm	light	normal	good
15	hot	heavy	low	good
16	warm	moderate	normal	poor
17	warm	moderate	high	average

that the decision {Yield} does not contingent on {Temperature, Rain, and Fertilizers}. Hence, Table 10.3 is inconsistent. To handle this inconsistency, identify the concepts where these conflicting cases belong. Such concepts are not clearly definable with the considered predictors. However, the concept can be approximated using the equivalence classes generated from indiscernibility relation of the predictors. For approximation, lower and upper approximations of these concepts are computed and then rules are generated from these approximations. For any rough concept *X*, certain and possible rules generated respectively from lower and upper approximations represent the decision rules corresponding to *X*. In a similar fashion, using the same algorithm, it is possible to generate all decision rules representing various concepts of {Yield}.

In Table 10.3, for the concept $X = \{2, 5, 7, 8, 9, 11, 12, 13, 17\}$ representing the farmers with Yield 'average', the lower approximation of X, , is {2, 5, 7, 8, 9, 11, 12, 13} and the upper approximation of X, $\bar{X}$, is {2, 5, 7, 8, 9, 10, 11, 12, 13, 17}. Consequently, the accuracy of this approximation is 0.8. Similarly, for the concept {3, 4, 10, 14, 15}, where the farmers with Yield, 'good' the lower approximation seemed to be {3, 4, 14, 15} and the upper approximation is {3, 4, 10, 14, 15, 17}. Approximation accuracy obtained is 0.6667. The concept {1, 6, 16} representing farmers with Yield 'poor', the lower and upper approximations are the same and are given by {1, 6, 16}. This indicates that the concept {1, 6, 16} is definable with the predictors {Temperature, Rain, Fertilizers} and hence the accuracy of approximation is 1. The other two concepts are roughly definable or rough sets. The equivalence classes produced by the indiscernibility relation after this fuzzy discretization come up with an overall classification accuracy of 0.8824.

After computing the lower and upper approximations of the concepts, LEM2 algorithm select those predictor-value pairs (a, v) from the decision table satisfying the condition $[(a, v)] \cap X' \neq \emptyset$, where the X' set represents the lower or upper approximation of the selected concept. Then, the selected predictor-value pairs are used by the algorithm to generate a local covering expounding the set X. Each member of the generated local covering constitutes a minimal complex which will corresponds to a single rule. Hence, local covering furnishes a set of decision rules to take the decision regarding the concept representing X. By assigning lower and upper approximations of the remaining concepts to the set X', decision rules of the various other concepts can be generated separately from the same LEM2 algorithm.

If the set is taken as {2, 5, 7, 8, 9, 11, 12, 13} (i.e., the lower approximation of the concept {2, 5, 7, 8, 9, 11, 12, 13, 17} representing the Yield = 'average'), the local covering given by the algorithm is {{(Rain, light) & (Fertilizers, high)}, {(Fertilizers, low) & (Temperature, warm)}, {(Rain, moderate) & (Fertilizers, low)}, { (Temperature, hot) & (Fertilizers, high)}, {(Temperature, cold) & (Fertilizers, high)}, {(Temperature, hot) & (Rain, moderate)} }. This local covering generates the following certain rules describing the 'average' Yield. The rules are presented in LERS (Learning from Examples based on Rough Sets) format.

2, 2, 2
(Rain, light) & (Fertilizers, high) → (Yield, average)
2, 2, 2
(Fertilizers, low) & (Temperature, warm) → (Yield, average)
2, 1, 1

(Rain, moderate) & (Fertilizers, low) → (Yield, average)
2, 1, 1
(Temperature, hot) & (Fertilizers, high) → (Yield, average)
2, 2, 2
(Temperature, cold) & (Fertilizers, high) → (Yield, average)
2, 1, 1
(Temperature, hot) & (Rain, moderate) → (Yield, average)

Certain rules describing the 'good' Yield (i.e., when = {3, 4, 14, 15}, the lower approximation of the concept {3, 4, 10, 14, 15, 17}) are:

2, 1, 1
(Temperature, warm) & (Rain, heavy) → (Yield, good)
2, 1, 1
(Fertilizers, normal) & (Rain, light) → (Yield, good)
2, 1, 1
(Temperature, cold) & (Fertilizers, normal) → (Yield, good)
2, 1, 1
(Temperature, hot) & (Fertilizers, low) → (Yield, good)

Certain rules describing the 'poor' Yield (i.e., when = {1, 6, 16}, the lower approximation of the concept {1, 6, 16}) are:

3, 2, 2
(Temperature, warm) & (Rain, moderate) & (Fertilizers, normal) → (Yield, poor)
3, 1, 1
(Temperature, cold) & (Rain, heavy) & (Fertilizers, low) → (Yield, poor)

Since, the concept representing Yield = poor is a crisp set, the lower and upper approximations are same and hence all the decision rules generated from this concept are certain rules.

If we select the set X′ as {2, 5, 7, 8, 9, 10, 11, 12, 13, 17}, (the upper approximation of the concept {2, 5, 7, 8, 9, 11, 12, 13, 17}, i.e., for [(Yield = average)]), the local covering generated by the algorithm is {(Fertilizers, high) & (Rain, moderate)}. Hence the only possible rule describing 'average' yield is:

2, 1, 2
(Fertilizers, high) & (Rain, moderate) → (Yield, average)

When = {3, 4, 10, 14, 15, 17}, the upper approximation of the concept {3, 4, 10, 14, 15}), the possible rule describing 'good' Yield is:

2, 1, 2
(Rain, moderate) & (Fertilizers, high)→ (Yield, good)

10.7 Evaluation Result

In most of the data mining applications, for data discretization, crisp intervals are normally used. In this paper, to compare the performance of the proposed rule mining approach, two different strategies are employed. In the first strategy, the generated rules are compared with the rules obtained from the same LEM2 algorithm but by mapping the collected data values of the variables onto non-overlapping (crisp) intervals as shown in Table 10.4. These crisp intervals are selected as intervals fully contained in the overlapping intervals defined for the fuzzification of the data values in the proposed approach. As an outcome of this mapping, a new decision table is generated which is given in Table 10.5.

Decision Table 10.5 shows three concepts representing the classification of farmers with Yield 'poor', 'average', and 'good'. These concepts are respectively {1, 6, 7, 8, 12, 13, 16, 17}, {2, 5, 9, 11} and {3, 4, 10, 14, 15}. From Table 10.5, the decision {Yield} does not completely depend on the predictors {Temperature, Rain, Fertilizers} because Table 10.5 is inconsistent. But in RST, inconsistencies are handled directly by constructing the lower and upper approximations of various concepts and then generating rules from these approximations.

For the concept {1, 6, 7, 8, 12, 13, 16, 17} representing farmers with Yield 'poor', the lower approximation is {1, 6, 7, 12, 16} and upper approximation is {1, 2, 4, 6, 7, 8, 10, 12, 13, 14, 16, 17}. In this case, the accuracy of approximation is 0.42. Similarly, the lower approximation of the concept representing 'average' Yield is {5, 9, 11} and upper approximation is {2, 5, 8, 9, 11, 14}. The approximation accuracy in this case is 0.5. For the concept representing Yield 'good', the lower and upper approximations are respectively {3, 15} and {2, 3, 4, 8, 10, 13, 14, 15, 17}. This approximation has an accuracy of 0.22. This shows that all the three concepts are rough sets. As mentioned in the proposed approach, certain rules are then generated using the implemented LEM2 algorithm by passing the lower approximation as an argument. The process is repeated for all the lower approximations. Similarly for generating the possible rules upper approximations are used instead of lower approximations.

When the lower approximation {1, 6, 7, 12, 16} of the concept representing 'poor' Yield is presented to the algorithm as the argument, the following certain rules are generated by the algorithm.

Table 10.4 Crisp intervals defined for the variables selected.

Temperature [Celsius]			**Rain [mm]**			**Fertilizers [Kg/Acre]**			**Yield [in 1000/Acre]**		
< 31	31-33	>33	<111	111-115	>115	<199	119-121	>121	< 6500	6500-7500	>7500
cold	warm	hot	light	moderate	heavy	low	normal	high	poor	average	good

Table 10.5 Decision table after crisp discretization.

Objects	Temperature	Rain	Fertilizers	Yield
1	warm	heavy	low	poor
2	hot	light	low	average
3	hot	heavy	high	good
4	cold	moderate	low	good
5	hot	heavy	normal	average
6	cold	heavy	low	poor
7	hot	moderate	low	poor
8	hot	light	low	poor
9	cold	light	normal	average
10	warm	moderate	normal	good
11	hot	light	normal	average
12	cold	heavy	high	poor
13	cold	moderate	low	poor
14	hot	light	low	good
15	hot	heavy	low	good
16	warm	moderate	low	poor
17	warm	moderate	normal	poor

2, 2, 2
(Rain, heavy) & (Temperature, cold) → (Yield, poor)
2, 2, 2
(Fertilizers, low) & (Temperature, warm) → (Yield, poor)
2, 1, 1
(Temperature, hot) & (Rain, moderate) → (Yield, poor)

Following are the certain rules describing 'average' Yield.

2, 2, 2
(Fertilizers, normal) & (Rain, light) → (Yield, average)

2, 1, 1
(Rain, heavy) & (Fertilizers, normal) → (Yield, average)

Induced certain rules describing 'good' Yield are:

3, 1, 1
(Temperature, hot) & (Rain, heavy) & (Fertilizers, low) → (Yield, good)
2, 1, 1
(Temperature, hot) & (Fertilizers, high) → (Yield, good)

Possible rules generated by the algorithm when the upper approximations of various concepts passed to the algorithm are:

1, 4, 6
(Rain, moderate) → (Yield, poor)
2, 1, 3
(Fertilizers, low) & (Rain, light) → (Yield, poor)
2, 2, 2
(Rain, heavy) & (Temperature, cold) → (Yield, poor)
1, 3, 4
(Temperature, warm) → (Yield, poor)
1, 3, 5
(Rain, light) → (Yield, average)
2, 1, 1
(Rain, heavy) & (Fertilizers, normal) → (Yield, average)
2, 1, 3
(Fertilizers, low) & (Rain, light) → (Yield, good)
2, 1, 2
(Rain, moderate) & (Temperature, cold) → (Yield, good)
2, 1, 2
(Rain, moderate) & (Fertilizer, normal) → (Yield, good)
3, 1, 1
(Temperature, hot) & (Rain, heavy) & (Fertilizers, low) → (Yield, good)
2, 1, 1
(Temperature, hot) & (Fertilizers, high) → (Yield, good)

In the second strategy, from the data values discretized using both fuzzy and crisp discretization approaches, decision rules are generated using the conventional Apriori rule mining algorithm. When the fuzzy discretized data values are used for rule generation, Apriori algorithm produce the following six certain rules.

2, 2, 2
(Temperature, cold) & (Fertilizers, high) → (Yield, average)
2, 2, 2
(Temperature, warm) & (Fertilizers, low) → (Yield, average)
2, 2, 2
(Rain, light) & (Fertilizers, low) → (Yield, average)
2, 2, 2
(Rain, light) & (Fertilizers, high) → (Yield, average)
3, 2, 2
(Temperature, warm) & (Rain, moderate) & (Fertilizers, normal) → (Yield, poor)
3, 2, 2
(Temperature, warm) & (Rain, light) & (Fertilizers, low) → (Yield, average)

If we set the minimum confidence level in Apriori algorithm as 0.5, the following 12 possible rules are also generated by the algorithm.

1, 5, 4
(Rain, light) → (Yield, average)
2, 4, 3
(Temperature, warm) & (Rain, light) → (Yield, average)
1, 7, 5
(Fertilizers, high) → (Yield, average)
1, 3, 2
(Temperature, hot) → (Yield, average)
2, 3, 2
(Temperature, warm) & (Fertilizers, normal) → (Yield, poor)
2, 3, 2
(Rain, heavy) & (Fertilizers, high) → (Yield, average)
1, 5, 3
(Temperature, cold) → (Yield, average)
1, 5, 3
(Fertilizers, low) → (Yield, average)
2, 4, 2
(Temperature, warm) & (Rain, moderate) → (Yield, poor)
2, 4, 2
(Temperature, warm) & (Fertilizers, high) → (Yield, average)
2, 4, 2
(Temperature, warm) & (Fertilizers, high) → (Yield, good)
2, 4, 2
(Rain, moderate) & (Fertilizers, normal) → (Yield, poor)

The following four certain rules are generated by the Apriori algorithm when data values discretized using crisp intervals are used for rule generation.

2, 2, 2
(Temperature, cold) & (Rain, heavy) → (Yield, poor)
2, 2, 2
(Temperature, warm) & (Fertilizers, low) → (Yield, poor)
2, 2, 2
(Temperature, hot) & (Fertilizers, normal) → (Yield, average)
2, 2, 2
(Rain, light) & (Fertilizers, normal) → (Yield, average)

When the same discretized dataset is used in the algorithm with a minimum confidence level of 0.5, the possible rules generated are:

1, 4, 3
(Temperature, warm) → (Yield, poor)
2, 4, 3
(Rain, moderate) & (Fertilizers, low) → (Yield, poor)
1, 6, 4
(Rain, moderate) → (Yield, poor)
2, 3, 2
(Temperature, cold) & (Fertilizers, low) → (Yield, poor)
2, 3, 2
(Temperature, warm) & (Rain, moderate) → (Yield, poor)
2, 3, 2
(Temperature, hot) & (Rain, heavy) → (Yield, good)
2, 3, 2
(Rain, heavy) & (Fertilizers, low) → (Yield, poor)
1, 10, 6
(Fertilizers, low) → (Yield, poor)
1, 5, 3
(Temperature, cold) → (Yield, poor)
1, 5, 3
(Rain, light) → (Yield, average)
1, 5, 3
(Fertilizers, normal) → (Yield, average)
1, 6, 3
(Rain, heavy) → (Yield, poor)
2, 4, 2
(Temperature, hot) & (Rain, light) → (Yield, average)

10.8 Discussion

Data discretization is very important to implement many data mining operations efficiently. In this work, a fuzzy set-based data discretization approach is employed to study the influence of FST in decision making. For evaluation purposes, in the same dataset a conventional set-based data discretization is also applied and the performances of both the approaches are studied. For set-based data discretization, suitable crisp intervals are selected to map the feature values of the original decision table. In the proposed work, instead of crisp intervals, overlapping intervals are considered for discretization. These overlapping fuzzy intervals help to handle abrupt shifting of data values from one crisp interval to another. The evaluation results indicate the effectiveness of fuzzy discretization together with rough approximation in handling vagueness present in the data.

The approximation accuracies of the concepts of the response variable {Yield} obtained after the fuzzy discretization and the usual crisp discretization are presented in Table 10.6. Figure 10.1 compares the accuracies obtained from both these approaches.

The information presented in Table 10.6 shows that accuracy of concepts approximation after fuzzy discretization is better than that of the accuracy obtained in the case of crisp discretization. Figure 10.1 explains the effectiveness of fuzzy-based data discretization in handling the vagueness present in data compared to the conventional set-based data discretization.

To study the efficiency of the proposed method of rule induction process, in addition to the accuracy of approximation, the total number of rules generated, and the confidence of these generated rules are also considered. Table 10.7 shows the total number of certain and possible rules generated by LEM2 algorithm together with total number of rules generated after fuzzy and crisp discretization.

Table 10.6 Accuracy of rough approximation after fuzzy and crisp discretization.

Concepts	Accuracy after fuzzy discretization	Accuracy after crisp discretization
poor	1	0.42
average	0.8	0.5
good	0.6667	0.22

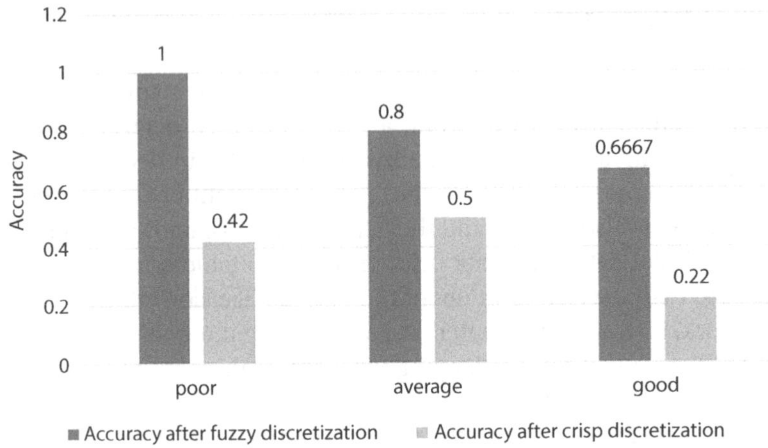

Figure 10.1 Approximation accuracy after fuzzy discretization and crisp discretization.

Table 10.7 Number of rules generated after fuzzy based and set-based discretization.

Data transformation applied	Total number of rules generated	Number of certain rules	Number of possible rules
Fuzzy based	14	12	2
Set based	18	7	11

From Table 10.7, it is important to note that the number of certain rules generated by LEM2 after fuzzy discretization is significantly greater than that of the rules generated after the set-based data discretization. Another important contribution of the proposed fuzzy-rough rule induction approach is the reduced set of possible rules. When rule mining is performed after fuzzy discretization combined with rough approximation, number of possible rules generated is only two. If the same rule mining process is performed after crisp discretization, the number of possible rules increases to 11. This shows that vagueness in taking decisions can be reduced considerably when rough set based rule induction is combined with fuzzy discretization.

From the certain rules generated using Apriori algorithm, in both cases, the decision rules representing the decision (Yield, good) is absent. So, when we apply Apriori algorithm, to generate a decision leading to "good"

Table 10.8 Number of certain rules generated by LEM2 and Apriori algorithms.

Discretization \ Algorithm	LEM2	Apriori
Fuzzy	12	6
Crisp	7	4

Yield with 100% confidence is impossible. But when we use LEM2 algorithm, the number of certain rules for (Yield, good) are four in the case of fuzzy discretization and two when crisp discretization is used. Another inference obtained from the rule generation using Apriori algorithm is that there is a possibility for generating more inconsistent rules. Table 10.8 lists the number of certain rules generated by LEM2 and Apriori algorithms from the dataset pre-processed using fuzzy and crisp discretization techniques.

From Table 10.8, we conclude that fuzzy discretization with LEM2 algorithm gives better scope for accurate and efficient decision making.

Conclusion

Rough Set Theory and Fuzzy Set Theory provide sound mathematical tools to handle vagueness present in information systems. This study focusses on a novel hybrid approach for decision rule induction combining fuzzy discretization, rough set approximation and the idea of local covering. The proposed algorithm is applied on a small dataset related to farming and rules are generated. The success of the proposed approach in constructing the rule base by selecting induced rules are proved in the experimental results. A crisp discretization is applied on the dataset instead of the fuzzy discretization to prove the efficiency of the proposed method. Then performance of both the approaches is compared in terms of both accuracy of approximation of various decision concepts and the number of certain rules generated. The Apriori algorithm is also used to evaluate the performance of the proposed model. Better accuracy in approximating various concepts and the ability to induce a greater number of certain rules, attest the proposed approach is efficient in handling vagueness present in data as well as in decision making.

References

1. Zdzislaw Pawlak, Rough sets and intelligent data analysis. *Information Sciences,* vol. 147, pp. 1-12, Elsevier, 2002.
2. Timothy J. Ross, *Fuzzy Logic with Engineering Applications*, Third Edition, John Wiley & Sons, 2010.
3. K. Srinivas, G. Raghavendra Rao, A. Govardhan, Rough-Fuzzy Classifier: A System to Predict the Heart Disease by Blending Two Different Set Theories. *Arabian Journal for Science and Engineering*, Vol. 39, Issue 4, pp 2857-2868, 2014.
4. Amitava Roy, Sankar K Pal, Fuzzy discretization of feature space for a rough set classifier, *Pattern Recognition Letters*, Vol. 24, pp 895-902, (2003) Elsevier.
5. Richard Jensen and Qiang Shen, New approaches to Fuzzy–Rough Feature Selection. *IEEE Transactions on Fuzzy Systems*, Vol. 17, No. 4, pp. 824-838, 2009b.
6. Songbo Tan, Yuefen Wang and Xueqi Cheng, Text Feature Ranking Based on Rough-set Theory, In *Proceedings of the IEEE/WIC/ACM International Conference on Web Intelligence*, pp. 659-662, 2007.
7. Rajeesh E, Anupama M and Shijo M Joseph, Use of Data Mining Techniques for Sustainable Growth in Coconut Farming, *Proceeding of 19th Swadesi Science Congress, 2009.*
8. Jiawei Han and Micheline Kamber, *Data Mining: Concepts and Techniques*, 2nd Edition, Morgan Kaufmann Publishers, 2006.
9. Jerzy W. Grzymala-Busse, Rough Set Theory with Applications to Data Mining. In *Encyclopedia of Data Warehousing and Mining*, Ed. by J. Wang, Information Science Publishing, pp. 973-977, 2006.
10. Pawlak Z, *Rough Sets: Theoretical aspects of Reasoning about Data*, Kluwer Academic Publishing, Dordrecht. 1991.
11. Jiye Li and Nick Cercone, Discovering and ranking important rules, *IEEE International Conference on Granular Computing*, Vol. 2, pp. 506-511, 2005.
12. Mahsa Farjoo, Hamed Fazlollahtabar, Minoo Talebi Ashoori, Design of a recommender system for online shopping using decision tree and Apriori algorithm, *Journal of Software Engineering and Intelligent Systems*, Volume 3, Issue 3, pp. 236-244, December 2018.
13. Alzoubi, W. A., Cluster Based Association Rule Mining for Courses Recommendation System, *International Journal of Computer Science & Information Technology (IJCSIT)* Vol. 11, No 6, December 2019.
14. Rabiatul A. A., Rashid, Puteri N. E. Nohuddin1, Zuraini Zainol, Analyzing Climate Variability in Malaysia Using Association Rule Mining. *International Journal of Engineering & Technology*, 7 (4.34) 394-397(2018).
15. Saloni Ambasana1, V.J. Shukla1 and Mahesh Vyas, The Perspicacity of Association Rule Mining in the Multifaceted Environment of Ayurveda, *Current Traditional Medicine*, Vol 3, No. 3, pp. 195-205, 2017, DOI: 10.2174/2215083803666170706145352.

16. Robert G Smith, Non-Manual Articulators in Sign Language Verbs: An Analysis with Data Mining Association Rules. 2018, *Proceedings of 1st Irish Deaf Research Conference (IDR2018).*
17. Demiriz A., Ekizoglu, B., & Kula, U., Clustering products under pairwise positive and negative association constraints in retailing. In *Proceedings of ANNIE, 2010.*
18. Danelys Cabrera, Association Rule Mining for Customer Segmentation in the SMEs Sector Using the Apriori Algorithm, Association Rule Mining for Customer Segmentation, 2019.
19. Sreelekha. S, D. Malathi, Neuro-Symbolic Integration with Association Rule Mining. *International Journal of Pure and Applied Mathematics* Volume 114 No. 12, pp. 769-776, 2017.
20. Hemant Kumar Soni. Peculiarly Approaches for Association Rule Mining using Genetic Algorithm, *4th International Conference on Advances in Electrical, Electronics, Information, Communication and Bio-Informatics (AEEICB-18).*
21. Shawkat, M., Badawy, M., and Eldesouky, A.I. (2021). A Novel Approach of Frequent Itemsets Mining for Coronavirus Disease (COVID-19), *EJECE, European Journal of Electrical Engineering and Computer Science*, Vol. 5, No. 2, March 2021.
22. Tandan, M., Acharya, Y., Pokharel, S., & Timilsina, M. (2021). Discovering symptom patterns of COVID-19 patients using association rule mining. *Computers in Biology and Medicine*, 131, 104249. https://doi.org/10.1016/j.compbiomed.2021.104249
23. K. R. Swetha, N. M, A. M. P and M. Y. M, "Prediction of Pneumonia Using Big Data, Deep Learning and Machine Learning Techniques," *2021 6th International Conference on Communication and Electronics Systems (ICCES)*, 2021, pp. 1697-1700, doi: 10.1109/ICCES51350.2021.9489188.

11

An Effective Machine Learning Approach to Model Healthcare Data

Shaila H. Koppad[1]*, S. Anupama Kumar[2] and Mohan Kumar[3]

[1]*Manager, Capgemini Technology Services India Limited, Bengaluru, Karnataka, India*
[2]*Department of MCA, RV College of Engineering, Bengaluru, Karnataka, India*
[3]*Department of Pulmonary, Sapthagiri Institute of Medical Sciences and Research Centre, Bengaluru, Karnataka, India*

Abstract

India ranks second in the world population, and a growing population sends an alarming signal to healthcare professionals. As population grows people tend to acquire complications related to health, and technological developments have to provide solutions to overcome challenges. The health updates published by the Government of India in 2017 focus on a rise in non-communicable diseases. This research work focuses on non-communicable Chronic Obstructive Pulmonary Disease (COPD) which is the major cause of death worldwide. The major challenge faced by the healthcare industry is gathering and integrating the clinical data in various formats and analysing them for the betterment of patients. The COPD data was collected from heterogeneous sources in diversified formats. The various data storing formats were evaluated depending on the different manipulations and operations; ORC file format was found efficient. The preprocessed data is stored in ORC format in Hive Component and it is found to be very efficient in terms of scalability and high availability. The different machine learning algorithms like Linear regression, Random forest and Decision table were found to be efficient in predicting the presence of COPD. The algorithms were analysed for the accuracy and performance, The Decision table algorithm was found to be 99% accurate for the dataset.

***Keywords*:** Healthcare, big data, visualization, data, machine learning

**Corresponding author*: shaila.koppad@gmail.com

Kukatlapalli Pradeep Kumar, Aynur Unal, Vinay Jha Pillai, Hari Murthy, and M. Niranjanamurthy (eds.) Data Engineering and Data Science: Concepts and Applications, (279–302) © 2023 Scrivener Publishing LLC

11.1 Introduction

India contributes 17.71% to the world population and stands at second position in global rank. The population of India is increasing by nearly 1% every year. Increased population causes poverty, lack of education, unemployment, child labor, food problems, a lowering of the standard of living and many more factors. Poverty and Unhealthy living conditions can be the reason for increase in death rates and health issues. Malnutrition and Communicable and non-communicable diseases are the major causes of health issues in India. Communicable diseases are dengue fever, hepatitis, tuberculosis, malaria, pneumonia and many more. The non-communicable diseases are diabetes, cancer, cardiovascular, chronic obstructive pulmonary, injuries and mental health disorders.

In India healthcare is one of the promising sectors which is growing immensely [1]. Today diseases are increasing with the increase in the population, leading to an increase in deaths and a public health challenge, as well as causing an effect on the economic growth of the country [2]. People are getting prone to various diseases. The causes may be a busy life, living style, poor maintenance and pollution [3].

In recent years, people are working round the clock by keeping themselves busy with high level of stress, causing health issues like stomach pain, chest pain, insomnia, headaches and high blood pressure due to chaotic nutrition and unhealthy lifestyles. In a busy life, people also tend to have restlessness, and insufficient sleep can be the major cause which would affect creativity and concentration. Diseases caused by germs in communities due to poor maintenance of the region are signs of poor hygiene [4]. These can be due to uncleaned toilets and surroundings, not washing clothes and bedding frequently and food poisoning. Diseases like food poisoning, diarrhea, skin infections, pneumonia, trachoma and gastroenteritis are caused by bacteria; hepatitis A, cold and flu are caused by viral infection; giardiasis, pediculosis, threadworm, hookworm, threadworm and roundworm infections are caused by Parasitic. Living style diseases are associated with people's lifestyle, and these are non-communicable diseases which are mainly caused by unhealthy eating habits, smoking, drug, alcohol, and lack of physical activity. Obesity, Type II diabetes, Atherosclerosis, Heart Disease, High Blood Pressure, Cancer, Stroke, Chronic Obstructive Pulmonary Disease (COPD), Cirrhosis, Arthritis, Metabolic 3 Syndrome, Colitis, Depression, Vascular Dementia and Nephritis are some of the disease caused due to people's lifestyles [5]. Pollution is of different types, such as water, soil, noise and air. Toxins from human and industrial waste are introduced into water supplies which are used by people. This water when consumed by people and animals causes various diseases like diarrhea, vomiting, diphtheria, skin lesions, kidney damage, hepatitis, and typhoid [6]. Humans produce waste materials which are gathered on the surface of the earth, and from these waste chemicals such as fertilizers and pesticides are absorbed into soil. When plants are grown from this soil and consumed by humans diseases like cancer, brain and nerve damage, kidney and liver diseases result. Noise pollution is created by humans when they produce a huge amount of loud noise from vehicle sources and industries. The diseases caused due to noise pollution decrease efficiency of people due to lack of concentration, fatigue, deafness, blood pressure and anxiety. Burning of waste material, fuel and other by-products released by human modern life are generating harmful gases that are released into the air and become the cause of air pollution. These pollutants, when inhaled by humans and animals, cause various diseases. The various diseases caused due to air pollution are lung cancer, stroke, heart disease, acute lower respiratory and COPD. In this regard the Healthcare system generates huge data pertaining to various healthcare associates [7]. This data is generated from various sources like Electronic Medical Record (EMR) [8], Clinical History, Electronic Health Records (EHR) [9], Laboratory Tests and doctors' perceptions. Healthcare is one of the promising sectors which is immense in India with inequalities [10]. This system includes various stakeholders like patients who are key stakeholders [11]. Doctors treat the patients and they are named as referral doctor, visiting doctor and resident doctor depending on the service they provide. Health organizations help in connecting doctor and patients by sharing resources. Government takes lots of responsibilities in improving the safety and quality of public health and also provides different schemes under which patients get treated at lower cost. Insurance companies are also part of the healthcare industry by providing health insurance to patients for their medical clams.

11.2 Types of Data in Healthcare

As the technology is improving, an enormous amount of data gets generated day by day in healthcare [12]. The source of data can be various departments such as patients, health providers, health plans, medical claims, supply chain, charts, finance, lab tests and adverse event [13]. The data from these departments can be in diversified formats which are generated by Electronic Medical Records (EMR), Research, Devices, Digital Hospitals and social media as shown in Figure 11.1.

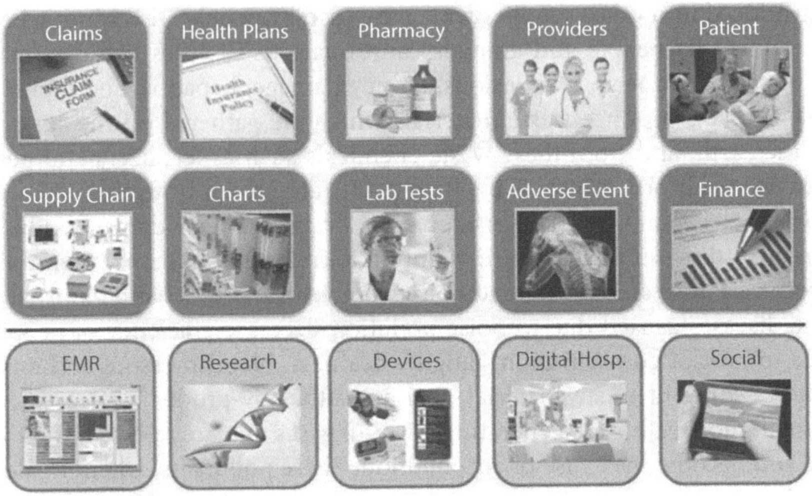

Figure 11.1 Variety of data generated in healthcare industry.

EMR data is generated by the health providers or health organizations which contain the patients' clinical history. Data generated from all the departments can be used for research purposes for developing efficient models. Heterogeneous health equipments are used in the treatment of patients but data generated from these devices are in diversified formats which is difficult for analysis [14]. Digital Hospitals use different Information and Communication Technology (ICT) for ease of communication and advance technologies for better patient treatment [15]. Social Media is another of the latest technologies which includes patient connections, exchange of ideas and the facilitation of services that do not involve selling [16]. It improves personal health management and promotes a healthy lifestyle [17].

Although the healthcare system in India is advanced there are certain gaps which cannot be neglected. The following are some of the gaps identified in this work:

- Data is not recorded or stored properly in hospital which is leading to loss of clinical history of the patients.
- Patients are not telling doctors their symptoms, which may be because they are not educated about the diseases or afraid of revealing the symptoms.
- Patients visit multiple doctors for treatment and hide medical history from them.

- During the visit to the hospital patients do not carry previous prescriptions or reports.
- Due to lack of medical records patients might be under-treated for the disease or over treated for a small illness.
- Huge data is getting generated from heterogeneous sources in diversified formats. It is difficult to maintain the data in diversified formats.
- For applying learning algorithms the data has to be in single format, which is not available.

11.3 Big Data in Healthcare

Healthcare Analytics is used to analyze health activities of patients, functionalities of the health care organizations, drug discovery, cost benefit analysis, health insurance, etc. [18]. By applying Big Data Analytics, the clinic-epidemiological data can be logically structured and analyzed. Big data analytics can be implemented efficiently in healthcare because it can examine huge data sets comprehending diversified data types. Big Data can be realistic to forecast epidemics, improve quality of life, avoid preventable deaths and cure disease [19]. Analytics is used to uncover hidden patterns,

Table 11.1 Big data in healthcare.

Sl. no.	Case study	Description
1	Diagnostics	Process of identifying sources of illness by applying data mining and analysis
2	Preventative Medicine	To prevent disease, predictive analytics and data analysis of genetics can be performed
3	Precision Medicine	Is used to drive hyper-personalized care from the aggregate of data
4	Medical Research	Research carried over to discover new treatments, cure disease and new medicines
6	Cost Reduction	Identifying the values for better patient care and long-term savings
7	Population Health	Recognizing health strategies based on demographics, socioeconomic and disease trends

unknown associations, market tendency, customer choices and other useful business information [20]. The challenges of handling data include search, transfer, analysis, capture, visualization and storage of unstructured data [21]. Big Data can be built as a centralized repository where all the diversified data maintained by different hospitals can be preprocessed and structured in a single format which helps in easy healthcare analytics. The health-related data is integrated and accumulated continuously, resulting in an incredible volume of data for each versatile patient [22].

Table 11.1 Different case studies which can be considered for research work under Big Data in healthcare.

11.4 Different V's of Big Data

Initially Big data analytics was basically described by three primary characteristics: Volume, Velocity and Variety [23]. As the complexity of data is increasing and the technology is improving the Big data V's are also increasing. Now there are eight V's, called volume, value, veracity, visualization, variety, velocity, viscosity and virality which describe the characteristics of Big data in healthcare [24].

The 8 V's [25] help to handle healthcare data efficiently as follows:

1. **Volume –** The volume of the data generated in healthcare is increasing day by day because of new disease, hospitals, admission of patients to the hospital and insurance companies and their plans, etc. Handling this huge data is critical issue so care has to be taken to reduce the size without loss of the data.
2. **Value –** The clinical history of the patients plays a vital role. When the patient is sick and has to be treated his clinical history provides value information regarding the previous treatments that the patient has undergone.
3. **Veracity –** In healthcare data patients may be treated on their clinical history. So it is necessary and challenging to keep the data clean to deal with information.
4. **Variety –** The data generated in healthcare is available in different formats: structured, semi-structured or unstructured due to various sources which generate these data. There are numerous applications used by stakeholders to capture, store and process the data which results in variety of data. Different hospital departments will have different

applications to hold the patient data, and laboratory tests conducted on the patient will also generate reports in different formats.

5. **Velocity** – The frequency and speed of data generated is very high in healthcare due to telemedicine, social media, use of smartphone for fitness apps and also to the reports of any test that are captured and shared to different doctors to take a second opinion before taking medication. The data is transformed in fractions of seconds from various regions. This will enable patients to get the analysis of their medical report in real time.
6. **Viscosity** – As the patients get admitted to the hospital information is available with the health organization. Now they can analyze and call for action to take precautionary measures to prevent the disease.
7. **Virality** – When any disease is spreading across the state or country, if there is a centralized data warehouse the analytics help to identify in which region the number of patient admissions is increasing, and this can be used to send more doctors to that particular region.
8. **Visualization** – Visualization helps to easily read and understand the data available. It helps to unlock the data and benefits from patients to providers to payers and suppliers. It helps the patients to understand their health condition, the effect of the disease and its impact. It also helps the hospital management to provide better facilities to patients.

11.5 About COPD

Chronic Obstructive Pulmonary Disease (COPD) is a non-communicable disease which is the fourth leading chronic disease that is causing deaths in the world [25]. According to the World Health Organization (WHO) projection, by 2020 this disease will be the third leading disease to cause deaths. Due to COPD, more than 3 million people died in 2012, accounting for 6% of all deaths globally [26]. The following Figure 11.2 depicts the types of diseases that lead to death worldwide.

As per the statistics data released by WHO every year, it is clearly understood that COPD stands in fourth position as a non-communicable disease causing death, which is depicted in Figure 11.3. Figure 11.4 shows COPD stands in third position causing more deaths worldwide for 2016 [27].

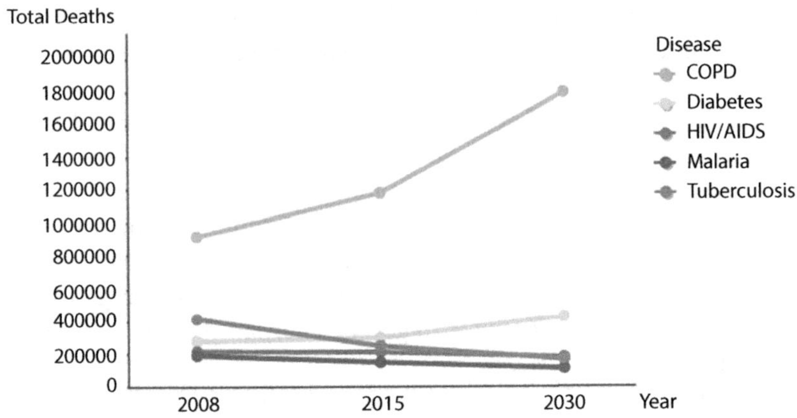

Figure 11.2 Diseases causing deaths [26].

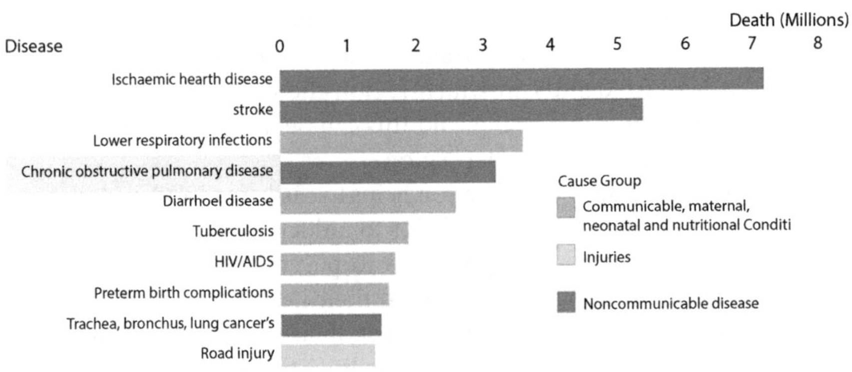

Figure 11.3 Top 10 global causes of deaths in 2000 [28].

Researchers confirmed the reason for the increase may be lifestyle habits which include smoking and consuming, air pollution, bio mass, occupation exposure, etc. Some of these parameters are measurable and can be directly analyzed.

The risk factors are the characteristics of the diseases that are the reasons people are prone to get the disease. The person might be knowingly or unknowingly involved in the risk factors that lead to the disease. The major reason for COPD patients [28] is smoking, which is considered as part of high level style of living. The other factors are biomass fuel which was used for cooking food to meet the basic daily need; occupational exposure is another cause for the disease, for example, people working in the bricks industry. Another factor is air pollution; the reason for that is

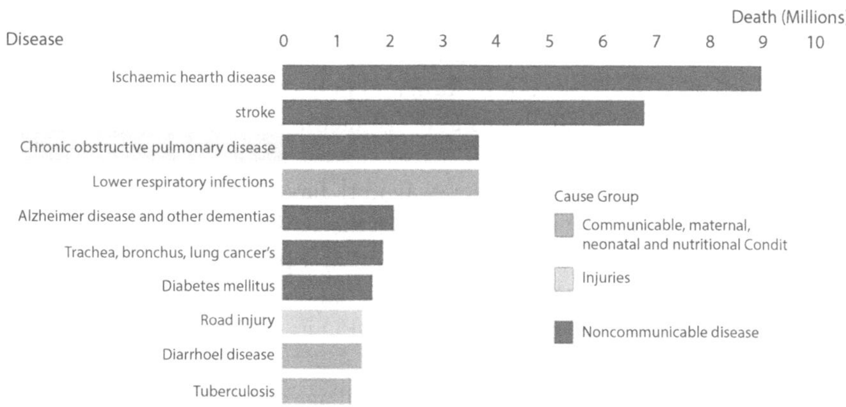

Figure 11.4 Top 10 global causes of deaths in 2006 [28].

people consume the harmful gases that are released in air and lungs are damaged, causing pulmonary disorder. A few of the factors may be genetic abnormalities, anomalous lung development and accelerated aging. To identify that the person is suffering from the disease or not, symptoms play a vital role. These symptoms are the key indicators of the disease. The symptoms among COPD and asthma are also identical, due to which it is difficult for doctors to identify that the patient is suffering from asthma or COPD. According to the Global Initiative for Chronic Obstructive Lung Disease (GOLD), which is used as a global strategy for COPD [29], the key symptoms among COPD patients are dyspnea, cough, sputum production, wheezing, and chest tightness. Additional symptoms are fatigue, weight loss and anorexia.

Arvind B. Bhome [30] has given an outline of the epidemiology COPD in India, the morality of the disease and the etiological determinants of COPD in India. He has also discussed the severity of the disease in various parts of India and concluded that there was an increase in the growth of disease from 1994 to 2010.

The *European Research Journal* (ERJ) [31] has identified the growth of COPD in India and the prevalence of COPD >57 million people suffering from obstructive airway diseases (OADs) by the end of 2016. The multiple causes for the spread of the disease would be biomass combustion fuels inside the home, non-conventional forms of tobacco products such as bidi, hookah, and chillum, which deliver relatively greater amounts of combustion by products, are excessively consumed in rural areas, etc.

The researchers in [32, 33] have applied different statistical techniques to understand the risk of COPD using various clinical tests like Arterial

Blood Gas (ABG) analysis, a complete hemogram, blood sugar levels, serum albumin levels, blood urea nitrogen levels (BUN), serum creatinine levels, SGPT, SGOT and serum sodium and potassium levels, etc. The relationship between these attributes and age factor was also analyzed using different statistical tools.

Sundeep Salvi and Anurag Agrawal [34] have discussed the various causes of the disease in India and provided the statistical analysis of COPD in India. They have also discussed the need of a preventive center for the disease in India. Further they have also pointed out that the various factors needed for predicting the prevalence of COPD in India is ignored and has to be taken care of.

The authors in [35] have studied eight villages each from Mysore and Nanjangud randomly by taking biomass smoke exposure and chronic bronchitis for the female population, and the researchers in [36] analyzed the people by conducting a spirometry test. Statistical analysis has been done over the data to estimate the prevalence of the disease and the output of the research. Work [17] states that the biomass smoke and poor lifestyle will lead to severity of the disease in Mysore and Mumbai.

Negi *et al.* [37] studied 126 patients (73.81% male) in North India using sampling and [38] studied the patients in Northwest India using Wright's mini peak flow meter. Socioeconomic status, COPD severity, health status, dyspnea and anthropometric measures of eligible patients were assessed using St George Respiratory Questionnaire (SGRQ) in Hindi Version [39]. To examine the relationship between Health-Related Quality of Life (HRQL) and risk factors Linear regression model was used and the values were adjusted for age and gender. For COPD patients the HRQL was significantly decreased across stages. The researcher in [40] concluded that prevalence of chronic bronchitis, bronchial asthma, and COPD was found to be 3.36%, 1.18%, and 4.21%, in rural Northwest India.

In [41] related from quadriceps weakness present in significant proportions in patients belonging to younger age groups and suffering from mild disease. These findings indicate that onset of muscle weakness in COPD may precede the onset of symptoms.

The Patients' details can be accessed with reference to unique ID, from the clinical centralized data repository. This method helps to know the treatments given to patients by which doctor [42]. They implemented J48 decision tree on COPD dataset with 10 cross validation. The results provided several statistics to make a prediction on given dataset. The algorithm's accuracy is 91.2%, ROC area is 1 and to build this model 0.03 seconds is the time taken.

The authors in [43] have considered the symptoms of COPD including Dyspnea, Smoking, Proliferation cough, Wheezing, Body Mass Index (BMI), fever and frequent chest infection. They say that symptoms in patients served as a predictive factor in COPD patient care. In this paper they had used big data and its techniques to collect COPD data from divergent format, manipulate data according to logics and predict the likelihood of occurring COPD patients through Logistic regression model. The prediction accuracy COPD dataset is 0.97 and area under curve is 0.98 which is closer to the ideal value 1.

Although many researchers [44, 45] investigated influencing individual parameters on COPD using statistical tools but as per investigator no research work was investigated considering all the parameters which has a major impact on COPD analysis.

The various research gaps which are been identified from the work are as follows:

i. The various parameters like clinical data sets, age and sex of the population are considered by the international researchers and this is not handled effectively by the Indian researchers. There are no standard attributes used by the Indian researchers to predict the severity of the disease.
ii. The variety of the data set used by the Indian researchers is very low since the storage and usage of medical records/ EHR has not been handled efficiently.
iii. Statistical analysis has been used by most of the researchers and the accuracy of the method is not measured.
iv. Computational techniques applied by the researches are very limited and need novel methods.

Due to technological developments, data is maintained in various formats which can be handled efficiently. Electronic health records are substantial and complex; they are challenging to deal with conventional programming and are not able to be effectively dealt with customary administration techniques. The huge data created by the healthcare industry is vast and versatile. The existing statistical and computational techniques are not sufficient to handle the versatile variety and huge volume of data.

But as challenges are increasing technologies are also getting advanced to find solutions for these challenges. One challenge is handling data with different characteristic called as Big Data. Every problem has solution; if big data is a problem, Hadoop is the solution to handle data with various characteristics.

Big data is one of the promising solutions in the current scenario that can handle all the challenges mentioned in the survey and analyze the data available in various formats. Technology can be used efficiently to understand the research work and organize, structure and find out the presence/absence of COPD disease among the patients using the reports and symptoms collected from them.

11.6 Methodology Implemented

The framework has been designed as six phases. The following Figure 11.5 illustrates work flow of framework structure of the work carried. The different phases are data collection, data structuring, data storing, data preprocessing, applying learning algorithms and data visualization.

i) Data Sources
The facts which are collected from different sources can be defined as data sources. The data generated from these heterogeneous sources can be categoriesed as structured, semi-structured and unstructured data. The data which are collected from these sources are available in different formats.

The dataset pertaining to the details of the COPD patients complaining pulmonary diseases is collected in semi-structured and structured format. The symptoms and the test for the COPD disease are according to

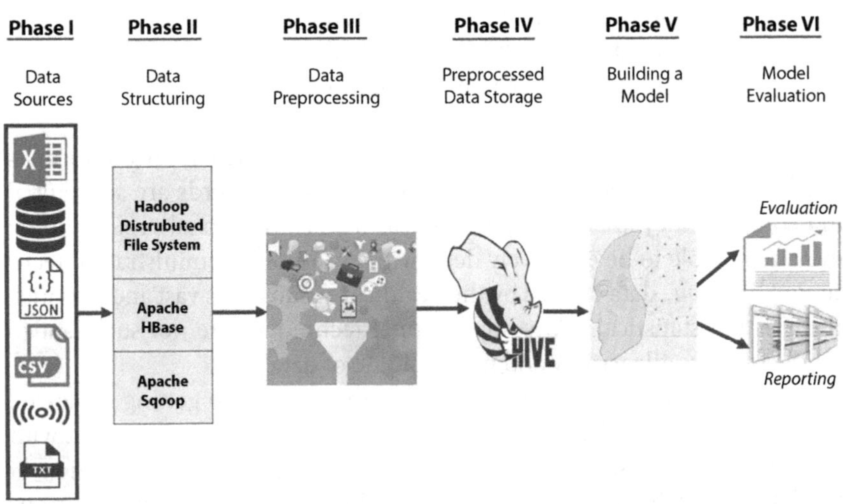

Figure 11.5 Phases of the work carried.

GOLD [46]. A total 20 different attributes are considered in this research which includes personal details of the patient's symptoms of the disease, test report details, etc.

ii) Data Structuring

When different data types are used, the challenges are more due to their type and variety of formats. Big data provides the way to systematically extract information, analyze heterogeneous data types and place into single format for analysis.

iii) Data Preprocessing

Data Preprocessing is an important phase in data analysis as they impact success rate of the implementation. This reduces complexity of the data under analysis as data in the real world is unclean. This process comprises data cleaning, transformation, feature selection and reduction.

In this work, subset of the original dataset by selecting only relevant columns is used. When the dataset was considered, some data was missing and corrupted as shown in Figure 11.6, so the dataset was prepared for analysis.

The dataset contains instances with missing values and handling these values was the challenge since wrong predications may lead to ending one's life in the healthcare domain. Such instances with too many missing values for attributes were avoided.

It is not always that missing values should be avoided but these values can be replaced. In this chapter the missing values are replaced by the mean of the existing values. For example: if the age of the patient was missing

PatientID	AGE	GENDER	Patient Type	COUGH	EXPECTO RATION	DYSPNEA GRADE	WHEEZE	FEVER	HEMOPT YSIS	PACKS	NOYs	PY	HEIGHT (IN CM)	WEIGHT (KG)	BMI	PULSE	HB	CHEST X RAY	FEV1%
AC01.0003013448	43	F	IP	2	1	2	1	1	1	1	28	28	158	71	22.31	72	15.9	2,3	63.43
AC01.0003052488	59	M	IP	1	2	0	1	2	2	1.1	27	29.7	152	48	23.34	94	13.6	1	84.55
AC01.0002998591	0	M	IP	1	1	2	2	2	2	1.5	32	48	149	53	22.67	200	15.8	1	24.37
AC01.0002992495	50	F	IP	2	2	0	2	R	1	1.1	14	15.4	166	48	25.6	108	14.7	2	86.36
AC01.0003036307	45	M	OP	1	1	2	2	1	2	1.9	6	11.4	160	80	24.29	107	15.4	2	23.27
AC01.0002993614	67	F	OP	1	2	0	2	2	2	1.3	11	14.3	152	52	23.14	107	11.5	1	81.57
AC01.0003032448	60	F	OP	N	2	3	1	2	2	1.2	13	15.6	164	72	21.7	102	12.5	1	26.44
AC01.0003020628	56	G	IP	2	1	1	2	2	2	1.5	20	30	159	70	25.14	95	12.1	2	24.38
AC01.0003011326	69	F	IP	1	1	4	2	1	1	1.1	69	20.9		64	24.69	79	16.1	2,3	63.5
AC01.0003031901	61	M	IP	2	2	4	2	2	1	1	32	32	156	70	24.54	74	16.1	2,3	64.36
AC01.0003016270	65	F	OP	2	1	1	1	2	1	1.5	8	12	168	76	80	70	13.5	1	27.59
AC01.0003032801	57	M	IP	1	F	0	2	2	2	1.6	2	3.2	147	64	21.58	108	16.2	2	84.52
AC01.0003014751	52	F	OP	2	1	2	2	2	1	1.5	27	40.5	150	80	21.68	84	14.7	2,3	27.46
AC01.0002995541	50	F	IP	1	1	4	1	1	2	1.3	5	6.5	169	54	25.53	78	16.6	1	23.58
AC01.0003032250	68	3	IP	1	2	0	2	1	1	1.8	35	63	152		22.4	74	13.8	2	24.45
AC01.0003002034	75	F	OP	2	2	4	2	N	2	1.4	17	23.8	159	74	21.37	107	13.1	2,3	28.36
AC01.0002989142	51	M	IP	2	1	3	1	2	1	1.7	23	39.1	151	47	22.47	94	13.8	2,3	68.42
AC01.0003035928	72	F	IP	1	1	0	1	1	2	1.5	7	10.5	157	55	22.2	74	16.6	1	83.4
AC01.0003004832	75	F	IP	2	1	4	2	2	1	1.3	24	31.2		60	21.23	108	14.6	2,3	65.42
AC01.0003018943	58	F	OP	2	2	5	2	1	1	1.8	15	27	166	49	22.62	98	12.5	2	50.37
AC01.0003050091	42	F	OP	1	1	0	1	2	2	1.2	12	14.4	165	77	25.71	99	16.2	1	86.36
AC01.0002996719	63	F	IP	2	1	0	2	2	1	0.9	12	10.8	170	66	22.27	88	16.4	1	84.54
AC01.0003001888	47	F	IP	1	1	3	2	1	-2	1.4	13	18.2	155	47	23.41	87	11.3	2,3	77.52
AC01.0003037761	66	M	IP	1	2	2	1	2	2	1.8	26	46.8	162	46	25.21	108	16.9	1	29.27
AC01.0003008105	187	M	IP	2	1	3	2	2	2	0.9	33	29.7	147	53	23.36	73	13.1	2,3	64.27
AC01.0003040940	62	M	IP	2	1	0	2	2	2	1.9	5	9.5	166	78	24.27	94	15.1	2,3	68.44
AC01.0003026584	56	F	OP	1	1	C	1	2	2	1.7	24	40.8	158	61	23.69	83	11.6	2,3	29.4
AC01.0003044989	62	M	IP	2	1	2	1	1	2	1.6	11	17.6	169	57	22.65	95	12.6	2,3	50.36

Figure 11.6 Missing and corrupted data.

then this value was replaced by taking the mean of the age attribute of the patients. Since patient age would not lead to major harm it was replaced.

It also has to be taken into consideration that missing values for all the attributes cannot be replaced. Suppose the symptoms of the patient like fever attribute value is missing than the value cannot be replaced for analysis. But the analysis can also be considering other attributes as feature selection.

iv) Data Storage

Hadoop provides better ways to store and format data on files. Closest to the relational way of storing data is row-based format. SequenceFile and Avro belong to this category, of which the former is the most primitive row-based format provided to users. Other specialized row-based formats that use SequenceFile at their base are also available in Hadoop. In most big data scenarios, column-based file formats are known to perform better than their row-based counterparts, as most queries require retrieval of a few columns for multiple rows [47]. The various file formats that were demonstrated to create, describe and provide information about table and time taken to perform these tasks is also mentioned.

Figure 11.7 illustrates sunburnt chart for summary of tasks performed on table available in different formats files and time taken to perform these various tasks. From the table it can be deliberated that ORCFile format is efficient in all tasks which takes less time to perform the given tasks. From the sunburnt chart it is clear that ORCFile format is efficient for storing COPD data.

v) Building Model

For building classification, a model systematic approach is carried which is called as classification technique built from input dataset. The association between class label and attribute set can be best suited using the classification techniques. These generated models should be capable of predicting target variable from input data provided means class labels of unknown records. The model should be trained with the help of the training set and should be applied for testing on the remaining data.

The data is divided into training data and test data. The learning algorithm is provided with training set then algorithm learns and model is built using the given data. Further to receive appropriate output the model is applied on the test data.

The presence of a supervisor in the learning process as teacher or mentor is called as supervisor learning. In this process the machine is trained using data which is well labeled or already identified with the correct

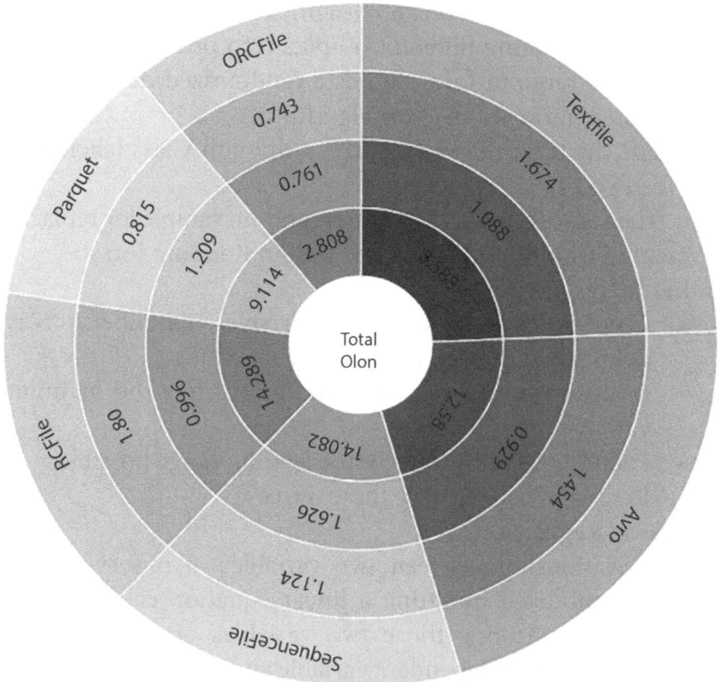

Figure 11.7 Sunburnt Chart representing task performed on different file formats.

answer. Later the same machine is provided with a new set of data; subsequently supervised learning algorithm analyses training data and generates the correct outcome.

The absence of a supervisor in the learning process as teacher or mentor is called as unsupervised learning. In this process the input data is neither classified nor labeled, but grouping of the unsorted information is carried depending on the similarities, differences and patterns without the trained data. The machines are restricted to search the hidden patterns in unlabeled data.

The different classification algorithms which are implemented in this work are Random Forest, Linear Regression and Decision Table.

- **Random Forest**
  Random forests or random decision forests are an ensemble learning method for prediction. It combines random selection of attributes. Each decision tree is produced based on random selection of samples/attributes. This algorithm

uses the randomization and bagging for prediction. Bagging means combining different outputs into one single prediction. The random forest builds a randomized decision tree in each iteration of bagging algorithm.

For the training set D = d1, ..., dn with Class labels Y = y1, ..., yn,.

The algorithm selects N times random sample for replacement in training set. Steps involved in Random forest:
In the training set D,

- Consider N number of input variables, a number n<N is deliberated, at each node
- n attributes are selected randomly from the N input variables
- Each tree is full-grown to the leading range possible
- Predict by aggregating these n trees

- **Linear Regression**
  The relationship between two variables of observed data can be modeled by fitting a linear equation called Linear Regression. Among those two variables one is called as explanatory or independent variable and the other as dependent varaible. There is significant association between these two variables but this does not imply one variable is caused by another.
  The designed liner model should basically determine that the relationship between variables of interest exists or not of observed data. From the two variables explanatory variable can be named as 'X', dependent variable as 'Y', 'b' as alope of line and the intercept as 'a' in the equation Y=a+bX of linear regression.
- **Decision Table**
  Decision tables are a concise visual representation for specifying which actions to perform depending on given conditions. They are algorithms whose output is a set of actions. The information expressed in decision tables could also be represented as decision trees or in a programming language as a series of if-then-else and switch-case statements. Every variable corresponds to decision and action to a procedure or operation to perform.

vi) Model Evaluation

In this work Machine learning, Big Data technique is used for training the system for the prediction of disease on COPD dataset. The Linear Regression, Decision Table and Random Forest algorithms are used in this work on COPD dataset.

Figure 11.8 illustrates the results of the Random Forest Algorithm. The time taken to build the model of COPD dataset is 0.8 seconds; the values are compared for 10 Cross-validation, Correlation coefficient is 0.995 which is closer value to 1 and the mean absolute error is 0.0084 and root mean squared error is 0.017.

Figure 11.9 illustrates the results of the Linear Regression Algorithm. The time taken to build the model of COPD dataset is 5.29 seconds, which is more compared to Random Forest; the values are compared for 10 Cross-validation; Correlation coefficient is 0.9983 which is closer value to 1 and the mean absolute error is 0.0015 and root mean squared error is 0.0274.

Figure 11.10 discuss the results of Decision Table Algorithm. The time taken to build the model of COPD dataset is 0.56 seconds, which is less compared to Random Forest and Linear Regression, the values are compared for 10 Cross-validation, Correlation coefficient is 0.9992 which is closer value to 1 and the mean absolute error is 0.0011 and root mean squared error is 0.0188.

Table 11.2 depicts the summary of efficiency of the algorithms in predicting the prevalence of the disease.

```
=== Classifier model (full training set) ===

RandomForest

Bagging with 100 iterations and base learner

weka.classifiers.trees.RandomTree -K 0 -M 1.0 -V 0.001 -S 1 -do-not-check-capabilities

Time taken to build model: 0.8 seconds

=== Cross-validation ===
=== Summary ===

Correlation coefficient                 0.9995
Mean absolute error                     0.0084
Root mean squared error                 0.017
Relative absolute error                 1.8869 %
Root relative squared error             3.6082 %
```

Figure 11.8 Results of random forest algorithm.

```
Time taken to build model: 5.29 seconds

=== Cross-validation ===
=== Summary ===

Correlation coefficient                  0.9983
Mean absolute error                      0.0015
Root mean squared error                  0.0274
Relative absolute error                  0.3453 $
Root relative squared error              5.8091 $
```

Figure 11.9 Result of linear regression algorithm.

```
Time taken to build model: 0.56 seconds

=== Cross-validation ===
=== Summary ===

Correlation coefficient                  0.9992
Mean absolute error                      0.0011
Root mean squared error                  0.0188
Relative absolute error                  0.2365 $
Root relative squared error              3.9793 $
```

Figure 11.10 Result of decision table algorithm.

Table 11.2 Performance analysis.

Algorithm applied	Time taken	Accuracy
Random Forest	0.8 sec	99.95 %
Linear Regression	5.29 sec	99.83%
Decision Table	0.56 sec	99.92%

Figure 11.11 describes the performarnce analysis of the algorithms implemented in this work. Decision Table is the best suited for the analyzing the prevelance of COPD among patients.

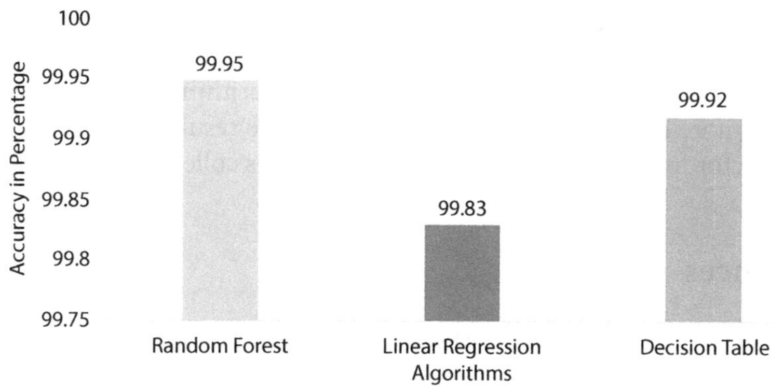

Figure 11.11 Result of decision table algorithm.

Conclusion

The health care industry is one of the major domains for researchers. Recent technological advancements have brought in new techniques and methods to understand the data and bring in new knowledge about it. This work is carried out for an effective machine learning approach to model healthcare data.

The detailed survey on the Healthcare data and COPD has given awareness about the symptoms of the disease and the parameters that should be used to classify COPD and asthma. The GOLD guidelines also helped to understand the standards that should be used to understand and diagnose the presence and level of COPD. The recent technological advancement helps in understanding the effect of using big data tools and machine learning techniques for diagnosing the prevalence and severity of the disease. The details of these findings were discussed in the literature survey.

The COPD patient data is collected from heterogeneous sources that were available in diversified formats. The major challenge is to handle the variety of data available as structure, semi-structured and unstructured formats. The methodology includes building a framework to collect, structure, preprocess, store, and build a model using machine learning algorithms and evaluate the performance of the model.

Machine learning algorithms are implemented to inspect the effects of the disease. Various classification algorithms are applied on different COPD datasets. These Machine Leaning algorithms helped to predict the presence and severity of COPD among patients. The implementation of Random Forest, Linear Regression and Decision Table over the data collected. The Decision Table Algorithm was efficient on the COPD dataset in predicting the prevalence of the disease. The time taken by this algorithm is 0.56 which is less compared to other two Random Forest and Linear Regression Algorithms. The Coefficient of the algorithm is 0.9992 near to 1 proved more efficient than other algorithms. The results are interpreted by the doctor in the hospital from which the data is collected.

References

1. Nilaish, A Review of Indian Healthcare Sector (April 13, 2017). Available at SSRN: https://ssrn.com/abstract=2952286 or http://dx.doi.org/10.2139/ssrn.2952286
2. Narain J. P. (2016). Public Health Challenges in India: Seizing the Opportunities. *Indian Journal of Community Medicine: official publication of Indian Association of Preventive & Social Medicine, 41*(2), 85–88.
3. Kelly, F. J., & Fussell, J. C. (2015). Air pollution and public health: emerging hazards and improved understanding of risk. *Environmental Geochemistry and Health, 37*(4), 631–649.
4. Joanna Esteves Mills & Oliver Cumming, The Impact of Water, Sanitation and Hygiene on Key Health and Social Outcomes: Review of Evidence June 2016.
5. Booth, F. W., Roberts, C. K., & Laye, M. J. (2012), Lack of exercise is a major cause of chronic diseases. *Comprehensive Physiology, 2*(2), 1143–1211.
6. Buttar, H. S., Li, T., & Ravi, N. (2005), Prevention of cardiovascular diseases: Role of exercise, dietary interventions, obesity and smoking cessation. *Experimental and Clinical Cardiology, 10*(4), 229–249.
7. Dash, S., Shakyawar, S.K., Sharma, M. *et al.*, Big data in healthcare: management, analysis and future prospects. *J Big Data* **6,** 54 (2019).
8. Siang Edmund, L. C., Ramaiah, C. K., & Gulla, S. P. (1). Electronic Medical Records Management Systems: An Overview. *DESIDOC Journal of Library & Information Technology*, 29(6), 3-12.
9. Evans R. S. (2016). Electronic Health Records: Then, Now, and in the Future. *Yearbook of Medical Informatics, Suppl 1* (Suppl 1), S48–S61.
10. Amritah Sandhu, The need to address structural inequality in the provision of healthcare in India https://timesofindia.indiatimes.com/blogs/voices/the-need-to-address-structural-inequality-in-the-provision-of-healthcare-in-india/ 2020,
11. Tarkkanen, Kimmo & Reijonen, Pekka & Harkke, Ville. (2013). Roles and Identification of Stakeholders in Health Care IS Development.
12. Hammer, J., Aiyar, Y., & Samji, S. (2007). Understanding Government Failure in Public Health Services. *Economic and Political Weekly*, 42(40), 4049–4057.
13. Balarajan, Y., Selvaraj, S., & Subramanian, S. V. (2011). Health care and equity in India. *Lancet* (London, England), *377*(9764), 505–515.

14. Tariq RA, Vashisht R, Sinha A, *et al.* Medication Dispensing Errors and Prevention. [Updated 2021 Jul 25]. In: StatPearls [Internet]. Treasure Island (FL): StatPearls Publishing; 2021 Jan-. Available from: https://www.ncbi.nlm.nih.gov/books/NBK519065/
15. Abhinav Sharma, Robert A. Harrington, *et al.*, Using Digital Health Technology to Better Generate Evidence and Deliver Evidence-Based Care, *Journal of the American College of Cardiology*, Volume 71, Issue 23, 2018, pp. 2680-2690, ISSN 0735-1097, https://doi.org/10.1016/j.jacc.2018.03.523.
16. Samal, Janmejaya. (2016). A brief assessment of Ayurveda, Yoga and Naturopathy, Unani, Siddha, and Homoeopathy health system based on Five-year Plans of India. 10. S91-S95. (review paper)
17. Raghupathi, W., & Raghupathi, V. (2014). Big data analytics in healthcare: promise and potential. *Health Information Science and Systems.*
18. Institute of Medicine (US) Committee on Regional Health Data Networks; Donaldson MS, Lohr KN, editors. Health Data in the Information Age: Use, Disclosure, and Privacy. Washington (DC): National Academies Press (US); 1994. 2, Health Databases and Health Database Organizations: Uses, Benefits, and Concerns. Available from: https://www.ncbi.nlm.nih.gov/books/NBK236556/
19. Dash, S., Shakyawar, S.K., Sharma, M. *et al.* Big data in healthcare: management, analysis and future prospects. *J Big Data* **6,** 54 (2019).
20. Aceto, Giuseppe & Persico, Valerio & Pescapè, Antonio. (2018). The role of Information and Communication Technologies in Healthcare: Taxonomies, Perspectives, and Challenges. *Journal of Network and Computer Applications.* 107. 10.1016/j.jnca.2018.02.008.
21. Ventola C. L. (2014). Social media and health care professionals: benefits, risks, and best practices. *P & T: A Peer-Reviewed Journal for Formulary Management, 39*(7), 491–520.
22. Jane, M., Hagger, M., Foster, J. *et al.* Social media for health promotion and weight management: a critical debate. *BMC Public Health* **18,** 932 (2018).
23. Ward, M. J., Marsolo, K. A., & Froehle, C. M. (2014). Applications of Business Analytics in Healthcare. *Business Horizons, 57*(5), 571–582.
24. Hong, L., Luo, M., Wang, R., Lu, P., Lu, W., & Lu, L. (2018). Big Data in Health Care: Applications and Challenges, *Data and Information Management, 2*(3), 175-197.
25. Mehta, Nishita & Pandit, Anil. (2018). Concurrence of Big Data Analytics and Healthcare: A Systematic Review. *International Journal of Medical Informatics.* 114. 10.1016/j.ijmedinf.2018.03.013.
26. Uthayasankar Sivarajah, Muhammad Mustafa Kamal, Zahir Irani, Vishanth Weerakkody, Critical analysis of Big Data challenges and analytical methods, *Journal of Business Research*, Volume 70, 2017, pp. 263-286, ISSN 0148-2963.
27. Md. TareqHasan, Amathulhadishakara,"Versatile Aspects of IoT in Medical Science" *American Journal of Engineering Research* (AJER) 2018 *American*

Journal of Engineering Research (AJER) e-ISSN: 2320-0847 p-ISSN: 2320-0936 Volume 7, Issue 4, pp. 89-96.

28. Volume, velocity, and variety: Understanding the three V's of big data, https://www.zdnet.com/article/volume-velocity-and-variety-understanding-the-three-vs-of-big-data/ 2018
29. Arockia Panimalar. S, Varnekha Shree. S, Veneshia Kathrine. A "The 17 V's of Big Data", *International Research Journal of Engineering and Technology (IRJET)* e-ISSN: 2395-0056 Volume: 04 Issue: 09, Sep. 2017, www.irjet.net p-ISSN: 2395-0072.
30. Jennifer Bresnick, Understanding the Many V's of Healthcare Big Data Analytics, https://healthitanalytics.com/news/understanding-the-many-vs-of-healthcare-big-data-analytics, 2017
31. Boutayeb, Abdesslam. (2010). The Burden of Communicable and Non-Communicable Diseases in Developing Countries. 10.1007/978-0-387-78665-0_32.
32. World Health Organization. http://www.WorldHealthOrganization.int/respiratory/copd/burden/en, 2018
33. The Global Burden of Disease, WHO October 2008. www.who.int/healthinfo/global_burden_disease/projections/en/index.html. Accessed 22 Dec. 2011.
34. Eisner, M.D., Anthonisen, N., Coultas, D., Kuenzli, N., Perez-Padilla, R., Postma, D., Romieu, I., Silverman, E.K., Balmes, J.R.: An Official American Thoracic Society Public Policy Statement: Novel Risk Factors and the Global Burden of Chronic Obstructive Pulmonary Disease, PubMed: 20802169.
35. WorldHealthOrganization. "http://www.who.int/news-room/factsheets/detail/the-top-10-causes-of-death" on 24 May 2018.
36. TO, P. G. (2018). Global initiative for chronic obstructive lung disease. *Am J Respir Crit Care Med, 197*(1).
37. Paraskeva, M. A., Borg, B. M., & Naughton, M. T. (2011). Spirometry. *Australian Family Physician, 40*(4), 216-219.
38. Ranu, H., Wilde, M., & Madden, B. (2011). Pulmonary function tests. *The Ulster Medical Journal, 80*(2), 84.
39. Arvind B. Bhome, COPD in India: Iceberg or volcano? *J Thorac Dis* 2012;4(3):298- 309. DOI: 10.3978/j.issn.2072-1439.2012.03.15
40. Toshita Kalkana, Saibal Moitra, Surinder K. Jindal, Subhabrata Moitra, "Increasing burden of COPD in rural India: an example why India warrants primary healthcare reforms", *ERJ Open Res* 2016; 2: 00032-2016 — DOI: 10.1183/23120541.00032-2016
41. Anup N. Nillawar, Kedar B. Joshi, Sandip Bharat Patil, Jayshree S. Bardapurkar, Suhas J. Bardapurkar, "Evaluation of HS-CRP and Lipid Profile in *COPD, Journal of Clinical and Diagnostic Research*, 2013 May Vol-7(5): 801-803.
42. Deepali J. Kamdar, Dharmeshkumar Patel, "A prospective study of the predictors of mortality in acute exacerbation of COPD: an Indian perspective",

International Journal of Advances in Medicine, Kamdar DJ *et al. Int J Adv Med.* 2017 Apr;4(2):362-366, ISSN 2349-3925 — eISSN 2349-3933
43. Sundeep Salvi, Anurag Agrawal, "India Needs a National COPD Prevention and Control Programme", Supplement to JAPI, February 2012, Vol. 60.
44. P.A. Mahesh, B.S. Jayaraj, A.K. Prabhakar*, S.K. Chaya and R. Vijaysimha, "Identification of a threshold for biomass exposure index for chronic bronchitis in rural women of Mysore district, Karnataka, India", *Indian Journal of Medical Research* 137, January 2013, pp. 87-94.
45. Lancelot Pinto; Radhika Banka; Matthew Shorofsky; Jean Bourbeau," The Prevalence of Undiagnosed COPD at an Annual Health Screening in Mumbai, India", *Chest.* 2015;148(4 MeetingAbstracts):703A. doi:10.1378chest.2279391
46. Patel, A. R., Patel, A. R., Singh, S., Singh, S., & Khawaja, I. (2019). Global initiative for chronic obstructive lung disease: the changes made. *Cureus, 11*(6).
47. Jiang, H., & Elmore, A. J., Boosting data filtering on columnar encoding with SIMD. In *Proceedings of the 14th International Workshop on Data Management on New Hardware* (p. 6). ACM. 2018.

12

Recommendation Engine for Retail Domain Using Machine Learning Techniques

Chandrashekhara K. T.[1]*, Gireesh Babu C. N.[1] and Thungamani M.[2]

[1]*BMS Institute of Technology and Management, Bangalore, Karnataka, India*
[2]*University of Horticulture Science, Bangalore, Karnataka, India*

Abstract

A recommendation system is highly beneficial to any business in the retail industry; not only does it increase revenue but it also enables retailers to provide their customers with the products they require. In this paper we are building a recommendation engine for a retailer that analyzes available products data to make predictions of which products will gain more revenue to retailer and which products are from which supplier. This can be done by performing clustering of the products based on the transaction history of the product purchases and then dividing the products into low-selling products, medium-selling products and high-selling products using canopy K-means clustering technique. RFM analysis is performed to classify customers and suppliers. Recency Frequency Monetary technique is a Marketing strategy used for analyzing customer behavior such as how recently a customer has purchased (recency), how repeatedly the customer purchased (frequency) and how much customer spends (monetary). It also classifies the suppliers into Low Gain, High Gain and Medium Gain which depends upon the product transactions of each of the suppliers. Which products will gain more revenue in the future and from which supplier is determined, and how many must be purchased on an average is predicted using ARIMA model.

***Keywords*:** K-means, canopy K-means, RFM, ARIMA, AR model

**Corresponding author*: chandru@bmsit.in

Kukatlapalli Pradeep Kumar, Aynur Unal, Vinay Jha Pillai, Hari Murthy, and M. Niranjanamurthy (eds.) Data Engineering and Data Science: Concepts and Applications, (303–316) © 2023 Scrivener Publishing LLC

12.1 Introduction

Formidable challenges and exciting opportunities are present in e-commerce due to the marketplace. The National Bureau of Statistics of China has collected data and it indicates that there were 4,667 million e-shoppers by the end of 2016. The e-retailing market has generated a huge market potential and it has influenced firms to grow their market presence. Online merchants provide their consumers with a variety of product alternatives and this leads to the overflow of information, and because of this the purchase decision process becomes difficult. If consumers compromise on their choice satisfaction, their purchases will be limited as well [7, 8].

The e-merchants' establishment or maintenance of long-term relationships with customers will be adversely affected because the results produced lack positive qualities. To overcome these challenges a recommender system was developed. These systems were basically designed to make personalized purchase suggestions. The suggestions are based on the filtered information about product features, demographics and customers' preferences. These recommendation systems are widely used in Business to customer outlets [2, 17].

Important applications of recommendation system include search engines like Yahoo, Google and Baidu. In these the recommendation system is implemented as an information filtering tool. It is also used by various e-commerce websites to make product suggestions [8, 16].

12.2 Proposed System

The proposed project performs clustering of the products based on the transaction history of the product purchase. It also classifies the suppliers depending upon the product transactions on each of the suppliers [4]. Which products will gain more revenue in the future and from which supplier is determined, and how many must be purchased on an average is done using ARMA. There are three modules in this project. They are:

1. Classification of Products – Canopy K-Means Clustering Algorithm.
2. Classification of Suppliers – RFM (Recency, Frequency and Monetary) Model.

3. Recommendation for Buyer – ARMA (Autoregressive Integrated Moving Average) model.

Apart from this, the project will contain a Product and Order Log Viewer.

This module performs clustering of the products based on the transaction history of the product purchases. It divides the products into Low-Selling Products, Medium-Selling Products and High-Selling Products. To perform this we use the Canopy K-Means Clustering Technique. In this we choose optimal centers and then perform k-means clustering [6].

As shown in Figs. 12.1 and 12.2, Classification of products is performed using Canopy K-means clustering algorithm. In Canopy K-means we use optimized centers for clusters calculation. The products are classified as low-selling, medium-selling and High-selling products based on the quantity and frequency of sale. Fig. 12.3 shows products are classified into different clusters. The algorithm begins with an initial set of clusters chosen at random phases. In each iteration, particular instance is moved to its nearest cluster Centre. The cluster centers are recalculated at each iteration [18]. The Centre of every cluster is computed as the average of all the instances belonging to a particular cluster. This module is responsible for classifying the items into three different categories, namely low-selling products, medium-selling products and high-selling products [1, 3].

Fig. 12.1 GUI for performing KMeans.

View K Means Clustering Output

Product II	Product Name	Frequency	No Of Days	Min Distance	Cluster No	Distance C1	Distance C2	Distance C3
1	Friday Night Lights	13	7	7.38241...	3	11.0736...	9.84321...	7.3824115301167
2	Money Ball	2	2	1.27475...	1	1.27475...	2.35702...	4.74341649025257
3	Paper Lion	2	2	1.27475...	1	1.27475...	2.35702...	4.74341649025257
4	Bollywood Glitz	1	1	2.37170...	1	2.37170...	3.59010...	6.04152298679729
5	Flim Fare Take The Heat	0	0	3.69120...	1	3.69120...	4.92160...	7.3824115301167
6	Decoding Rahul Gandhi	0	0	3.69120...	1	3.69120...	4.92160...	7.3824115301167
7	Anna Hazare	0	0	3.69120...	1	3.69120...	4.92160...	7.3824115301167
8	ISRO Satellites	0	0	3.69120...	1	3.69120...	4.92160...	7.3824115301167
9	Cave of Winds	0	0	3.69120...	1	3.69120...	4.92160...	7.3824115301167
10	ISRO Dreams	0	0	3.69120...	1	3.69120...	4.92160...	7.3824115301167
11	Java Programming Complete ...	1	1	2.37170...	1	2.37170...	3.59010...	6.04152298679729
12	Salman Dreams	0	0	3.69120...	1	3.69120...	4.92160...	7.3824115301167

Fig. 12.2 K-means clustering output.

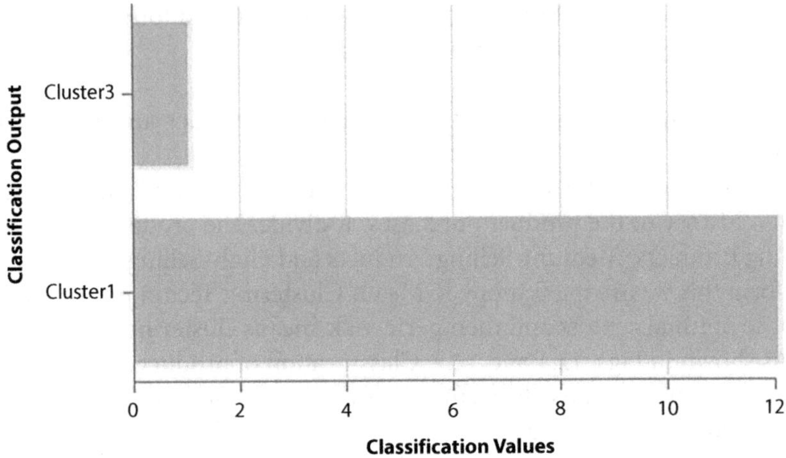

Fig. 12.3 Classification of products into different clusters.

12.2.1 Classification of Suppliers

This module performs segmentation of the suppliers based on the transaction history of the product purchases. It divides the Suppliers into Low-Profit Suppliers, Medium-Profit Suppliers and High-Profit Suppliers. To perform this, we use RFM (Recency, Frequency and Monetary) Segmentation. It segregates the suppliers based on how recently, how frequently and how profitably their supplies are sold.

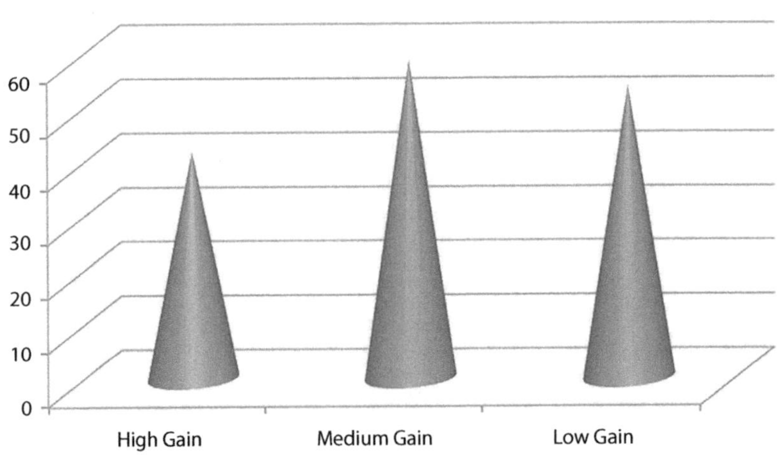

Fig. 12.4 Classification of suppliers using RFM model.

Fig. 12.4 shows the output of the RFM unit for supplier classification in the form of a columns graph. The classification of suppliers is performed using Recency, Frequency and Monetary segmentation. The suppliers are classified into high-gain, medium-gain and low-gain suppliers. There are over 50 medium-gain suppliers, over 30 high-gain suppliers and about 45 low-gain suppliers. It is suggested that the retailers procure products from the suppliers offering high gain and try to avoid the low-gain suppliers [9].

12.2.2 Recommendation for Buyer

This module uses Autoregressive Integrated Moving Average (ARMA) model. ARMA is a popular model in Statistics and Econometrics. The given transaction data is analyzed by the ARMA model and it provides a prediction for the procurement of stock in the subsequent months. This prediction gives the most profitable range of the most profitable product and suppliers based on the previous sales pattern.

The data set consists of the past data of the book order. The retail market is the main focus of our study. Dataset solely records the retail transactions. Our strategy requires totally different aspects of analytics to get accuracy in our sales prediction [11].

As shown in Fig. 12.5 transaction logs acts as an input. Two processes are executed, namely K-Means and RFM Computation, Forecasting which is a sequence of data mining algorithms. The Output is classification of users and Classification of Products and Forecasting [6].

As shown in Fig. 12.6, DFD Level 1 which has input namely transaction logs. The process list are Perform K Means Algorithm – which divides products into low-selling products, medium-selling products and high-selling products Perform RFM Analysis which is used to find loyal, medium loyal and then no loyal and Classification of Customers and Products. The output is a List of Recommended customers as per clustering, list of products as per clustering and forecasting analysis for products [5].

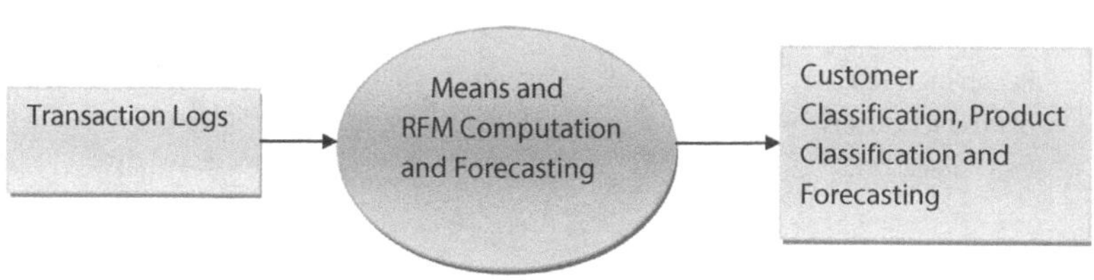

Fig. 12.5 DFD level 0.

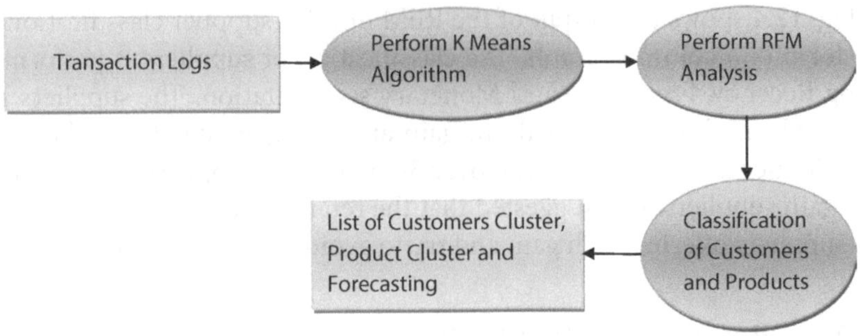

Fig. 12.6 DFD level 1.

As shown in Fig. 12.7, DFD Level 2, which is a detailed description of DFD level 2. The process Performed are K Means Algorithm – which divides products into low-selling products, medium-selling products and high-selling products then Perform RFM Analysis which is used to find

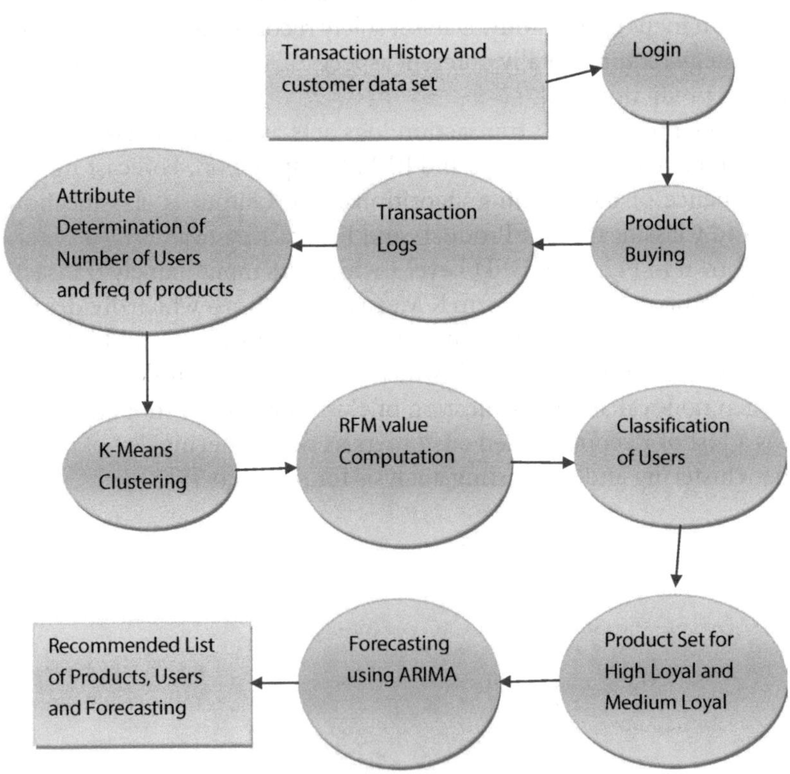

Fig. 12.7 DFD level 2.

loyal, medium loyal and then no loyal and Classification of Customers and Products. The output is a List of Recommended customers as per clustering, list of products as per clustering and forecasting analysis for products.

12.2.3 Forecasting Using ARIMA Model

Time Series analysis using ARIMA Model can be described as below.

1. IDENTIFY statement is used in the identification stage to denote the response series and choose ARIMA models for it. Time series used in later statements possibly differencing them, and calculates autocorrelations, inverse autocorrelations, partial autocorrelations, and cross correlations are read by IDENTIFY statement. ARIMA model that could be fit are suggested by IDENTIFY statement output. Many options are there to test stationary and tentative ARIMA identification [10].
2. ESTIMATE statement is used to specify the ARIMA model to check estimation and diagnostic checking stage. Diagnostic statistics are produced by the ESTIMATE statement. Significance test is performed to determine the parameters indicate the terms are unnecessary in the model. Goodness-of-fit statistics help in comparing this model with others. White noise tests are performed which indicate the white noise residuals. This test shows whether additional information and it can be utilized for complex model. If some problems are encountered then another model can be implemented and repeat the same steps [13, 15].
3. FORECAST statement is used in the forecast phase and it forecasts values of the time series and generate confidence intervals. Confidence intervals in the ARIMA model is generated using ESTIMATE statement

The Steps Involved for ARMA Model can be defined as follows:

1. Find the Unique Products from Order Log
2. Find the set of values in terms of the monthly purchase
3. Compute the difference between consecutive values
4. Find the average of the difference across the months
5. Provide a rounded integer on approximation of purchase

Detailed Steps of ARMA (Auto Regression Moving Average)

1. It is a forecasting method.
2. Uses information from variables to forecast the trend.
3. The regression is done based on past values.
4. Unvariate analysis is performed.
5. Probabilistic and stochastic variables are computed.
6. Designed for forecasting future movement.

ARIMA

1) The BJ Type of time series models Yj to be expanded by past or lagged values
2) Explain itself from history
3) The model will have two sections

Autoregressive (AR) model
Moving Average (MA) model
gdp(t)=a+bgdp(t-1))+ut
gdp(t)=a+b(Lgdp)+ut
gdp(t) = current value of time t
gdp(t-1) = previous value for time t-1

AR
Ar(2)=a+b1 gdp(t-1) +b2 gdp(t-2)+ut
Ar(3)=a+b1 gdp(t-1) +b2 gdp(t-2)+ b3 gdp(t-3)+ut

MA
gdp(t)= gamma+do *u(t) +d1 *u(t-1)
U(t)=error term
U(t-1)=previous error term

MA(2)
gdp(t)=gamma +do ut +d1 u(t-1) + d2 u(t-2)

Combining AR and MA
ARMA(p,q) =gdp(t)=a+b1 gdp(t-1)+do ut+ d1 ut-1

This work makes use of ARIMA.

a) ARIMA makes use of integration component

b) In most cases the sales values are not stationary hence it has to be transformed by making use of differencing approach, and this is called as integration

ARIMA p,q,d

1) Provide list of previous sales of various days
2) forecastsize -- it is number of days
3) int p = 3;int d = 0;int q = 3;int P = 1;int D = 1;int Q = 0;int m = 0;
4) forecast length = forecaststartindex - forecastendindex where
   forecaststartindex=length of data
   forecastendindex=length of data +1
5) generate a training data =salesdata (startindex:endindex)
6) Performdiffrentiate process

   Input

   a) arima parameters
   b) TrainingData

   Output
   diffrentiate data

7) Perform Mean of diffrentiate data
8) Executing shifting process of teh wave

   Input - mean value
   stationary data
9) Executing of Forecasting

Differentiate Process

1) Convert sales data into array
2) Obtain array with a one-timediffrence
3) for (int j = d, k = 0; j <src_len; ++j, ++k) {

   dst[k] = src[j] - src[k];
   }

Integration Process
Input - differentiation process array, mean value
Output - array with each value added with mean

Forecasting Process
HananRicean

1) Integrated data
2) Length of data
3) Maximum Iterations =5
4) Estimate using AR

```
for (int j = 0; j < _offsets.length; ++j) {
linearSum += timeseries[tsOffset - _offsets[j]] * _coeffs[j]; }
          offset is total time records available
          coeffs[j]=error values
```

5) Estimate using MA

```
final double error = data[j] - forecast;
or (int j = 0; j < _offsets.length; ++j) {
linearSum += timeseries[tsOffset - error[j]] * _coeffs[j];
}
```

6) Adding AR and MA value will give ARMA

12.3 Results

12.3.1 ARIMA Forecasting

As shown in Fig. 12.8 Input data provided to the GUI. Fig. 12.9 depicts the forecasting output: Product ID, Product Name, Average Forecast value, Total Sales which can happen and then list of day sales in the last column. Here we use Autoregressive integrated Moving average model to analyze the given transaction data; it provides prediction for procurement of stock

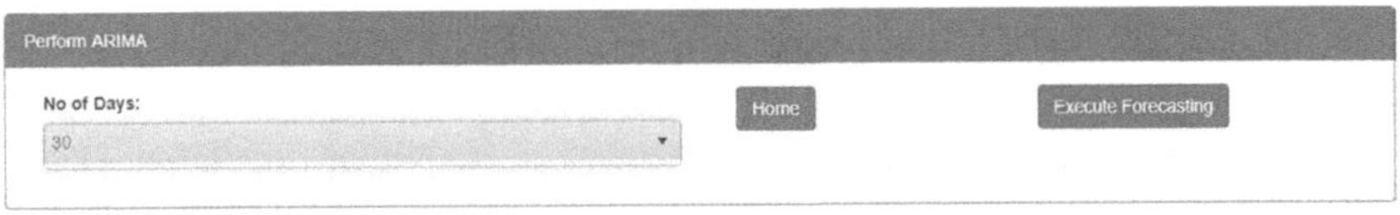

Fig. 12.8 Forecasting input.

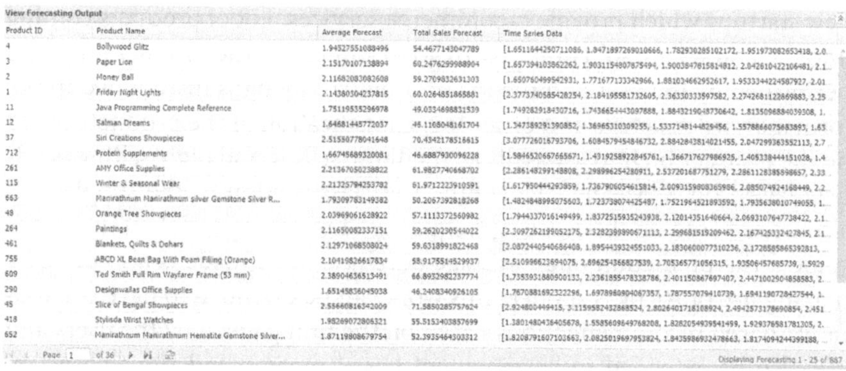

Fig. 12.9 Forecasting output.

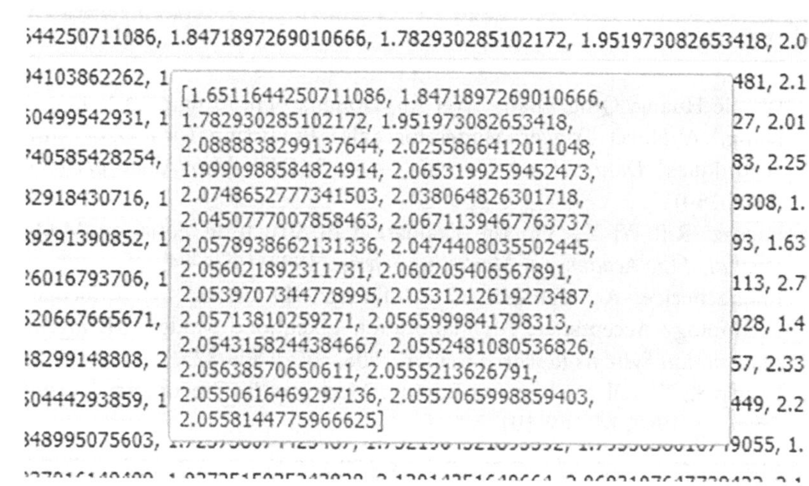

Fig. 12.10 Complete forecast sales details using ARIMA model.

in subsequent months. This prediction gives the most profitable range of products and suppliers based on the previous sales pattern [12, 14].

Fig. 12.10 shows the complete forecast details of the various days sales details using ARIMA model.

12.4 Conclusion

The e-commerce system is using the entire modern way of shopping in the consumer market. There are a lot of well-established companies and new startups which provide e-commerce services. Lot of products coming into the market are sold online with good strategies. The Product Buying is performed by the user and the log of product buying is maintained in the format of order details and order log. Classification of the products based on the transaction into low-selling, medium-selling and high-selling products based on K Means Algorithm. The classification of the suppliers is also done based on the recency, frequency of purchase and then monetary values. The prediction of an average purchase to be made is also done by making use of ARIMA model. This work can be extended to have collaborative-based recommendations based on the rating given by the users and can also be extended to have content-based recommendations based on transaction logs of the user.

References

1. Wenjie Huang, Qing Zhang, Wei Xu, Hongjiao Fu, Mingming Wang, Xun Liang, "A Novel Trigger Model for Sales Prediction with Data Mining Techniques" *Data Science Journal*, 14, p.15. DOI: http://doi.org/10.5334/dsj-2015-015.
2. Bagozzi R.P., Yi Y., "On the Evaluation of Structural Equation Models." *Journal of the Academy of Marketing Science*, 1988, 16:74-94.
3. Bhattacherjee A., Sanford C. "Influence Processes for Information Technology Acceptance: An Elaboration Likelihood Model." *Management Information Systems Research Center*, 2006, 30(4): 805-825.
4. Beatty S., Ferrell M. Impulse Buying: Modeling Its Precursors. *Journal of Retailing*, 1998, 74: 169-191.
5. Bruner GCII., Kumar A. "Explaining Consumer Acceptance of Handheld Internet Devices." *Journal of Business Research*, 2005,58(5): 553-558.
6. Cai S., Xu Y. "Effects of Outcome, Process and Shopping Enjoyment on Online Consumer Behavior." *Electronic Commence Research andApplications*, 2006, 5, 272-281.
7. Chen Y., Chen C., Lin Y., Yeh R. Predicting College Student' Use of ELearning Systems: An Attempt to Extend Technology AcceptanceModel. *PACIS 2007 Proceedings*, 121.
8. Chen H., Tseng H. "Factors That Influence Acceptance of Web-based E-learning Systems for The In-service Education of Junior HighSchool Teachers in Taiwan." *Evaluation and Program Planning*, 2012, 35(3): 398-406.
9. Chen LY. "The Quality of Mobile Shopping System and Its Impact on Purchase Intention and Performance." *International Journal of Managing Information Technology*, 2013, 5(2): 23-32. 28.
10. Chen L., Pu P. "Experiments on User Experiences with Recommender Interfaces." *Behaviour & Information Technology*, 2014, 33(4): 327-394.

11. Chin WW., Gopal A., Salisbury W. D. "Advancing the Theory of Adaptive Structuration: The Development of a Scale to Measure Faithfulness of Appropriation." *Information Systems Research*, 1997, 8(4): 342–367.
12. Cho YH., Kim JK. Kim SH. "A Personalized Recommender System Based on Web Usage Mining and Decision Tree Induction." *Expert Systems with Applications*, 2002, 23(3): 329-342.
13. Chung N., Koo C., Kim J. "Extrinsic and Intrinsic Motivation for Using a Booth Recommender System Service on Exhibition Attendees' Unplanned Visit Behavior." *Computers in Human Behavior*, 2014, 30: 59-68.
14. Cosley D., Lam SK., Albert I., Konstan JA., Riedl, J. "Is Seeing Believing? How Recommender System Interfaces Affect Users' Opinions. In *Proceedings of the SIGCHI conference on Human factors in computing systems*. ACM, 585-592, 2003.
15. Dabrowski M., Acton T. "The Performance of Recommender Systems in Online Shopping: A User-centric Study." *Expert Systems with Applications*, 2013, 40(14): 5551-5562.
16. Anand Shanker Tewari; Kumari Priyanka, "Book recommendation system based on collaborative filtering and association rule mining for college students", *2014 International Conference on Contemporary Computing and Informatics (IC3I)*, 2014, pp. 135–138.
17. RA Achana, RS Hegadi, TN Manjunath, "A novel data security framework using E-MOD for big data", *2015 IEEE International WIE Conference on Electrical and Computer Engineering (WIECON-ECE)*, pp. 546-551, IEEE.
18. M Divya, TN Manjunath, RS Hegadi, "A study on developing analytical model for groundnut pest management using data mining techniques", *2014 International Conference on Computational Intelligence and Communication Networks*, pp. 691-696, IEEE.

13

Mining Heterogeneous Lung Cancer from Computer Tomography (CT) Scan with the Confusion Matrix

Denny Dominic* **and Krishnan Balachandran**

School of Computer Science and Engineering, Christ (Deemed to be University), Bengaluru, India

Abstract

Early detection of any sort of cancer, particularly lung cancer, which is one of the world's most lethal illnesses, can save many lives. Life expectancy can be improved and the degree of mortality reduced by adopting the early forecast. While there are different methods like X-ray and CT scans to detect lung cancer cells, CT images resulted as more favored. The 2D images are used for more accurate medical results, such as CT scans. The proposed approach here will address how to interpret the CT images for the Mining Heterogeneous Lung Cancer from Computer Tomography (CT) Scan with the Confusion Matrix. This research will explore how the image conversion can be achieved through different methods of image processing to obtain better results from CT images. The Confusion Matrix helps to estimate inequality in a picture pattern. After the evaluation of the processed images by Confusion Matrix, a final accuracy with a result of 93% is obtained.

Keywords: Computer tomography (CT), LBP features, preprocessing, processing, smoothening

13.1 Introduction

Mining Heterogeneous Lung Cancer from Computer Tomography (CT) Scan with the Confusion Matrix is a part of research work carried on

**Corresponding author*: denny.cmi@gmail.com

Kukatlapalli Pradeep Kumar, Aynur Unal, Vinay Jha Pillai, Hari Murthy, and M. Niranjanamurthy (eds.) Data Engineering and Data Science: Concepts and Applications, (317–334) © 2023 Scrivener Publishing LLC

to detect lung cancer in the early stage. Here, multiple parameters are taken into account like Image scans, Symptoms and Risk factors, thus to say, Heterogeneous data on lung cancer. Due to inadequate prognostic approaches that are now available, early diagnosis of lung cancer continues to be difficult. Cancer survival is higher if it is detected early and adequate care is administered [1]. The lymph moves through the lymph vessels that flow into the lymph nodes in the lungs and the middle of the chest. A CT scan for the detection of lung cancer is better than an X-ray. Smoking is said to be the primary cause of lung cancer. Therefore, in order to construct an intelligent computer-assisted diagnostic system, a framework must be introduced.

In order to improve the grading of CT images, this method must be quick and reliable. A typical attribute would not be adequate to effectively classify all classes [1]. The high-resolution chest computed tomography (CT) Scan is commonly accepted to evaluate lung cancer. In the early stages, the small cell seems to be very difficult to identify. The prediction of lung cancer in the earlier stage is a difficult task by all the means. Since lung cancer is one of the most dreaded illnesses, to save people's lives, early diagnosis is strongly encouraged. We can cut mortality rates and improve life expectancy accuracy by focusing on early detection in 13% of all new cancer diagnoses and 24% of all cancer deaths. Lung cancer may be detected using a variety of ways, including X-rays and CT imaging, although CT pictures are recommended. To acquire better outcomes on the same ailment, medical photographs are always preferable. The suggested approach will describe how Watershed segmentation may be used to the pre-processing of CT images in this paper.

The data are fed into deep learning algorithms in order to increase the accuracy of picture scans. The results are then utilized to verify the veracity of the results by the medical examiner.

The Median Filter, Gabor Filter, and Watershed segmentation are used for the majority of the pre-processing.

This research paper will explore how image alteration may be done to get better outcomes from CT scans using various image processing approaches. Smoothing of pictures with median filters, image enhancement, and picture segmentation will all be part of the proposed method's design. Precision in pre-processing is an imperative issue in order to establish a more useful approach for doctors to detect the type of lung nodules identified on CT scans.

The lung, airway, chest wall, artery, and pulmonary fissure are the key places where the pulmonary nodules appear, and it becomes impossible for clinicians to access the exact location of the cancerous nodules. The main aim of this research is to gain insight into the inequality in the Lung Cancer nodules which are depicted in the CT images, so that cancer cells can be predicted early using Confusion Matrix to find the image inequality.

This research work has segmented lungs images for classification with malignant nodule and without malignant nodules. This experiment was carried out Using MATLAB. It presented inequality among the Lung images which is evaluated by confusion matrix. To evaluate a model for improved outcomes, the confusion matrix and Receiver Operating Characteristic (ROC) curve were used. Confusion matrix defines the performance and goal groups. The findings were derived using a Confusion matrix with a True positive value and a True negative value, indicating that all of the selected values were appropriate and no throwaway qualities existed.

13.2 Literature Review

The literature review below would include major work carried on during this research and the reference for the same. The below narration would give an idea how the research gap was found out and the solution for the same. Noise filtering is the initial step in identifying lung cancer. The two basic processes of pre-processing are denoising and Weiner filtering. One of the challenges in picture processing is white noise [3]. Ilya Levner [2] has implemented topographic feature and object marker watershed segmentation. The important operation in the image processing is pixel sorting and segmentation. Pixel grouping is needed if Objects belonging to the same specified class are near to each other. The better outcome in a collection of non-overlapping areas, the unsupervised environment may use the watershed algorithm.

In terms of similar nodule characteristics of similarity, speculation and clarification, the definition of quantification has become a major concern [7–9]. Therefore, the need for quantitative features to be gained by image segmentation has become an urgent need. Anywhere in the lung, anywhere, a pulmonary nodule may be present. In the extraction of the attribute, the final product is explained as if the processed image has some normality or abnormality [4] and segmentation of the image is the basis of classification. Ginneken [4] has broken the lung image extract classification into either rule-based or pixel-based groups. Most categories [5, 6] prefer the rule-based approach, with distinct steps, adding the rule for achieving a particular result.

The main research in the image scan are region growth, Thresholding, edge detection, ridge detection, fitting of geometric models and functions, morphological operations, and dynamic programming.

Acronyms	Descriptions
CT	Computer Tomography
2D	Two Dimensional
3D	Three Dimensional
DICOM	Digital Imaging and Communication in Medicine
ROC	Receiver Operating Characteristic
PACS	Picture Archiving and Communication Systems are examples of these types of devices
CLAHE	Limited Adaptive Histogram Equalization
LIDC	Lung Image Database Consortium
ROI	Receiver operating characteristic
CNN	Convolution Neural Network
ZCA	Zero component analysis
LBP	Local Binary Pattern
LESH	Local Energy-based Shape Histogram
RMSE	root-mean-square error
TP	Real Positive
TN	Real Negative
FP	False Positives
FN	False Disadvantages

13.3 Methodology

13.3.1 Description of the Data

The CT Scan is an expanded variant of the X-ray in which an X-ray machine is attached to the device. The computer processes images taken from angles and distances and shows them in three-dimensional, cross-sectional (tomographic) and fragments. Bones, muscles, blood vessels and organs are seen clearly in this way. For diagnosis, treatment and medical

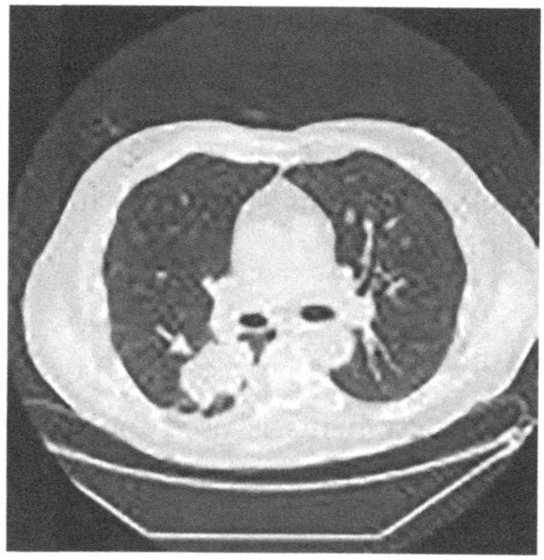

Figure 13.1 CT image.

development, image CT scanning is useful. In this study, CT images are collected as real-time data from different medical colleges and are standard for Digital Imaging and Communication in Medicine (DICOM).

The Digital Imaging and Medical Communication (DICOM) standard is for Information and associated data from medical imaging to be communicated and managed. DICOM is widely used by various manufacturers, such as Scanners, servers, workstations, printers, network gear, and Picture Archiving and Communication Systems are examples of these types of devices (PACS), to consolidate and distribute medical images. For example, the patient ID in the file must be included in the chest X-ray image file, so that the image is not inadvertently removed from that information. In order to define and view the file, this is the same way that image formats like JPEG embed tags shown in Figure 13.1.

13.3.2 Image Preprocessing

The images obtained from various medical colleges in India and from the LIDC (Lung image database consortium) were preprocessed. The images collected vary in their size and form, so they are resized to 128 * 128 pixels. The photos are transformed to grayscale pictures again. Image editing is carried out with the preprocessing and processing steps. With image smoothening, image enhancement and, finally, even segmentation, pre-processing is accomplished. Images of the lungs have a low contrast, making it difficult to spot nodules. That is why, for picture improvement, we used Contrast Limited Adaptive Histogram Equalization (CLAHE).

This approach separates photos into blocks, then calculates the histogram values for each block using a set number of gray levels. This approach separates photos into blocks, then calculates the histogram values for each block using a set number of gray levels. The LIDC (Lung Image Database Consortium) was employed in this study, which included 221 lung CT images from cancer patients.

13.3.3 Image Segmentation

The image with different segments is used for easy analysis and better understanding of any picture. They segment the picture into the regions or objects that make it up. The end consequence of segmentation is that whole images are filled by different segments or by a collection of contours derived from the image [5]. Segmenting grey matter, white matter, etc., is the primary use.

The contrast between the backdrop and the lung nodule is really high. As a result, obtaining ROI (Receiver operating characteristic) of lung nodules using threshold segmentation is quite difficult. CNN is used to eliminate the interference. The CNN (Convolution Neural Network) technique is used to segment images. Multiple perceptron layers, such as convolutional and Re-Lu layers, are present in a CNN. Normalized layers include dropout, convolutional, and pooling. For image combining, pooling and convolutional layers are commonly utilized in CNN architecture.

Mean pooling and maximum pooling are two actions performed by pooling layers. Mean-pooling calculates nearby pixels, whereas max pooling calculates the maximum number of feature points.

Size errors are reduced by mean pooling, while estimated errors are reduced by max pooling. For picture segmentation, CNN algorithms are utilized. It completes every CNN in this way.

13.3.4 Image Processing

Using different features, such as local processing of binary patterns, the preprocessed image is processed to improve the accuracy of lung scan that which will accurately distinguish cancer cells. The distribution of LBP in a similar surface picture among different regions can be extraordinary.

13.3.5 Zero Component Analysis (ZCA) Whitening

Whitening is a data transformation in such a way that its matrix of covariance is the matrix of identity. Whitening thus decorrelates characteristics.

It is utilized as a form of preprocessing. If you have *N* data points in *Rn*, then the matrix of covariance is *R n*n*n* This has estimation equation 13.1 as:

$$\sum jk^{(n)} = 1/(N+1)\sum(1=0)^n(x*ij - xj)(x*ik - xk) \quad (13.1)$$

where '*xj* denotes the jth component of the samples x estimated mean.

Any *W-Rn*n* matrix that satisfies the *WTW* = *C-1* condition whitens the results.

ZCA whitening is the *W* = *M-(1/2)* alternative.

13.3.6 Local Binary Pattern (LBP Feature)

The LBP system administrator calls a nearby example of a picture pixel, which is determined by comparing its dim value with that of its neighbors. To build an LBP image, the LBP rotation invariant is used. It evaluates every image pixel. The picture pixel is dubbed a close example by the LBP administrator, and its dim value is calculated by comparing it to that of its neighbors. To create the LBP picture, the rotation invariant of LBP is applied.

Every pixel in the picture is analyzed. The texture of a picture is represented using this way. The technique is used to depict the texture of the scene, as seen in Figure 13.2, also help estimate an image's local binary pattern distribution. The downside is a scarcity of local statistical texture data from the picture. This method ignores variations between small regions, but is used over the entire picture to calculate pattern distribution as shown in Figure 13.2.

Figure 13.2 LBP image after converting using LBP features.

13.3.7 LESH Vector

It helps to estimate the local binary pattern distribution of an image as well. A lack of local statistical texture data from the image is the downside. This approach ignores differences between small regions, but is used to quantify pattern distribution over the whole image as shown in Figure 13.3.

Owing to the barbaric lung environments, lung cancer detection devices are facing several problems. Similarity of the lung area's arterial width, ribs, bronchial veins, and bronchi, and a number of nodule types, such as cavities and nodules of ground glass [6]. These include the lung's cumulative functions, such as nodule dissection, extraction of features and classification [6]. Contributing to the performance of the device include the identification of lung abnormalities: device optimization using parameters for image acquisition and reconstruction, nodule location, size of data collection, and cross recognition [7].

13.3.8 Local Energy Map and Orientation Map

When designing an operation under the MAP (Maximum Priority) Estimation Probability Framework in combination with Markov Random Fields, energy is used to describe the measurement of 'information'.

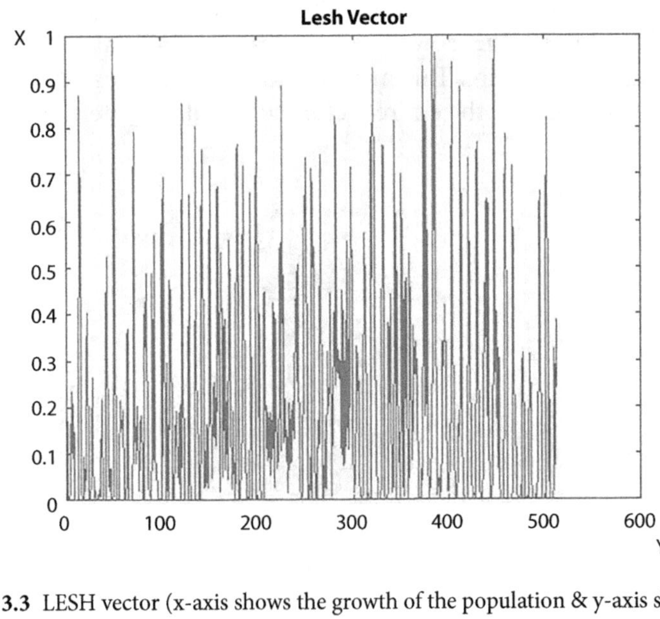

Figure 13.3 LESH vector (x-axis shows the growth of the population & y-axis shows the density).

Figure 13.4 Orientation map using LESH vector.

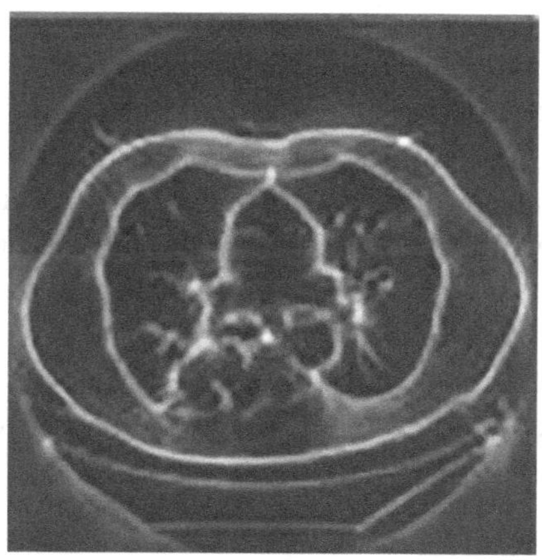

Figure 13.5 Local energy map using LESH vector.

Often there is a negative energy reduction measure and sometimes a positive energy reduction measure to increase it as shown in Figures 13.4 and 13.5.

13.3.9 Training with Deep Learning Methods

Convolutional neural networks (CNNs, or ConvNets) are important tools for deep learning, and they are particularly good at interpreting picture data. CNNs, for example, may be used to categorize photos. You may use a regression layer at the network's end to predict continuous data like angles and lengths. They blend the input data with the matching predicted output in the training data set. It gives their model "ground truth" data, which is normally created by a human or semi-automated process. The test data set provides information that will be used to test your model. This data, on the other hand, has no "anticipated" outcome. These data may be used to analyze how successfully the model is trained and the model properties during the machine learning testing phase. They use this model to train their model by mixing the inputs with the intended outputs during the testing phase. It creates a Convolutional neural network architecture which is used to anticipate the angles of rotated handwritten digits once it is trained. Optical character recognition can benefit from these predictions. The data collection includes synthetic pictures of handwritten numerals as well as the matching rotation angles (in degrees) for each picture. d*igitTrain4DArrayData* and *digitTest4DArrayData* are used to load the training and validation pictures as 4-D arrays. The rotation angles in degrees are outputs *YTrain* and *YValidation.* There are 5,000 photos in each of the training and validation data sets.

When training neural networks, it is important to ensure data is normalized throughout the process. Normalization aids in the stabilization and acceleration of gradient descent network training. If data is not scaled properly, the loss can become NaN, and network parameters can diverge during training. Rescaling data such that its range becomes [0,1] or such that it has a mean of zero and a standard deviation of one are two common methods of normalizing data.

The following data can be normalized.

Data should be entered as, before you feed the predictors into the network, normalize them.

13.4 Result

The result can be seen from the image below in Figure 13.6. If you utilize batch normalization layers at the end of the network to normalize the layer outputs, the network's predictions will be normalized before training begins. If the result differs significantly from the expectations, network training may not converge. If the answer is not scaled properly, try normalizing it and seeing if network training helps. The outputs of the layers. A batch normalization layer may be used to normalize the outputs of each convolutional and fully connected layer. Figure 13.6 shows the Training Progress with final outcome after Training the Processed data.

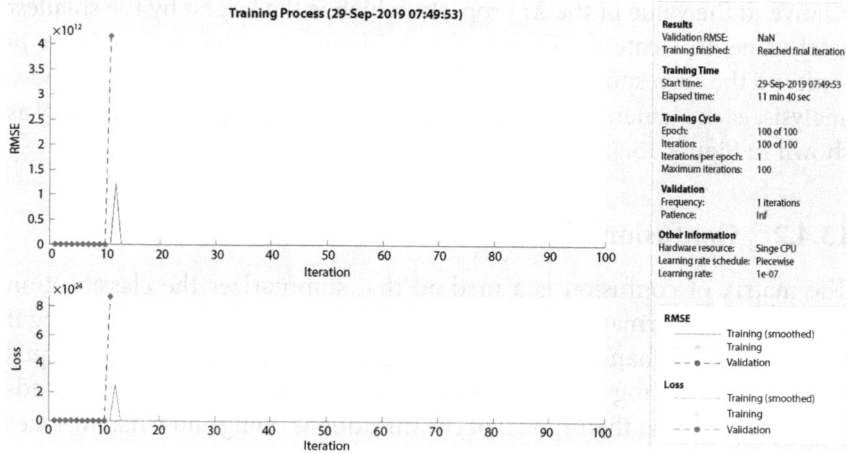

Figure 13.6 Training progress.

a. *Calculate the following to assess the model's performance:*
The proportion of forecasts that fall within a reasonable error range. The expected and actual angles of rotation's root-mean-square error (RMSE).
b. *Calculate the difference between the projected and actual rotation angles.*
From the true angles, calculate the number of forecasts within an acceptable error range. Set the temperature to 10 degrees. Calculate the % of forecasts that fall inside this range. Make a scatter plot of your guesses.
c. *Compare the expected and actual values on a graph.*
To align the numerals and show them together, you can utilize Image Processing Toolbox routines. Use imrotate to rotate 49 sample digits according to their projected degrees of rotation.
d. *Show the original digits with their rotations rectified.*
Montage is a technique that can be used.
e. *To calculate the discrepancies between the expected and actual angles of rotation.*

Use the root-mean-square error (RMSE).

13.4.1 Lorenz Curve

A graphical representation of the cumulative distribution of a given characteristic in a population is the Lorentz curve. The elements are sorted relative to the value of the *Xi* property, which is the largest by the smallest asset value, to create the Lorentz curve. Then, according to its relative *pi* rank and the corresponding cumulative *qi* percentage of the attribute under analysis, each element is graphed. For each, more specifically, I=1,2,3..*N* as shown in Figure 13.7.

13.4.2 Confusion Matrix

The matrix of confusion is a method that summarizes the classification algorithm's performance. The consistency of classification is misleading if they have more than two classes in each class, and they have an unequal number. Calculating the uncertainty matrix provides a clear understanding of exactly what their model of classification is doing and what mistakes it makes. Figure 13.8 shows what are the outcome of the training progress. It gives us:

Real Positive (TP): they have got the illness

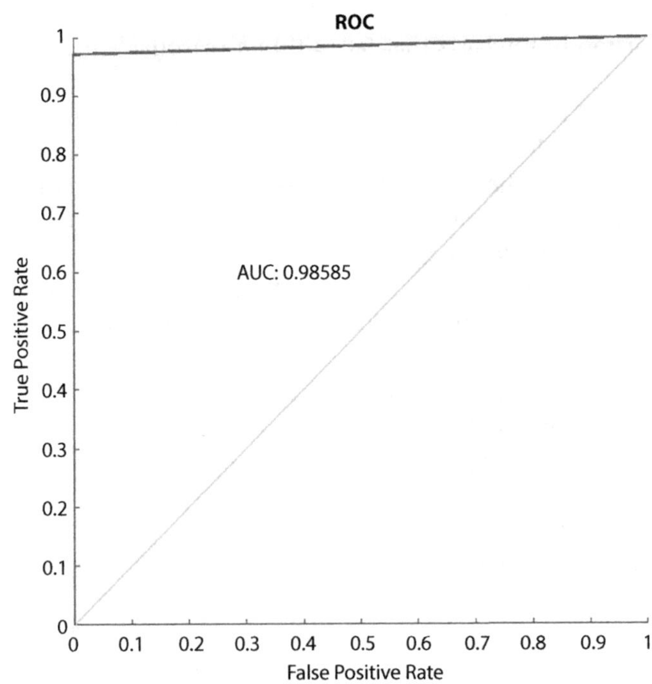

Figure 13.7 ROC curve.

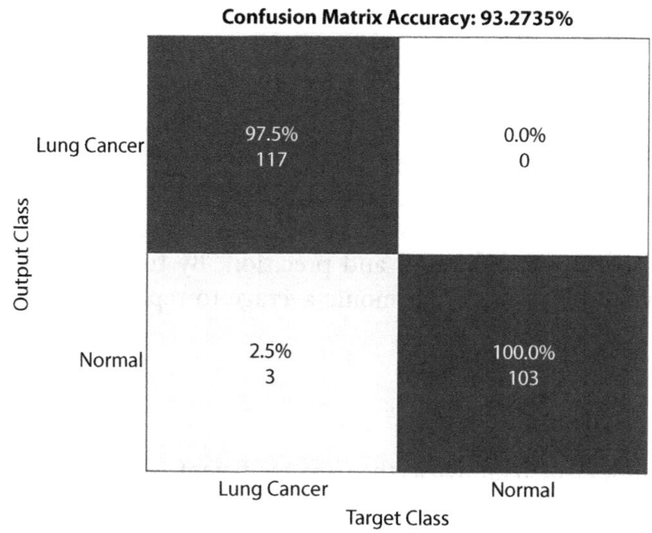

Figure 13.8 Confusion matrix.

Real Negative (TN): they do not have the disease.
False Positives (FP): yes, but they are not really a disease.
False Disadvantages (FN): no, but they actually have the disease as shown in Figure 13.8.

$$\textbf{Recall} = \frac{TP}{FN + TP} \tag{13.2}$$

It had an estimate of how precise it was from all the positive classes. This needs to be as high as possible.

$$\textbf{Precision} = \frac{TP}{FP + TP} \tag{13.3}$$

They are right on how positive the positives really are in all the positive groups.

Accuracy
Of all groups, in this situation, they have predicted exactly how it will be 4/7. This needs to be as high as possible.

F Measurement

$$\mathbf{F - Measure} = \frac{(2 * \mathbf{Recall}) * \mathbf{Precision}}{\mathbf{Precision} + \mathbf{Recall}} \tag{13.4}$$

With low precision and high recall, it is difficult to compare the two models. So it makes use of F-scores to compare them. The F-score helps to simultaneously assess recall and precision. By further punishing the extreme values, it uses the harmonic average to replace the arithmetical mean.

13.4.3 Gini Coefficient

The Gini index, often known as the Gini coefficient, is a measure of income distribution. It is frequently used to measure income distribution or, less usually, wealth distribution among a population as a measure of economic inequality. The Gini coefficient is used to find lung nodules that may be scanned using CT (Computerized Tomography). Over the years, the study of disparities in the distribution of a particular variable in a population has gotten a lot of attention. Because it is not the goal of this paper to enumerate all of these metrics, the suggested study focuses on the Lorentz curve and its coefficient, which is the Gini coefficient, which is more relevant to this paper. They analyzed the voxel density concentration in the pulmonary nodules using the Lorentz curve and the Gini coefficient in this study. A correlation-based criterion is used to estimate the function's ability. It uses decision tree splitting rules to look at how much impurity decreases when a function is chosen. The capacity of the feature to distinguish possible photos from superior outcomes is referred to as impurity. Figure 13.9 shows the reduced inequality/imbalance of the image as the value (0.19257) is closer to zero, which means this particular image does not carry a Lung Cancer. Figure 13.10 represents the Combined gini coefficient after processing all the images.

Because it adheres to the following criteria, the Gini coefficient is one of the most often used indices of economic inequality:

a. Confidentiality
   The identity of high- and low-income individuals in a population is not revealed by the coefficient.
b. Independency scale
   The Gini coefficient is calculated independently of the size of the economy, how it is assessed, or how affluent a country

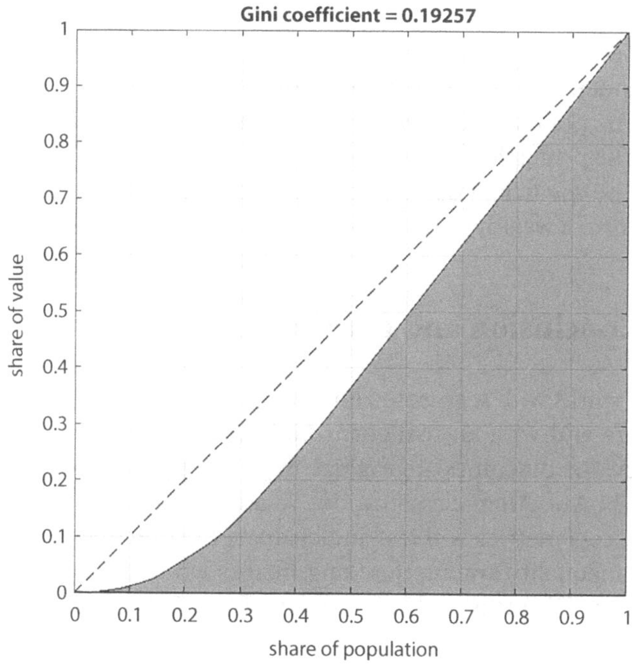

Figure 13.9 Gini coefficient.

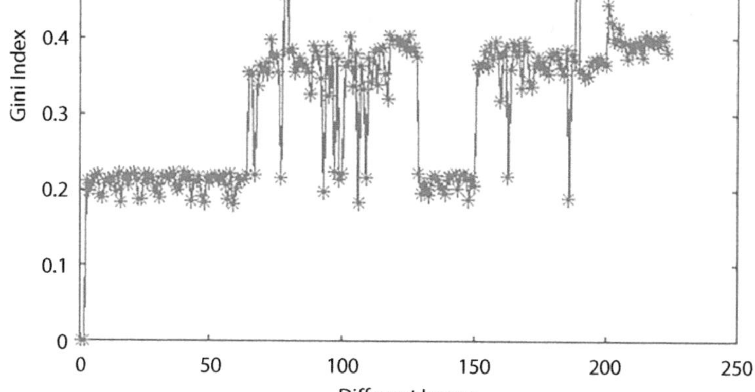

Figure 13.10 Combined gini coefficient after processing all the images.

is. Due to comparable income distribution, both rich and poor nations may have the same coefficient.

c. Independence of the population
The population size has no bearing on the coefficient.

d. The principle of transfer
The coefficient depicts scenarios in which money is moved from a wealthy person to a poor person.

13.5 Conclusion and Future Scope

This study works with segmented lung images for classification with malignant nodule and without malignant nodules. By using classification and consistency, the outcomes are evaluated. This experiment was carried out using MATLAB. After classification, X and Y-Validation are executed. Finally, success metrics will test and evaluate the outcomes. This research presented inequality among the Lung images which is evaluated by confusion matrix. We measured an accuracy of 93.2735% with the aid of that confusion matrix. To evaluate a model for improved outcomes, the confusion matrix and Receiver Operating Characteristic (ROC) curve were utilized.

Confusion-matrix defines the performance and goal groups. The false positive rate was 3.1%, indicating that only a few of the specified parameters were above 50% because biomarker data was insufficient due to the fact that most hospitals did not do this test and hence could not obtain sufficient data. The percentage of false negatives was 100%, indicating that all of the specified features were within the parameter. The ROC curve is used to calculate the false positive and true positive rates.

The findings were derived using a Confusion matrix with a True positive value of 96.9% and a True negative value of 0%, indicating that all of the selected values were appropriate and no throwaway qualities existed. Finally, we have an accuracy of 98.3051%, which is better than the literature that was studied.

References

1. A. Cruzroa, J.C. Caicedo, F.A. Gonzalez, Visual pattern mining in histology image collections using bag of features. *Artificial Intelligence in Medicine*. vol. 52, no. 2, pp. 91–106, 2011.

2. Ilya Levner, Hong Zhangm, Classification driven Watershed segmentation. *IEEE Transactions on Image Processing*. vol. 16. No. 5. May 2007.
3. Anita, Chaudhary, Sonit, Sukhraj, Singh, Lung Cancer Detection on CT Images Using Image Processing. *International Transaction on Computing Sciences*. vol. 4, 2012.
4. B.V. Ginneken, B.M, Romenyand, M.A, Viergever, Computer-aided diagnosis in chestradiography: a survey. *IEEE, Transactions on Medical Imaging*. vol. 20, no. 12. Dec. 2001.
5. Disha, Sharma, Gagandeep, Jindal, Identifying Lung Cancer Using Image Processing Techniques. *International Conference on Computational Techniques and Artificial Intelligence (ICCTAI'2011)*, vol. 17, pp. 872-880, 2011.
6. Nguyen, H.T., *et al.* Water snakes: Energy-Driven Watershed Segmentation. *IEEE Transactions on Pattern Analysis and Machine Intelligence*. vol. 25, no. 3, pp. 330-342. March 2003.
7. Farag, A., Ali, A., Graham, J., Elshazly, S., Falk, R., Evaluation of geometric feature descriptors for detection and classification of lung nodules in low dose CT scans of the chest. Paper presented at the Biomedical Imaging: From Nano to Macro. *2011 IEEE International Symposium, Chicago. IL. USA*. March 30 to April 2, 2011.
8. Lin, P-L., Huang, P-W., Lee, C-H., Wu, M-T., Automatic classification for solitary pulmonary nodule in CT image by fractal analysis based on fractional Brownian motion model. *Pattern Recognition*. 2013; 46 (12): 3279–3287.
9. Farag, A., Elhabian, S., Graham, J., Farag, A., Falk, R., Toward precise pulmonary nodule descriptors for nodule type classification. *Medical Image Computing and Computer-Assisted Intervention. MICCAI 2010*. Berlin. Germany. Springer. 2010.

14

ML Algorithms and Their Approach on COVID-19 Data Analysis

Kambaluru Ashok[1], **Penumalli Anvesh Reddy**[2*] **and Kukatlapalli Pradeep Kumar**[2]

[1]*Department of CS, Kyungpook National University, Daegu, South Korea*
[2]*Department of CSE, Christ University, Bangalore, Karnataka, India*

Abstract
This chapter begins with characterizing Supervised Learning and Unsupervised learning and investigates Machine Learning algorithms in every one of the sub domains of Regression, Classification, Clustering, and so forth. It also talks about the engineering of calculations like Linear Regression, Logistic Regression, K-Means, K Nearest Neighbors, Hierarchical, DB Scan, Decision Tree, Random Forest Regression, and Random Forest classifier. Utilization of every algorithm to investigate the dataset will be displayed by carrying out it on renowned dataset model, and output of each piece of code is displayed with their preview. This section likewise takes care of the issue of predicting the future number of COVID-19 cases and the precision behind each model or algorithm is shown and investigated utilizing different measurements dependent on situation or issue articulation, for example, either issue is on forecast or order. This chapter does not focus on the solution of COVID-19 data analysis or expectation, rather it will be followed and will task different models dependent on need with conclusive target being clear comprehension of the Machine Learning algorithms and its execution in Python.

***Keywords*:** Machine learning, covid analysis, covid data visualization, supervised learning, un-supervised learning, regression, classification

**Corresponding author*: penumalli.anvesh@btech.christuniversity.in

Kukatlapalli Pradeep Kumar, Aynur Unal, Vinay Jha Pillai, Hari Murthy, and M. Niranjanamurthy (eds.) Data Engineering and Data Science: Concepts and Applications, (335–350) © 2023 Scrivener Publishing LLC

14.1 Introduction

Machine Learning has arisen as a valuable apparatus for modelling issues that are generally hard to detail precisely [1]. Old-style computer programs are explicitly programmed by hand to play out an assignment. With Machine Learning, some portion of the human contribution is supplanted by a learning algorithm. As the accessibility of computational limits and information has expanded, Machine Learning has gotten even more practical throughout the years, to the point of being almost universal [2].

The main objective of Machine Learning is to provide actionable insights from an input dataset. "Actionable insights", in the sense, arrive at a decision, denoted as knowledge in a Machine Learning context.

In general, we are given a function that gives output (observed data) for the given input. The question is how to build (model) function from the given input and observed data. Once a function is modeled, we can use it for predicting output for future inputs. This is what typical machine learning algorithms address. Modeling is the process of approximating a function's parameters from provided input and output data to create a linear or nonlinear function. Curve fitting on a dataset is another name for it. Visualize a "mathematical function" that can fit on data points whenever the term "model" is used. This model helps to make a choice or comprehend a data pattern without having to recode everything. So, Machine Learning is the process of converting a dataset into a basic or complicated mathematical function that mimics the dataset's behavior. For a given input dataset, more than one model may be possible. We must experiment with various functions by fitting the dataset that results in maximum accuracy. Machine learning algorithms are built utilizing statistics, probability, calculus, and linear algebra, which assist in determining which function perfectly matches a dataset.

To understand the different types of Machine Learning model or algorithms, it is important to understand the types of datasets beforehand [3].

14.2 DataSet

Dataset is a collection of data which can have a number of features, also called variables in math, attributes in a database, and just columns in a table as shown below at Table 14.1. Similarly, it can have a number of records, also called as instance in programming, sample in probability, rows in a table. Labeled and unlabeled are the two types of datasets that are passed to Machine Learning algorithms as inputs.

Table 14.1 Diabetes symptoms and condition.

Glucose concentration test	**Blood pressure (mm Hg)**	**SkinFold thickness (mm)**	**Body mass index**	**Diabetes pedigree function**	**Age**	**Diabetes?**
148	72	35	33	0.627	50	yes
85	66	29	26.6	0.351	31	no

14.2.1 Labeled Datasets

- These datasets consist of all the features including the output feature at the very last. The values of the output feature are also called Observed values.
- It can be a numeric value (continuous) or a discrete number of categorical values (class labels). Example: Diabetes Feature is the Output Feature.

14.2.2 Unlabelled Datasets

- Unlabelled Datasets do not contain Output features in the dataset. It is simply a labeled Dataset without Output feature. Example: Without Diabetes Feature

14.2.3 COVID-19 Data

The dataset we use here is a labeled dataset which has 32 columns where in the last column total cases are dependent variables and which need to be predicted. The columns of dataset are [4]

(['total_cases','new_cases','total_deaths', 'new_deaths',
'total_cases_per_million','new_cases_per_million','date','total_deaths_per_million',
'new_deaths_per_million', 'total_tests','loc',
'total_tests_per_thousand', 'new_tests_per_thousand','iso_code',
'new_tests_smoothed', 'new_tests_smoothed_per_thousand', 'tests_units',
'stringency_index', 'population', 'population_density', 'median_age',
'aged_65_older', 'aged_70_older', 'gdp_per_capita', 'extreme_poverty',
'cvd_death_rate', 'diabetes_prevalence', 'female_smokers',
'new_tests','male_smokers', 'handwashing_facilities', 'hospital_beds_per_100k'])

This has the information of the total number of cases chronologically in various countries and we use the data of India which can be filtered using Python to perform a machine learning algorithm on it.

14.3 Types of Machine Learning Algorithms

Machine Learning Algorithms are categorized into four types based on the dataset they work on. Supervised learning, Unsupervised Learning, Semi-Supervised Learning, Reinforcement Learning are those four types as shown in Figure 14.1.

14.3.1 Supervised Learning

Supervised Learning is where an algorithm learns from the input dataset which has output data for each record in the data. The output feature of the dataset helps in building a model that can be made use of in predicting output for future inputs. These algorithms are Trained and Tested, where:

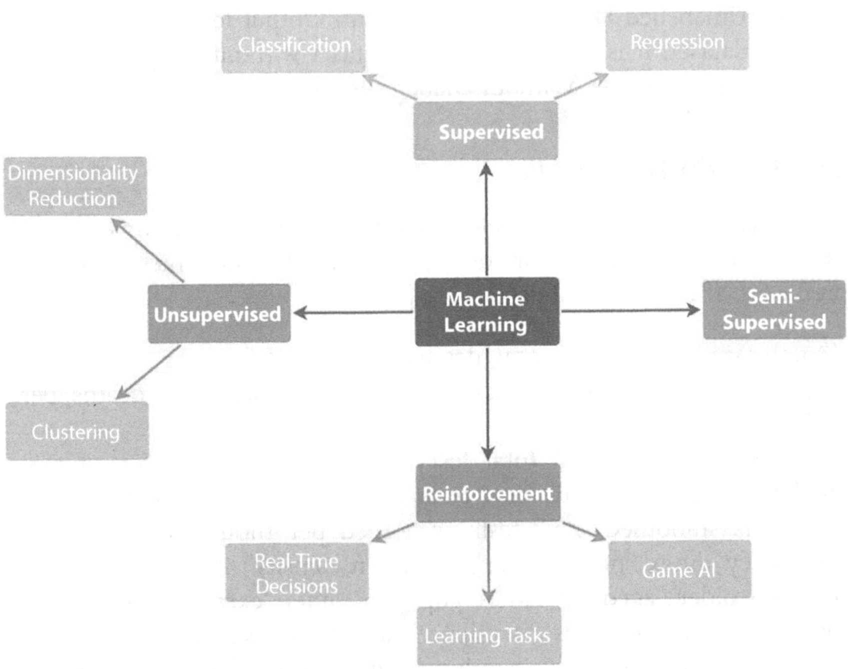

Figure 14.1 Machine learning categories.

Table 14.2 Diabetes symptoms.

Glucose concentration test	Blood pressure (mm Hg)	SkinFold thickness (mm)	Body mass index	Diabetes pedigree function	Age
148	72	35	33	0.627	50
85	66	29	26.6	0.351	31

- Training process involves building a model or mathematical function by approximating the values of the variables in the given input dataset as shown above in Table 14.2 for characteristics in diabetis prediction.
- Testing process is where the built model is evaluated and corrected.

The maximum efficiency of these algorithms is achieved by tuning the values of variables to fit the function on data.

Regression and Classification are the two types of Supervised Learning.

Regression – for predicting numerical value. Example: simple linear regression, gradient descent, regularization, polynomial regression.

a. Regression

Linear Regression [https://towardsdatascience.com/] is the first and basic model which gives an easy understanding of how algorithms work on relations between two data variables. Linear Regression is an algorithm which determines the statistical relation between variables which means dependent variable (y) is not accurately measured but it is a deterministic value obtained using independent variable (x). The relation between y and x is described as

$$Y = m(X) + c \tag{14.1}$$

Y is dependent variable and X is an independent variable that helps in predicting value of Y. m is slope of line which if positive indicates increase in X increase Y and negative value indicates decrease in X increase Y and c is constant which are chosen by algorithm during model training in such a way that error is minimized.

Squaring is done so that negative errors do not cancel out positive errors. This error is also termed as residue and analyzing the residual value pattern helps in increasing model performance.

b. Decision Tree Regressor [https://towardsdatascience.com/]

This is a famous and mostly used algorithm as it gives a better accuracy for most of the use cases and this works as both regression and classifier having basic idea as splitting the data given by continuously optimizing the objective function called information gain given as follows.

In the above function, N is total number of samples at specified node and D is dataset of parent and child, f is a feature to perform classification to iterate through tree and I is impurity measure.

This splits the entire data being stored in root node and splits in a way where leaf nodes (represent outcome) are finally reached by following if or else questions (Interior nodes) that represent the features of data set just like a tree structure and average value of category present at leaf node that is reached for test data after iterating through tree by asking splitting conditions is the final predicted value.

c. Random Forest Regressor [https://towardsdatascience.com/]

This is an extension to the decision tree regression algorithm implementing the idea of having more number of trees than a single decision tree, which is why it is called a forest of trees and random since it takes a random number of trees as shown below in Figure 14.2. It has the limitation of over-fitting problem of decision tree algorithm where in training data gets exactly fitted to single tree resulting in very high train accuracy but not good at test accuracy, and a second limitation being non-interpretability.

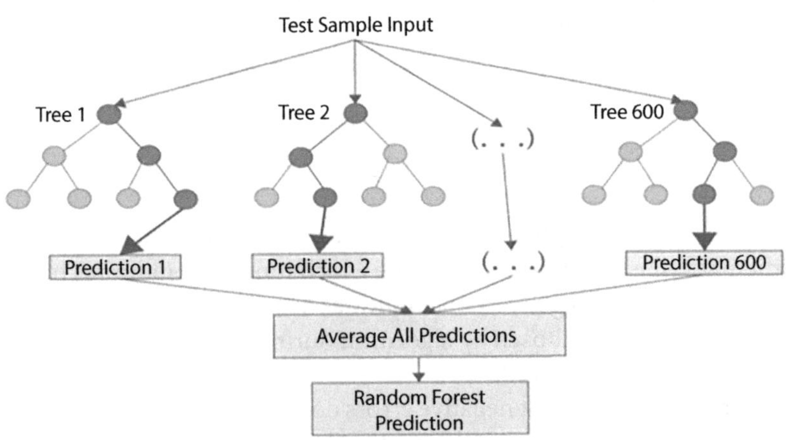

Figure 14.2 Random forest regressor.

d. XGBoost Regressor [https://towardsdatascience.com/]
XGBoost is one of those popularly used algorithms which is an extension to gradient boosting technique (this implements gradient descent technique to boost the model performance from previous model which is called boosting). The main advantage of xgboost is it removes problem of over-fitting by using regularization by penalizing models using the lasso and ridge regression and handles missing values on its own. It also uses parallel processing to run faster, and it has other advantages, such as faster tree pruning using depth first search where it uses max depth instead of negative loss criteria to stop splitting of nodes, and out of core computing ability since each thread is allocated with internal buffer to store stats, analysis.

e. Approach on Covid Dataset [https://towardsdatascience.com/]
Figure 14.3 is the visualization of relation between date and total cases reported in India. We can see some relation being present here where in some part of the graph has the linear relation between date and number_of_cases. When we build the Linear regression model on it, the mean square error resulted was 523149269830.4031 and mean squared error of xgboost was 528912311889.3791 and mean squared error of decision tree was 521839699774.9545 and random forest gave a mean squared error of 521521889450.35724 as shown below in Figure 14.3.

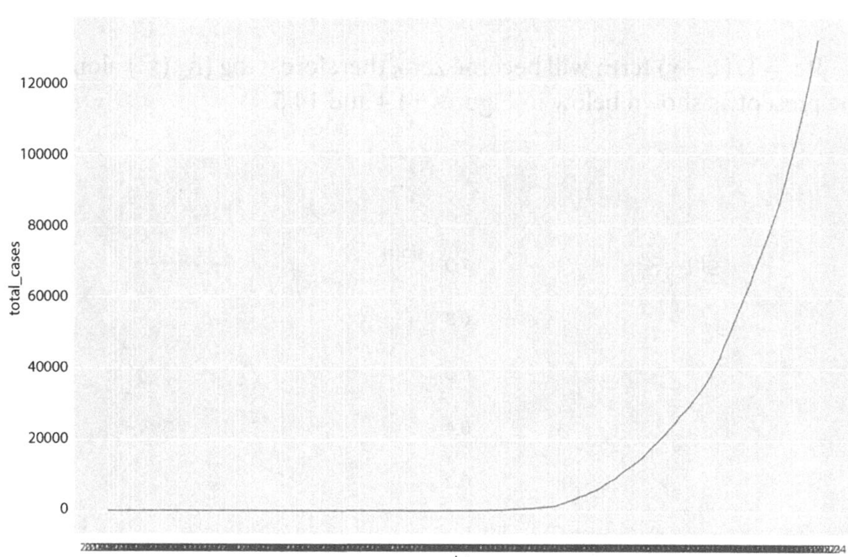

Figure 14.3 Visualization of relation between date and total number of cases.

Classification [https://towardsdatascience.com/] [5]
Classification or categorization – for predicting class labels. Example: decision tree, random forest, ANN, deep learning, SVM, KNN.

a. Logistic Regression for Classification: [https://towardsdatascience.com/]

This is one of those popular regression algorithms used and this mainly helps in knowing the probability or confidence of predicted value being actual value. In the case of logistic classification we simply use this probability set to a threshold value to decide whether the output is 0 or 1 where 0 and 1 are just two different categories of classification; for example, o being mail to classify as spam and 1 being not spam or even alternative.

In the Logistic Regression we use activation functions based on the type of classification we need (Binomial (2 classes), Multinomial (more than 2 classes)). In the case of binomial we use the sigmoid activation function which gives output always in the range of 0 and 1. In multinomial we use the soft-max function.

$$\text{Cost}(h_\Theta(x), \text{Y (actual)}) = -\log(h_\Theta(x)) \text{ if } y = 1$$

$$-\log(1 - h_\Theta(x)) \text{ if } y = 0 \qquad (14.2)$$

$$\text{Cost}(h_\Theta(x), y) = -y\log(h_\Theta(x)) - (1 - y)\log(1 - h_\Theta(x))$$

If $y = 1$, $(1 - y)$ term will become zero, therefore $-\log(h_\Theta(x))$ alone will be present as shown below in Figures 14.4 and 14.5.

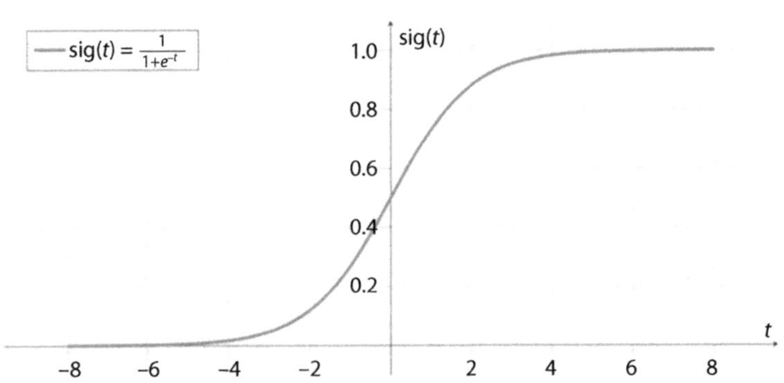

Figure 14.4 Sigmoid.

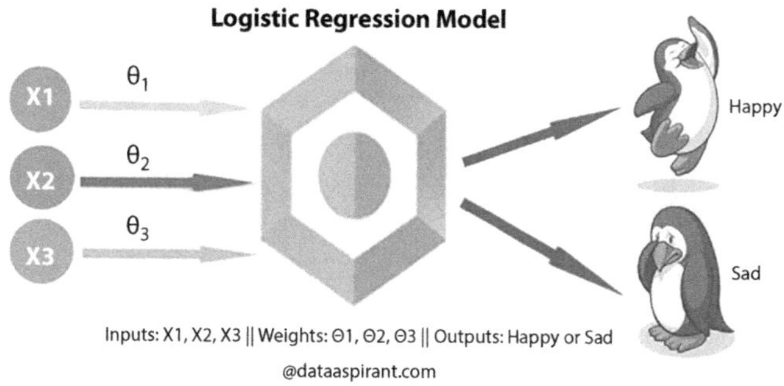

Figure 14.5 Linear regression model.

If y = 0, (y) term will become zero, therefore – log (1 – h_Θ (x)) alone will be present

The reason for not using mean squared error as cost function for logistic unlike used for linear regression is that. We know gradient descent converges into global minima (minimization results in maximum likelihood) only if its convex and logistic results convex gradient descent if used with mean squared error.

b. Decision tree classifier [https://towardsdatascience.com/]

Decision tree classifier is similar to decision tree regression algorithm which has root node with all data input, branch nodes to represent the decision rules and leaf nodes to give the output which do not have any more branches. Tree is built here using CART (classify and regression tree).

It begins with a tree at root node that has entire data and divides into subset or subtress by selecting the most important feature using Attribute Selection Measure technique and this process continues till there is no possible further division marking node as leaf node as shown below in Figure 14.6.

Attribute selection technique uses two approaches where in information gain, value of each feature is calculated for its relation to target class and one that has high IG value will be split out first.

$$\text{Information Gain} = \text{Entropy(S)- [(Weighted Avg)*Entropy (each feature)} \quad (14.3)$$

Entropy(s)= -P(yes)log2 P(yes)- P(no) log2 P(no)

S= Total number of samples, P (yes) = probability of yes, P (no) = probability of no

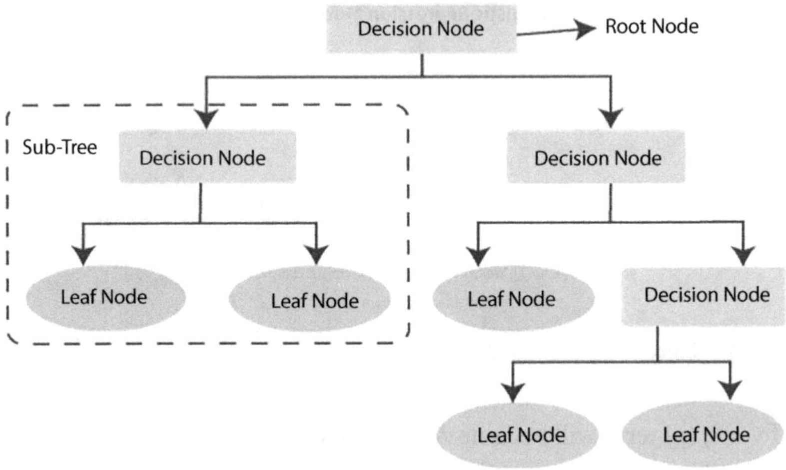

Figure 14.6 Decision tree classifier.

The other approach is Gini index which is measure of impurity used while creating CART using Gini Index = 1- ΣjPj2 which is used by cart algorithm to create binary splits.

The main disadvantage of decision tree classifier is over-fitting which is solved by random forest classifier.

c. Random Forest Classifier

Random Forest Classifier is similar to decision tree algorithm but instead of averaging the final outputs of various trees we just use maximum voting technique, where we take out the category that is specified by the most number of trees as shown in Figure 14.7. It solves the problem of over-fitting of decision tree and also handles well with data of more features. The only limitation is it does not perform well in regression task and there might be cases of other algorithms giving good accuracy.

d. Naïve-Bayes Classifier

Naïve-Bayes are classifier algorithms. The basic idea is implementing Bayes concept of probability to classify the category and idea of naïve which says one feature is independent of other in deciding the final target class. The main advantage of this is it can perform well for high dimensional data and also builds models faster. The main application of this is seen in text clarification as text classification data has more features and there might be multi-class classification; the disadvantage is that no consideration of correlation between features is considered.

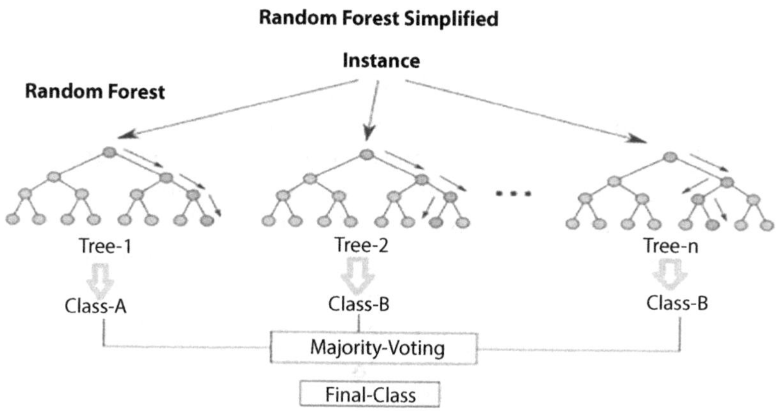

Figure 14.7 Random forest classifier.

$$P(A|B) = \frac{P(B|A)P(A)}{P(B)} \tag{14.4}$$

Above is the formula used to determine probability of event A on event B using likelihood probability P(B/A). This is a simple algorithm which has steps of making a table that has probability of each feature of data which is calculated using posterior probability and then class with more probability is output for given input feature.

14.3.2 Unsupervised Learning

Unlike Supervised Learning, these algorithms explore the unlabeled datasets and try to find the patterns in the data like cluster, association rule and dimensionality reduction. The absence of the output feature is the reason these algorithms are under Unsupervised Learning. These models are not trained and tested, but extract patterns which are also called Descriptive Modeling, Data Mining.

Clustering – grouping data points. Example: k-means, hierarchical, fuzzy c-means [6].

a. K-Means Clustering

K-means is a famous and popularly used unsupervised learning iterative algorithm used for clustering the unlabeled data points into needed numbers of clusters ensuring no data will be belonging to two clusters. It employs the technique of minimizing the sum distance between centroid of clusters and data points around it. Steps of algorithm include selecting k number of where k is number of clusters) centroid points at random position. It assigns each data point to nearest centroid and iteratively calculates the variance of each clusters by totaling the distance between data points and centroid, and reassigns the centroid position to ensure variance is minimal and stops till there is no repositioning of centroid.

The challenge with this is choosing optimal k value. This is solved by the elbow method which starts trailing from 1 and to specified count n and calculates wcss called within cluster sum of squares for each k value, and gives the graph where we consider the start point or bend from which is there is no significant drop of wcss and choose it as k.

b. Hierarchical Clustering

This removes the need of deciding the number of clusters and also it does not have an assumption of making similar size clusters as k means have. It is more like a tree structure called a dendrogram. This has a top-down approach called divisive where in entire data is assumed to be in single cluster and we split till we reach optimal number of clusters. Another approach agglomerative is called bottom-up approach, wherein we start with considering each data point as single cluster and build up till we reach one cluster by combing nearest clusters and calculating the linkage distance using metric given (e.g., Euclidean). And we the dendrogram plot and use the plot to know optimal number of clusters as shown below.

We see the distance at each cluster formation and can clearly visualize that maximum distance between clusters is present in 4th case and so number of clusters is equal to 5 since there are total of 5 vertical lines that are passing through horizontal line that is drawn for maximum distance.

c. Approach of Clustering on COVID-19 Data

Clustering is just about grouping similar data. The best approach of clustering on COVID-19 is grouping the articles written on COVID based on the keywords present in the article. This helps research teams find the articles with the same theme and study for further research rather than manually finding the related articles, which consumes more time since thousands of articles and papers are written every day.

The limitation here is dealing with multiple languages and the long time for training since there are many words to classify in many articles, making computation costly. This can be avoided by dimensionality reduction wherein the abstract of each paper can be significantly focused.

14.3.3 Semi-Supervised Learning

Semi-Supervised Learning is the middle ground between Supervised Learning and Unsupervised Learning. These algorithms work on input data that contain both the labeled and unlabeled records. This type of learning model is difficult to build since it is neither supervised nor unsupervised learning, yet algorithms can effectively use a mixture of labeled and untellable data. Similar to Supervised Learning, these algorithms train the model with the available output records in the data and also work on the rest of the unlabeled records to explore patterns like Unsupervised Learning.

14.3.4 Reinforcement Learning

This type of learning is basically independent of input dataset. It does not have an input dataset for training to build the model but it sequentially learns from the dynamic input environment. It is roughly trained using a reward-based method, but it is not tested because there is no output feature, which sometimes makes this type of learning come under Semi-Supervised. This is a feedback-based learning where it learns by hit and trial to make a decision in an unknown environment based on the previous results obtained through hit and trial.

Applications in different sectors:

- Robotics – robot navigation, robo-soccer, walking, juggling, etc.
- Adaptive control – factory processes, telecommunication, helicopter pilot, etc.
- Game playing – tic-tac-toe, chess, maze, PubG, counter strike, etc.
- Chemistry – optimizing the chemical reactions.
- Business – strategy planning.
- Manufacturing – deep reinforcement learning to pick goods and put in some containers.
- Finance sector – evaluating trading strategies.

In these applications, pre-built models are not effective as input data depends on other players in the game [7].

14.4 Conclusion

This chapter provides an analysis of the various Machine Learning Algorithms used for feature extraction and classification of COVID-19 data obtained from different reliable sources. All the Machine Learning algorithms discussed in this chapter have various applications based on the type of data they use and the procedure they follow. The Supervised Learning will predict future performance or outputs based on the input labeled dataset, whereas the Unsupervised Learning will traverse through the data to find a hidden pattern or a logic in the provided input unlabeled dataset. The Semi-Supervised Learning will try to predict the missing output feature values and also further output values based on the respective inputs after training with the limited labeled part of the dataset, mining through the rest. Reinforcement Learning is an imitation of Human behavior. It works in an unknown environment without any input data, trying to figure out the situation by hit and trial, and learning through the obtained results. The adaptation of these algorithms in different sectors have been very effective and creative in finding better and efficient solutions for the problems being faced now and those that will be faced in the future. As challenging as it is to apply machine learning algorithms to massive amounts of data being raised, the continuous research on the respective fields is providing new and innovative solutions to meet the requirements for the processing of these huge chunks of data.

References

1. V. Alcacer, V. Cruz-Machado, Scanning the Industry 4.0: A Literature Review on Technologies for Manufacturing Systems (2019), *Engineering Science and Technology, an International Journal*, vol. 22, issue 3, pp. 899-919, https://doi.org/10.1016/j.jestch.2019.01.006
2. Muhammad, L. J., Islam, M. M., Usman, S. S., & Ayon, S. I. (2020). Predictive data mining models for novel coronavirus (COVID-19) infected patients' recovery. *SN Computer Science, 1*(4), 1-7.
3. Yeşilkanat, C. M. (2020). Spatio-temporal estimation of the daily cases of COVID-19 in worldwide using random forest machine learning algorithm. *Chaos, Solitons & Fractals, 140*, 110210.
4. Arpaci, I., Alshehabi, S., Al-Emran, M., Khasawneh, M., Mahariq, I., Abdeljawad, T., &Hassanien, A. E. (2020). Analysis of twitter data using evolutionary clustering during the COVID-19 pandemic. *Computers, Materials & Continua, 65*(1), 193-204.

5. Öztürk, Ş., Özkaya, U., & Barstuğan, M. (2021). Classification of Coronavirus (COVID-19) from X-ray and CT images using shrunken features. *International Journal of Imaging Systems and Technology*, 31(1), 5-15.
6. Zarikas, V., Poulopoulos, S. G., Gareiou, Z., & Zervas, E. (2020). Clustering analysis of countries using the COVID-19 cases dataset. *Data in Brief, 31*, 105787.
7. Bellandi, V., Ceravolo, P., Maghool, S., & Siccardi, S. (2020, December). A Comparative Study of Clustering Techniques Applied on Covid-19 Scientific Literature. In *2020 7th International Conference on Internet of Things: Systems, Management and Security (IOTSMS)* (pp. 1-8). IEEE.

15

Analysis and Design for the Early Stage Detection of Lung Diseases Using Machine Learning Algorithms

Sindhu Madhuri*, Mahesh T. R., Vivek V., Shashikala H. K. and C. Saravanan

Department of Computer Science and Engineering, Faculty of Engineering and Technology, JAIN (Deemed-to-be University), Bangalore, Karnataka, India

Abstract

Machine learning (ML) is a fundamental method by creating complex, automated, and optimal methods by analyzing high-dimensional and multi-modal biomedical data. In medical systems, ML has importance to solve many health issues. This helps to diagnose infections faster and more effectively if they are identified earlier. This will save many lives and reduce the strain on the system. Computed Tomography (CT) photographs of the lungs include a variety of structures that are essential for diagnosing and analyzing pulmonary disease. Lung diseases like tuberculosis, pneumonia, repository diseases and lung cancer are all serious health risks. At present identification of lung cancers, lung diseases, pulmonary diseases, asthma, tuberculosis problems are major health issues and increasing rapidly. Many health issues are emerged in human body, lung diseases are most common and complex diseases leading to death in early stages. Lung diseases commonly occur due to loss of ability to breathe and lead to death due to late diagnosis. The lung diseases that occur in patients tend to play an important role but becoming difficult to detect the diseases in early stages in saving lives of patient in time. Physicians try to identify lung diseases by taking help of chest radiograph medical images which are one of the preliminary requirements for undergoing better diagnosis. Early detection of lung diseases can be helpful to patients for a better increasing in their health to cure and fast recovery. Due to rapid growth in the technology, the medial field has encounter dramatical changes and advancements in the way the diseases are simulated and analyzed efficiently. Researchers proposed various methods and approaches, but there is scope to improve the early detection of lung

**Corresponding author*: g.sindhumadhuri@jainuniversity.ac.in

Kukatlapalli Pradeep Kumar, Aynur Unal, Vinay Jha Pillai, Hari Murthy, and M. Niranjanamurthy (eds.) Data Engineering and Data Science: Concepts and Applications, (351–372) © 2023 Scrivener Publishing LLC

diseases via more efficient customized algorithms. In this approach, the objective is to improve the evaluation metrics and increase the diagnostic accuracy for early detection of lung diseases in patients. In this process, image segmentation methods are analyzed by applying various supervised and unsupervised machine learning algorithms and come up with better classification technique to design feature subsets stepwise. The features of lung medical images are extracted using statistical and geometrical methods for the detection of lung boundaries for better comparison of the results. Based on evaluation metrics like Mean Squared Error (MSE), Matthews Correlation Coefficient (MCC), Accuracy (A), Precision (P), Recall (R), Elapsed Time (ET), a comparison study is carried out by analyzing different ML techniques and finally results are demonstrated for the detection of lung diseases in early stages.

***Keywords*:** Medical Images (MI), image segmentation methods, Machine Learning (ML), Evaluation Metrics (EM), Matthews Correlation Coefficient (MCC)

15.1 Introduction

Day by day, human living environment drastically changes by rise in the preasure on human health, increase on public risk, human's lifestyle, climate and there is lots of change in world's envorinment. Human health problems are rapidly increasing every day because of infections, diseases which lead to million's of people falling ill and facing other risky health factors. Many people end their lives and few are slowly recovered due to the massive increase in health problems due to lung diseases, air pollution and breathing problems [1, 2]. Most of the disorders are caused due to lung diseases like repository diseases, Asthama, Lung cancer, Tuberculosis, Cronic Obstructive Pulmonary Disease (COPD), pneumonia, and coronavirus disease (COVID-19). Serious interstitial lung diseases, bacterial infections, breathing problems are caused due to coronavirus disease [3, 4]. Based on the medical records and physical observations of the patients, the repository system disorders need to be identified by the doctors or physicians during the diagnosis stage [5]. The most common chest radiography medical imaging used is X-rays, Computed Tomography (CT) of a patient with the detection of few lung diseases as shown in Figure 15.1.

X-rays is one of the most frequent, cost-effective and old radiological procedure used for lung disgnosis which provides enough information for

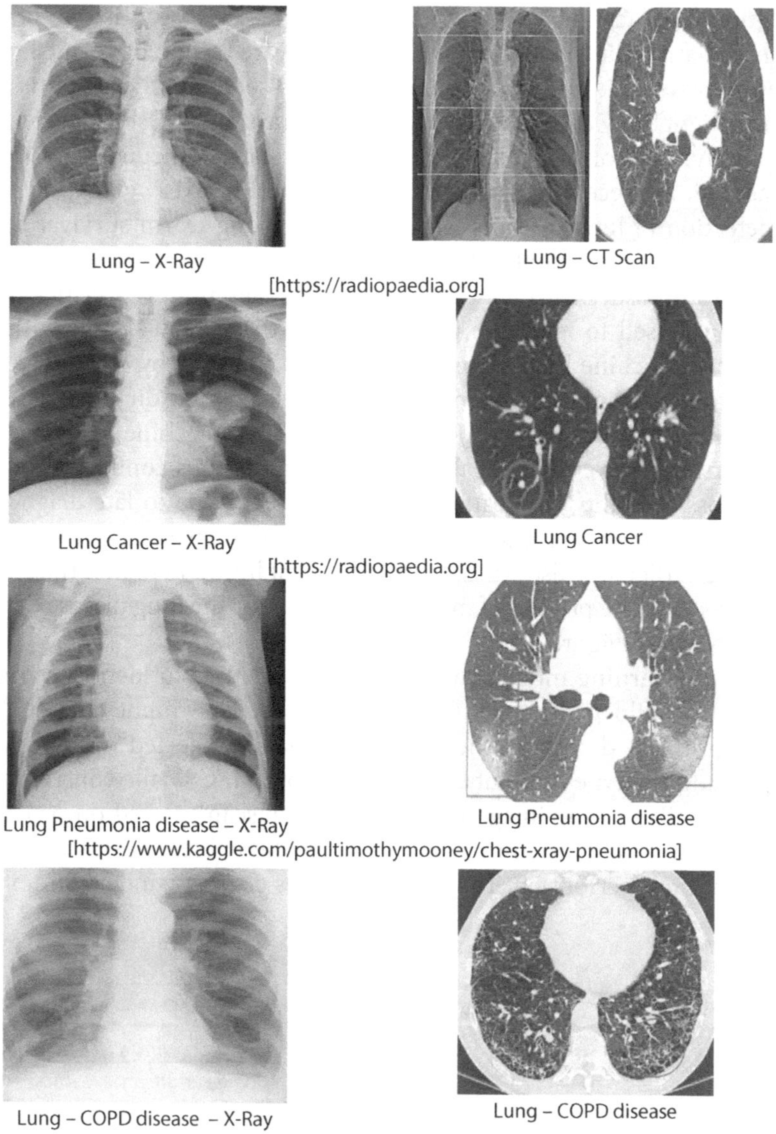

Figure 15.1 Chest radiography lung disease images – X-ray and CT.

detection of exact cause of lung disease, whereas CT scan is one of the costly radiation diagnostic tool with new software technology which provides more detail information to produce cross-sectional slices soft tissues of the patient body than plain X-ray image. X-rays and CT scan images are commonly used for disgnosing lung tumors, cancers, pneumonia, COPD diseases [6, 7]. Recently, Machine Learning (ML) has very great potential for detection of lung diseases which improves long-tearm survival diagnosis by analyzing patient lung medical images in solving many possible future health issues. In medical field, machine learning is a critical role and mainly used to fetch the information of the patient details with its symptoms. Machine learning is a fundamental method by creating complex, automated, and optimal methods by analysing high-dimensional and multimodal biomedical data. In medical systems, machine learning has importance to solve many health issues. Lung diseases commonly occur due to loss of ability to breathe and lead to death due to late diagnosis. This helps to diagnose infections faster and more effectively if they are identified earlier and also will save many patient lives and reduce the strain on the system. The process of machine learning for the lung disease detection is shown in Figure 15.2.

Machine learning models main objective is it tries to identify various patterns to build predictions of the future based on input dataset. For the lung disease detection, the input datasets are collected from various sources and the type of input can be in file format. CSV file consisting of patient details or lung image file (X-ray or CT scan). Based on problem statement and input, the process take place where feature extraction and classification is performed. The model provides us better understanding of

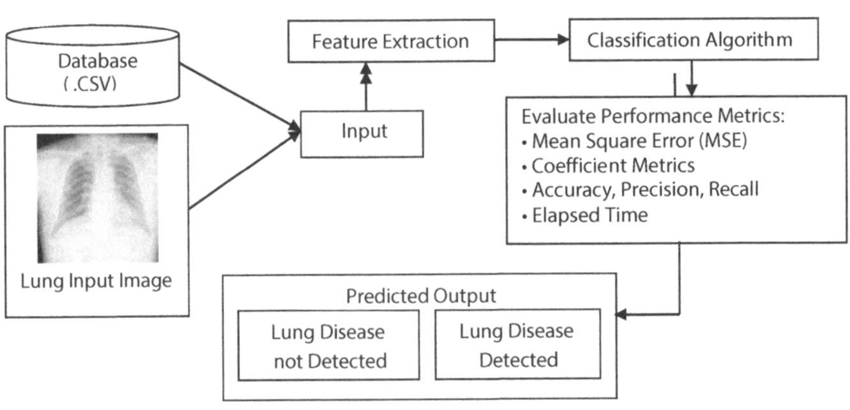

Figure 15.2 Process of machine learning.

features that are in the input dataset [8, 9]. Feature extraction is more time consuming and expensive when performed in manual process. But this can be solved using automated using machine learning classification algorithms like Decision Tree, K-means algorithm, Linear Regression, Support Vector Machines (SVM), etc., which help to classify the input dataset into features and label class attributes to predict the target data [10]. The algorithm needs to be tested correctly based on the input dataset is fed into machine learning algorithm for predicting output results correctly. If the output is not predicted correctly, then the algorithm needs to be re-training till the target output is achieved. This increases our accuracy results slowly over time by choosing our best machine learning algorithm. Doctors and researchers carried out lots of investigations for detection and prediction of lung diseases using machine learning algorithms for X-ray images [11, 12].

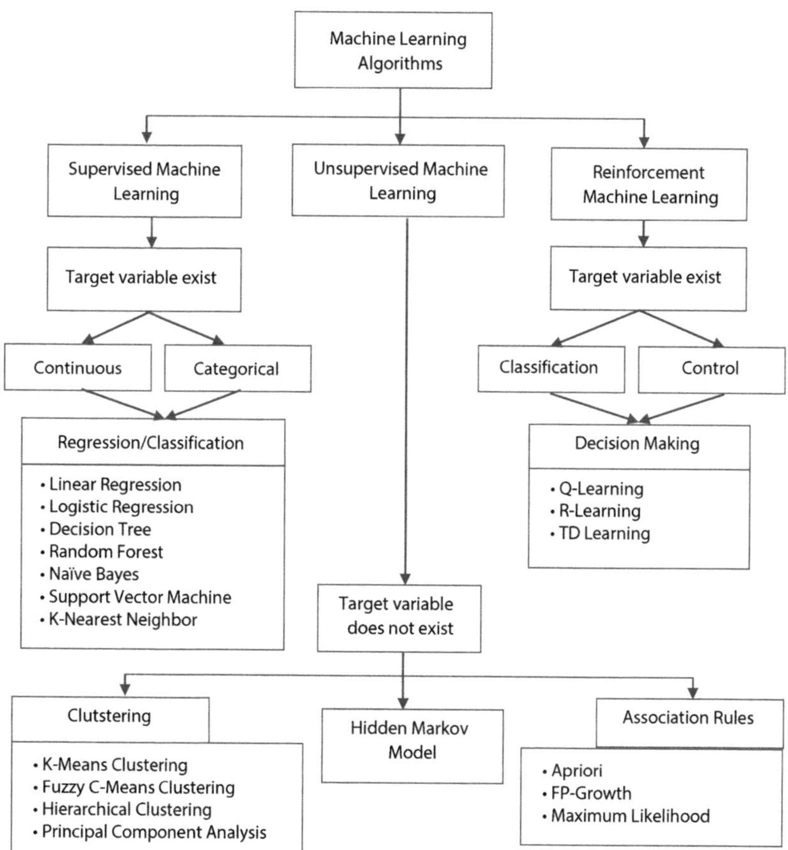

Figure 15.3 Classification of machine learning algorithms.

Machine learning algorithms are classified into different categories as shown in Figure 15.3.

For measuring the health care performance by exploring various patterns for the detection of lung diseases mainly uses supervised machine learning algorithms. As we collect the data from various sources and we know the type of the input dataset, the input dataset is sent through machine learning algorithm to train the model. In supervired machine learning algorithm, we mainly use the known data or labeled data for training dataset. Once the model is trained we can use unknown dataset into model to achieve predicted output [13]. Figure 15.4 shows the process of supervised machine learning for lung disease detection.

In unsupervised machine learning, the input dataset is unknown and training dataset is unlabelled. The unknown dataset is fed into machine

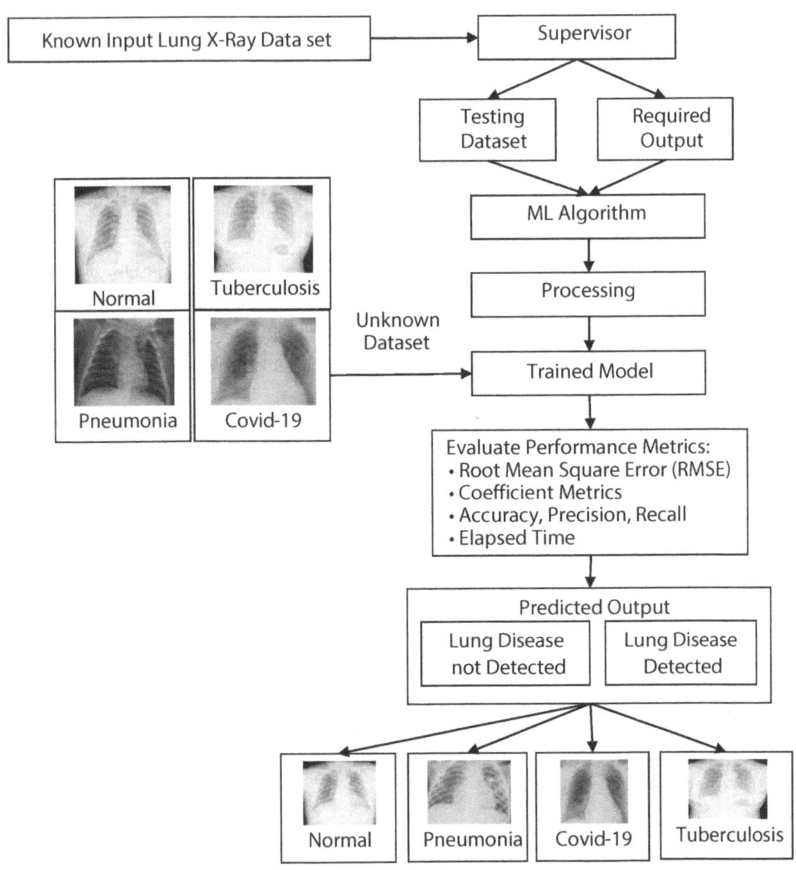

Figure 15.4 Process of supervised machine learning for lung disease detection.

learning algorithm to train the model for further processing. It mainly tries to search for patterns from the unknown input dataset and produces required output [13]. Figure 15.5 shows the process of unsupervised machine learning for lung disease detection.

Reinforcement learning process is different from supervised and unsupervised machine learning. It mainly consists of three components – agent, actions and environment. The agent is like a decision-maker to the system,

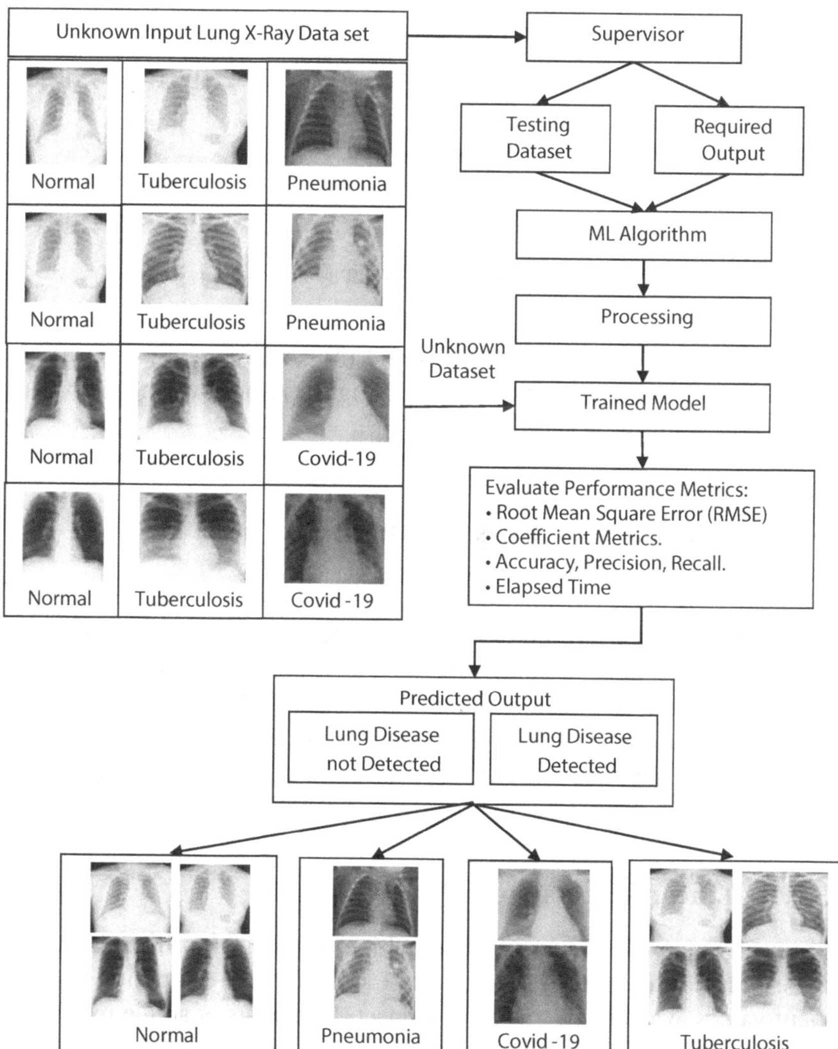

Figure 15.5 Process of unsupervised machine learning for lung disease detection.

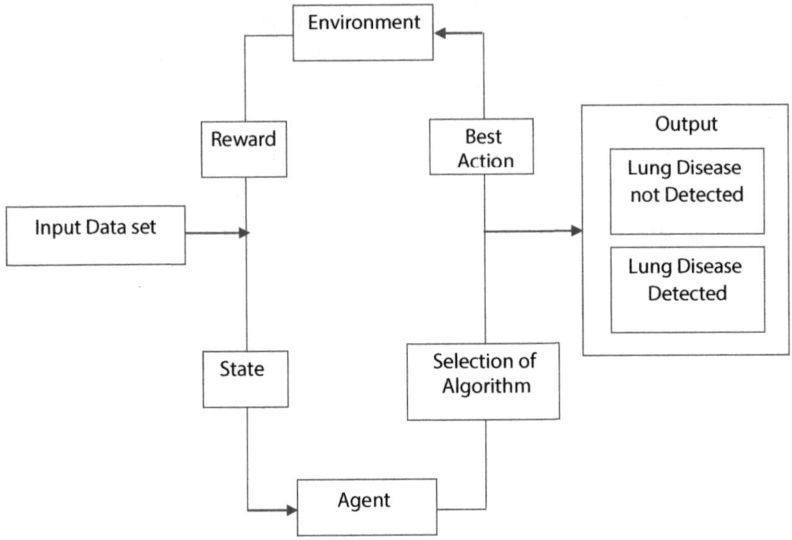

Figure 15.6 Process of reinforcement machine learning.

the action is what the agent performs and the environment is everything the agent performs and interacts [14, 15]. Figure 15.6 shows the process of reinforcement machine learning.

15.2 Machine Learning Algorithms

The analysis and design for the detection of lung diseases like asthma, pneumonia, lung cancer, and tuberculosis is carried out by taking the input datasets of .csv and lung images from kaggle website. Based on different machine learning algorithms, the analysis of lung disease is detected in early stage and investigated by measuring performance metrics. The supervised machine learning algorithms that are applied for the analysis are explained in this chapter.

15.2.1 Linear Regression

Linear regression [16] supervised machine learning algorithm is one of the best regression model used to represent the data by estimation of coefficient values available in the input dataset. It consists of single input and mathematical statistics is used to estimate the coefficient values. The linear regression model is calculated as equation below:

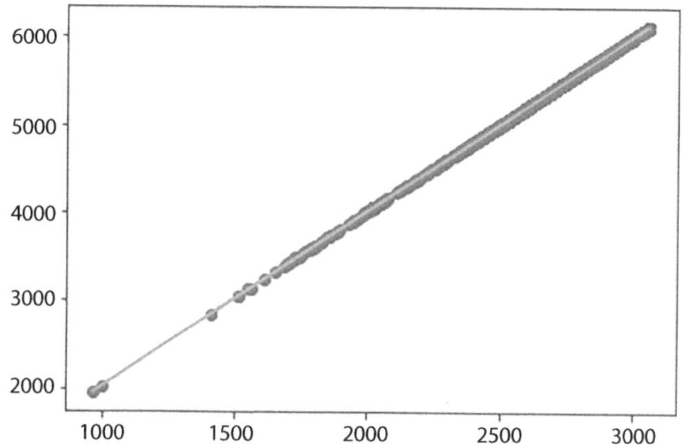

Figure 15.7 Outcome of linear regression algorithm for detection of lung diseases.

$$y = C0 + C1 * x \tag{15.1}$$

Where, the variable 'y' is dependent variable, 'x' is independent input line called plane or a hyper-plane and the coefficients are 'C0' and 'C1'. The outcome of this linear regression algorithm gives an straight line and identify the best fit values of the coefficients 'C0' and 'C1'. From the input dataset (.csv), we could able to analyse the training dataset of 70% and testing dataset of 30% and achieved that patient has no lung disease detected with best fit straight line using linear regression algorithm as shown below Figure 15.7.

15.2.2 Logistic Regression

Logistic Regression supervised machine learning model is to predict the probability of the desired output or target value. The aim of this algorithm is to minimize the risk factor that is available in initial sample space with a loss function by linearly as shown in below equation:

$$X^m = \{(x1, y1), (x2, y2), \ldots \ldots.., (xn, yn)\} \tag{15.2}$$

where, 'x' is object features in space; 'y' is finite dataset of numbers consisting of names or labels and 'X^m' is learning task of a linear classifier minimizing risk with a loss function [17, 18]. From the input dataset (.csv), we could able to detect the patients require disgnosis as '1' and no diagnosis as '0' for the lung disease detection with the accuracy results more than 77% as shown below Figure 15.8.

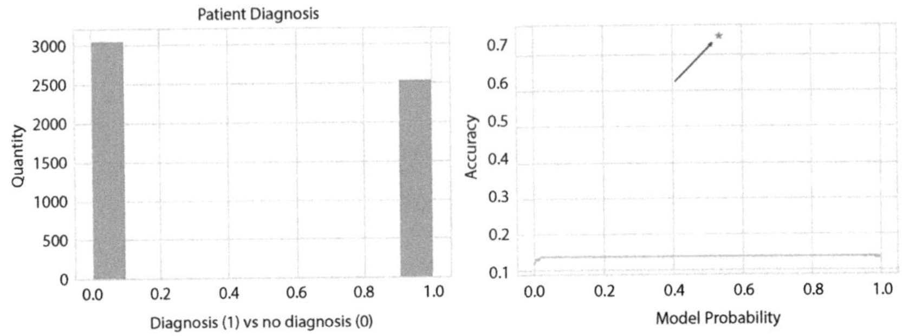

Figure 15.8 Outcome of logistic regression for detection of lung disease.

15.2.3 Decision Tree

Decision tree supervised machine learning algorithm is basically a tree-structure method used to solve classification problems. The decision tree algorithm consists of decision node – used to make decisions of features from the given input dataset and the leaf nodes – represent decision outputs of class labels. The objective of this algorithm is to predict the target variable value where the decision tree uses tree structure to solve a particular problem based on the leaf node. This algorithm is easy to understand but difficult with the huge dataset as it makes decisions in complicated manner [19]. Figure 15.9 shows the working of decision tree for lung disease detection.

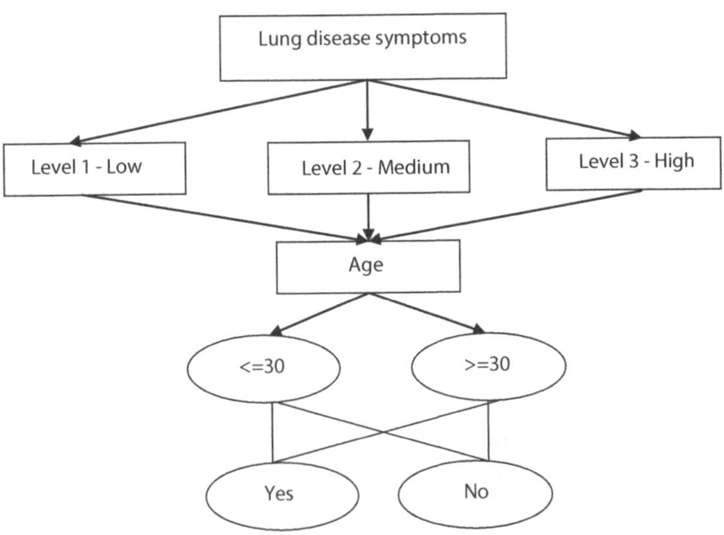

Figure 15.9 Working of decision tree for lung disease detection.

From the input dataset (.csv) lung disease patient data, we could able to detect the patient's type of the lung diseases like 'Atelectasis', 'cardiomegaly', 'tuberculosis', 'lung cancer', 'asthama', 'pneumonia', 'pneumothorax' as shown in below Figure 15.10.

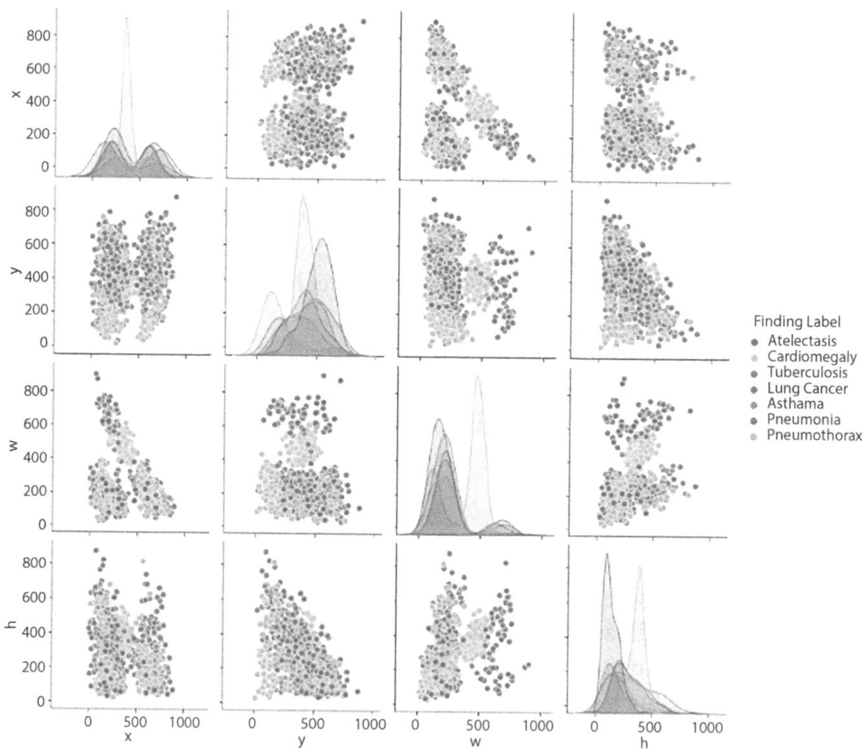

Figure 15.10 Outcome of decision tree algorithm for detection of lung disease.

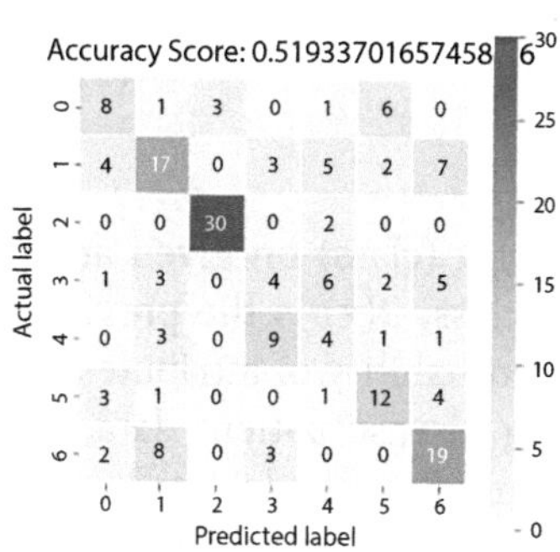


```
Classification report -
              precision    recall  f1-score   support

           0       0.44      0.42      0.43        19
           1       0.52      0.45      0.48        38
           2       0.91      0.94      0.92        32
           3       0.21      0.19      0.20        21
           4       0.21      0.22      0.22        18
           5       0.52      0.57      0.55        21
           6       0.53      0.59      0.56        32

    accuracy                           0.52       181
   macro avg       0.48      0.48      0.48       181
weighted avg       0.51      0.52      0.52       181
```

Figure 15.11 Confusion matrix and decision tree classification report.

Based on the outcome of the lung diseases detected from patient datset, the decision tree classification report with confusion matrix accuracy of 51% is shown in below Figure 15.11.

15.2.4 Random Forest

Random forest supervised machine learning algorithm creates 'n' number of decision trees mainly on various input data subsets. It calculates the average for better improvement of accuracy predicted on the given dataset. Random forest gives better results than decision tree algorithm as it performs average of the result which reduces over-fitting. The working of random forest uses two important concepts 'bagging' and 'random subspace' methods. Bagging means the collection of bags in a forest and each bag is represented as decision tree. Each decision tree takes input data randomly by replacing it with original dataset. Every tree node has features of random subset called random subset method [20–22]. From the input dataset (.csv) lung disease patient data, we could able to detect the patient's health condition by considering the parameters like 'poor health' and 'good health' and generated confusion matrix as shown in below Figure 15.12.

15.2.5 Naïve Bayes

Naïve Bayes supervised machine learning algorithm calculates the probability value of the dataset belongs to instance class. The class label with the final class calculating the highest probability value is considered. This algorithm best suits for two-fold or two-class, multi-class kind of problems. This algorithm identifies the features and characteristics of patient suffering from lung diseases by calculating the probability of attribute every state [23]. The probability is calculated as:

$$P(C/x) = \frac{\mathrm{P(x/C) * P(C)}}{\mathrm{P(x)}} \tag{15.3}$$

Where, P(C/x) is the post probability of target (class) on given attributes, P(C) is the previous probability of the class, P(x) is the previous probability of predicted output, P(x/C) is the probability of attributes on given class. From the input dataset (.csv) lung disease patient data, we could able to detect the patient's type of the lung diseases detected as '1' and not detected as '0' and achieved 88% accuracy than other algorithms.

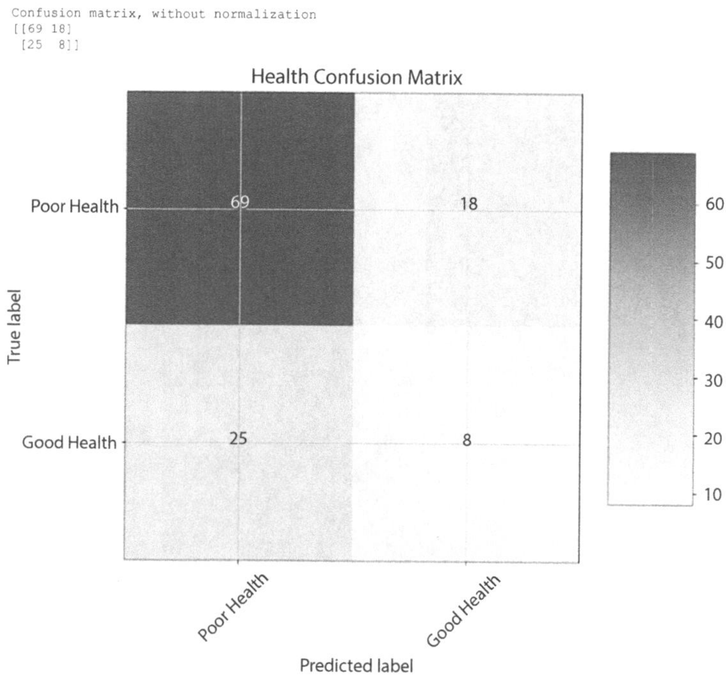

	Features	Importance
0	Patient ID	0.738940
1	OriginalImageWidth	0.094601
2	OriginalImageHeight	0.066183
5	Y	0.062468
4	OriginalImagePixelSpacing_y	0.033284

Figure 15.12 Confusion matrix and random forest classification report.

The naïve bayes classification report with confusion matrix is shown in below Figure 15.13.

15.2.6 Support Vector Machine (SVM)

Support vector supervised machine learning algorithm is used to analyse data by classifying input images as normal or consisting of disease images. This algorithm works by separating optimum linear hyper planes of data items in 'n' dimensional space where 'n' is number of features [24]. Support vector builds binary classifications and to find best hyperplane which seperates into two classes as -1 and +1. The objective of SVM is to find the minimum distance point from the two classes by detecting the maximized distances. SVM classification is calculated using the formula by setting Radial Basis Function (RBF) manually:

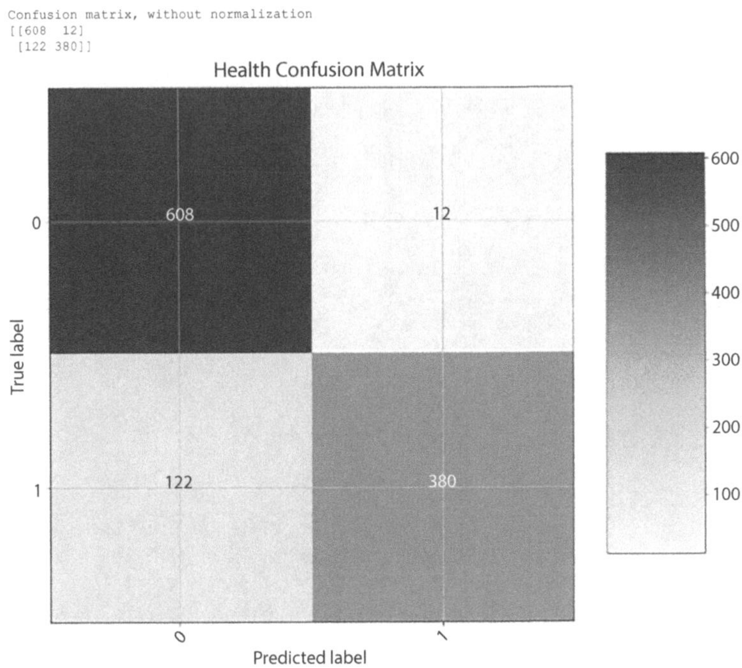

Figure 15.13 Confusion matrix using Naïve bayes algorithm.

$$K(x,x') = e^{-\tilde{a}\ ||x-x'||2}\ and\ \gamma = \frac{1}{n\ features * \sigma} \tag{15.4}$$

Where, 'γ' is a scalar as a single training dataset, '||x-x'|| [2]' is squared Euclidian distance between the features.

From the input dataset (.csv) lung disease patient data, we could able to detect the patient's lung diseases detected and not detected as shown in below Figure 15.14 and achieved 92% accuracy.

15.3 Evaluation Metrics and Comparative Results for Early Detection of Lung Diseases

In the binary classification metrics, lung disease detection is classified into two classes as positive class and negative class so called confusion matrix [25]. The confusion matrix representation is shown in below Table 15.1.

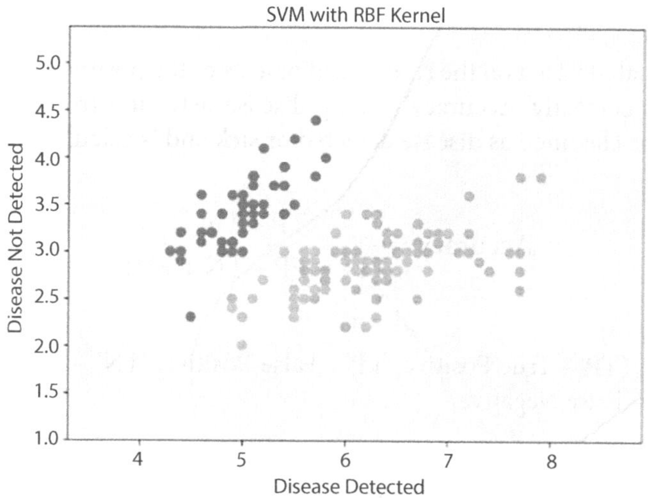

Figure 15.14 Detection of lung disease using SVM algorithm.

Table 15.1 Confusion matrix representation.

		True/actual	
		Positive	**Negative**
Predicted	Positive	True Positive (TP)	False Positive (FP)
	Negative	False Negative (FN)	True Negative (TN)

Regression and classification algorithm using machine learning algorithm is proposed to improve the performance evaluation metrics like accuracy, precision, recall, error etc. Further diagnosis of lung disease detection can be acquired based on outcome values: True Negative (TN) – which indicates no disease detected, and False Negative (FN) – indicates the no prediction of lung disease in patients, True positive (TP) – indicates the prediction of lung disease in patients and False positive (FP) – indicates no lung disease but predicts patient having lung disease. And based on the evaluation metrics, analysed the results by implementing in Python jupyter notebook.

15.3.1 Accuracy (A)

It is calculated based on the recognized objects of the positive class that are predicted correctly. Accuracy of lung disease detection from the patient database is classified as disease detected or sick and is calculated as:

$$\text{Accuracy} = \frac{\text{TP} + \text{TN}}{(\text{TP} + \text{FP} + \text{TN} + \text{FN})} \tag{15.5}$$

Where, 'TP' - True Positive, 'FP' - False Positive, 'TN' – True Negative and 'FN' – False Negative.

15.3.2 Precision (P) and Recall (R)

Precision and recall metrics are calculated based on number of suspected patients detected with lung diseases. The precision and recall are calculated as:

$$\text{Precision} = \frac{\text{TP}}{(\text{TP} + \text{FP})} \quad ; \quad \text{Recall} = \frac{\text{TP}}{(\text{TP} + \text{FN})} \tag{15.6}$$

15.3.3 Mean Squared Error (MSE)

The performance measures at every iteration of training set, the Mean Squared Error (MSE) is calculated between the actual and predicted values as shown in below equation:

$$\text{MSE} = \frac{1}{\text{n}} \sum_{\text{n}=1}^{\text{n}} (\text{X}_{\text{predicted}} - \text{X}_{\text{actual}})^2 \tag{15.7}$$

Where, 'n' is the number of samples.

15.3.4 Matthews Correlation Coefficient (MCC)

MCC is mainly applied when the classifiers value always lies between -1 and 1 for the lung disease detected. And if the MCC calue is '0' indicates no lung disease is detected. It is calculated using the formula:

$$\text{MCC} = \frac{\text{TP} * \text{TN} - \text{FP} * \text{FN}}{\sqrt{(\text{TP}+\text{FP})(\text{TP}+\text{FN})(\text{TN}+\text{FP})(\text{TN}+\text{FN})}} \tag{15.8}$$

Table 15.2 shows the performance metrics using supervised machine learning techniques:

Table 15.2 Performance metrics using machine learning techniques.

Supervised ML algorithm	Accuracy (%)	Precision (%)	Recall (%)	MCC (%)
Linear Regression	70.02	0.60	0.62	0.59
Logistic Regression	77.81	0.79	0.68	0.44
Decision Tree	52	0.44	0.42	0.37
Random Forest	69.01	0.62	0.67	0.54
Naïve Bayes	88.87	0.71	0.74	0.6
SVM	92.27	0.85	0.87	1.0

Figure 15.15 shows the performance metrics using graphical representation.

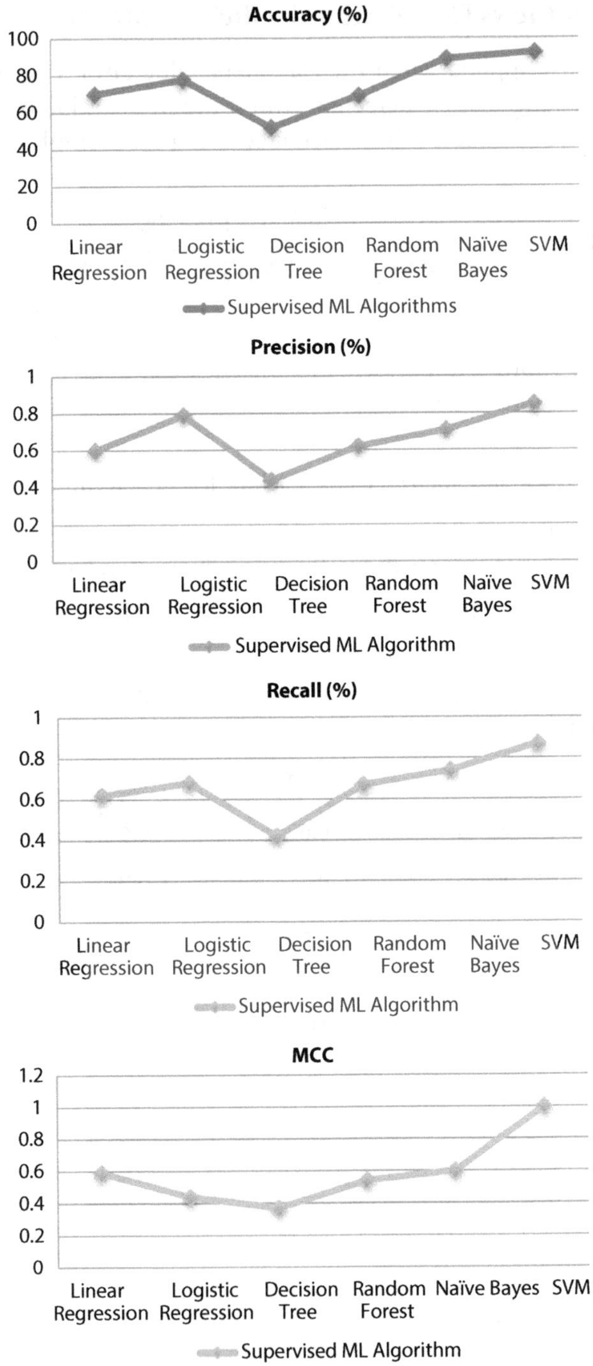

Figure 15.15 Performance metrics using graphical representation.

15.4 Conclusion

This chapter provides an analysis of the various techniques used for feature extraction and classification of early stage lung disease detection like asthma, pneumonia, lung cancer, tuberculosis. The machine learning algorithms plays important role in analysis and detection of lung diseases. By using supervised machine learning regression and classification algorithms like linear regression, logistic regression, decision tree, random forest, naïve bayes and svm algorithms the comparative analysis of accuracy mearures are compared and presented. And using SVM algorithm, the analysis and detection of lung diseases accuracy gives better results. This is time consuming process than patients undergoing check-ups or various testsfor the detection of lung diseases. This research work would be extended using unsupervised machine learning algorithms like K-means clustering techniques and using deep learning neural network techniques in addition with many pre-processing methods for better accuracy results for increasing the performance and evaluation metrics.

References

1. Bharati S, Podder P, Mondal R, Mahmood A, Raihan-Al-Masud M., Comparative performance analysis of different classification algorithm for the purpose of prediction of lung cancer. Advances in intelligent systems and computing, vol. 941. Springer; 2020. p. 447–57. https://doi.org/10.1007/978-3-030-16660-1_44.
2. Anuradha D Gunasinghe, Achala C Aponso and Harsha Thirimanna, Early prediction of Lung Diseases, Proceedings of International Conference for Convergence in Technology, 2019.
3. Anon., Two lung diseases killed 3.6 million in 2015: study., 2018. [Online] [Accessed 4 November 2018].
4. Anon., Understanding how your lungs work and how COPD affects your body., s.l.:s.n., 2007.
5. MedHelp., Available: http://www.edhelp.org/Medical-Dictionary/Terms/2/8964.htm., (2017, January 26). [Online].
6. American Thoracic Society., Diagnostic Standards and Classification of Tuberculosis in Adults and Children. Am. J. Respir. Crit. Care Med. 2000, 161, 1376–1395.
7. Setio, A.A.A.; Traverso, A.; de Bel, T.; Berens, M.S.; van den Bogaard, C.; Cerello, P.; Chen, H.; Dou, Q.; Fantacci, M.E.; Geurts, B.; *et al.*; Validation, comparison, and combination of algorithms for automatic detection of

pulmonary nodules in computed tomography images: The LUNA16 challenge,. Med. Image Anal, 42, 1–13, 2017.
8. U. Fayyad, G. P. Shapiro, and P. Smyth, The KDD process for extracting useful knowledge from volumes of data, Commun. ACM, vol. 39, pp. 27–34, 1996.
9. U Fayyad, G Piatetsky-Shapiro, P Smyth, Data Mining to Knowledge Discovery in Databases, AI Magazine, vol.17, no.3, pp. 37-54, 1996.
10. Balagatabi, Z. N., & Balagatabi, H. N., Comparison of Decision Tree and SVM Methods in Classification of Researcher's Cognitive Styles in Academic Environment, Indian Journal of Automation and Artificial Intelligence, 1(1), 31-43, 2013.
11. M.R.H. Mondal, S. Bharati, P. Podder, P. Podder, Data analytics for novel coronavirus disease, informatics in medicine unlocked, 20, Elsevier (2020).
12. S. Bharati, P. Podder, R. Mondal, A. Mahmood, M. Raihan-Al-Masud, Comparative performance analysis of different classification algorithm for the purpose of prediction of lung cancer, Advances in intelligent systems and computing, vol. 941, Springer (2020).
13. Ethem Alpaydin, Introduction to Machine Learning, 3rd ed. Cambridge.
14. Singh S, Agents and reinforcement learning [M], San Matco, CA, USA:Miller freeman publish Inc,1997.
15. Bush R R & Mosteller F., Stochastic Models for Learning [M]., New York: Wiley, 1955.
16. Data Mining - Techniques, Methods and Algorithms: A Review on Tools and their Validity, Mansi Gera and Shivani Goel, International Journal of Computer Applications (0975 – 8887) Volume 113 – No. 18, March 2015.
17. Hastie T., Tibshirani R., Friedman J., The Elements of Statistical Learning, Springer, 2009, ISBN: 978-0387848570.
18. Ajvazyan, S.A., Byhshtaber, V.M., Jenyukov, I.S., Meshalkin, L.D. Prikladnaya ststistika: klassifikasiya i snizheniye razmernosti [Applied statistics: classification and dimension reduction], Moscow Finansi a statistika, 1989.
19. S. Lee, A. Kouzani, and E. Hu, Random forest based lung nodule classification aided by clustering," Computerized Medical Imaging and Graphics, vol. 34, no. 7, pp. 535–542, 2010.
20. L. Breiman, Bagging predictors, Machine Learning, vol. 24, no. 2, pp. 123–140, 1996.
21. T. K. Ho, Random Decision Forests, in Proceedings of the 3rd International Conference on Document Analysis and Recognition, Montreal, QC, pp. 278–282, 1995.
22. Pattekari, S.A. and Parveen, A Prediction system for heart disease using Naïve Bayes., International Journal of Advanced Computer and Mathematical Sciences **3** (3) (2012) 290-294.
23. B. H. Boyle, Support vector machines: data analysis, machine learning, and applications, http://public.eblib.com/choice/publicfullrecord.aspx?p=3021500, 2011.

24. D. M. W. Powers, Evaluation: From Precision, Recall and FMeasure to ROC, Informedness, Markedness & Correlation, Journal of Machine Learning Technologies, vol. 2(1), pp. 37-63, 2011.
25. T. Fawcett, An Introduction to ROC Analysis, Pattern Recognition Letters, vol. 27(8), pp. 861-874, 2006 DOI:10.1016/j.patrec.2005.10.010.

16

Estimation of Cancer Risk through Artificial Neural Network

K. Aditya Shastry[1*], Sanjay H. A.[2], Balaji N.[3] and Karthik Pai B. H.[3]

[1]*Department of ISE, Nitte Meenakshi Institute of Technology, Bengaluru, Karnataka, India*
[2]*M S Ramaiah Institute of Technology, Benglaru, India*
[3]*Department of ISE NMAM Institute of Technology, Nitte, Karkala, Udupi, Karnataka, India*

Abstract
Cancer represents a widespread medical challenge that is being faced by the world. It is being regularly explored by scientists and researchers due to its high rate of mortality. In the preceding few years, deaths caused by cancer have been on the rise. Studies have shown that developed nations like the USA are also suffering due to cancer with more than 7 lakh deaths and 3 million new cancer patients every year. The Artificial Neural Network (ANN) represents one of the well-known Computational Intelligence (CI) techniques that can handle complex datasets and produced the desired outputs. ANNs are effective mechanisms that are widely employed for constructing prediction models for cancer. Keeping these points in mind, this chapter presents an exhaustive survey of ANN techniques used for cancer risk estimation.

***Keywords*:** Cancer, risk, Artificial Neural Network, prediction

16.1 Introduction

Cancer signifies a category of illnesses that are related to each other. In all categories of cancer, the cells of the body start splitting and spread rapidly to other tissues. Cancer can arise from any body part that consists of many cells. In normal conditions, the cells of humans develop and split to create

**Corresponding author*: adityashastry.k@nmit.ac.in

Kukatlapalli Pradeep Kumar, Aynur Unal, Vinay Jha Pillai, Hari Murthy, and M. Niranjanamurthy (eds.) Data Engineering and Data Science: Concepts and Applications, (373–408) © 2023 Scrivener Publishing LLC

new cells as and when the body requires them. The new cells replace the old or damaged cells [1]. Cancer varies from one person to another due to the different combinations of changes in genes. As cancer develops, further degradation of the body occurs. Even if two persons have the same tumour, the cells may undergo diverse changes in genes. Cells that are normal. Certain changes that occur maybe because of cancer instead of the reason for cancer. More than a hundred categories of cancer exist. They are named after the organs/tissues in which they form. For instance, lung cancer occurs in the lung cells, brain cancer occurs in the brain cells, and so on. The cell type that created cancer can also be used to describe cancer [2].

ANNs denote "computing systems" inspired by the "biological neural networks" that constitute animal brains. ANN is constituted by several neurons which are interconnected to each other. These neurons loosely resemble the neurons present in the human brain. Similar to the brain synapses, every connection in ANN can send signals to other neurons. The signal received by an artificial neuron is processed and transmitted to the next connected neuron. A real number is used to denote the signal and each neuron output is calculated by a non-linear function of the input sum. Edges are the names given to the connections. Weights are assigned to the connections/edges which are tuned during the process of learning. These weights increase/decrease the signal strength at a connection. The next neuron gets activated if the signal it receives is above a certain threshold. Normally, layers are formed from a collection of neurons. The inputs get transformed differently in diverse layers. The signals propagate from 1st layer (input layer) to the final layer (output layer), probably travelling through middle layers several times [3].

Recently, CI techniques along with machine learning (ML) methods are being widely employed in the prediction of medical-related data particularly for health-related diagnosis. The accurate diagnoses related to health is one of the critical challenges faced by medical professionals. Globally, numerous groups of researchers are putting efforts into devising ANNs for accurate health-related diagnoses. The ANNs have been found to enhance the accuracy of these diagnoses. The research in ANN has been going on for the past 50 years and continues to be one of the hot research topics in the health domain especially in the areas of urology, oncology, radiology, cardiology, etc. Its role in developing decision support systems is vital [4].

ANNs form vital tools for the identification and monitoring of cancer. If cancers can be detected at an early stage, then their growth can be stopped. This process can cure the patient of cancer. Hence, an early diagnosis is usually done based on the patient data (demographic and clinical data). It is observed that around 30% of casualties because of cancer can be avoided if they are detected during the initial stages. Developing effective treatments for cancer cure involves a lot of time and is quite expensive. Researchers are still in the process of discovering the main reasons for cancers and devise effective treatments for them. In this regard, ANNs represent a diverse technique for solving this issue. They are occasionally referred to as the 6th generation of computing. ANNs are being widely employed in different domains and have been shown to perform effectively in the health domain since they are feasible, adaptable, and robust with sound theoretical aid. Over a couple of decades, extensive research has been performed for the diagnosis of cancer in an automated way [5].

This chapter emphasizes the use of ANNs for cancer risk estimation. The remainder of the chapter is as follows. Section 16.2 elaborates some of the related work where ANNs are being utilized for cancer risk estimation. Section 16.3 provides the details of certain popular cancer datasets that are frequently used in experimentation. Section 16.4 discusses certain applications in which ANNs have been used for cancer risk estimation. Section 16.5 provides the comparison of ANN with other ML and CI models for cancer risk estimation. Section 16.6 is on the discussion of the results which is followed by conclusion, future scope, and references.

16.2 Case Studies Related to Cancer Risk Estimation Using ANN

This section discusses relevant case studies of ANN in the domain of cancer estimation.

The authors [6, 7] applied ANN for the classification of cancer. Here, the authors devised a model for categorizing small tumours of a round blue cell which leads to cancer during childhood. They were categorized into four categories, namely "neuroblastoma (NB), rhabdomyosarcoma (RMS), non-Hodgkin lymphoma (NHL), and the Ewing family of tumors (EWS)". The diagnosis accuracy varies from one category to another. The authors experimented on 80+ records of cDNA microarray data comprising of 6,000+ genes. They split the data into 60+ records for training and 20+ data for testing. Principal Component Analysis (PCA) was utilized for reducing the number of dimensions into 10 PCs which accounted for around 60% data variance. The ANN architected consisted of a 2-layer linear perceptron with 10 inputs and 4 components of a threshold. Gradient Descent (GD) was used for training the network with a learning rate =0.7 and momentum coefficient=0.3. The maximum number of iterations was set to 100. For validation 3-fold cross-validation (CV) was utilized which was iterated 1,000+ times which led to the training of 3,000+ ANNs. The majority voting concept was employed for classifying the test records. The authors observed accuracy of more than 90% when ANNs were employed for cancer classification.

In [8], ANN with fuzzy clustering was applied for detecting seven groups of genes ("FGFR4 - fibroblast growth factor receptor 4, AF1Q - transmembrane protein, NAB2 - NGFI-A binding protein 2 (ERG1 binding protein 2), CDH2 - cadherin 2, N-cadherin (neuronal), EHD1 - EH domain containing 1, LSP1 - lymphocyte-specific protein 1, FVT1 - follicular lymphoma variant translocation 1") that might be utilized for exact categorization of the four subdivisions of tumours. These "7 genes" carry distinctive features which are specific to cancer and together play a vital part in tumor biology. ANN with one hidden layer comprising 100+ nodes was employed. The clustering procedure led to the detection of 6 clusters from 20 genes. 100% test accuracy was achieved by the authors. Authors [9] employed ANN for detecting micro RNAs (miRNAs) for differentiating between stage II tumours and normal tissues. This was performed with the objective of providing better cancer therapy. The authors made use of 3-layer ANN with feedforward back-propagation GD that took miRNA as inputs. The output was a dual categorization (Stage 2 tumour or control). The number of hidden layer neurons was selected using cross-validation. After the miRNAs were detected, analysis was performed by comparing the normal tissues and the matched tumours by employing t-test ANOVA. Three miRNA signatures were detected for stage II tumour with 100% median accuracy.

The work by [10] utilized single layer ANN on a dataset comprising 60+ tumours for categorizing human astrocytic tumours into three classes. Inputs to the ANN were genes that were selected using the leave-one-out CV technique. The genes were ranked based on each iteration of the validation and the success rate was graded. These genes were collected into one set after removing around 60 genes due to redundancy. Hierarchical clustering identified a clear distinction between various forms of genes. The classes of ANN were utilized for survival analysis that resulted in the clustering of survival groups.

Autoencoder NN was used by [11] along with a method for reducing the number of dimensions for classifying cancer. The model was able to learn new types of cancer from unlabelled data without remodelling the features. As opposed to other techniques that need the training data to match the specific type of cancer for accurate labelling, the information for this method was found by integrating gene expression from diverse types of tumours so long as the information was produced through a similar technique for assessment of gene expression.

The method devised by the researchers [12] consists of two modules. One module is dedicated to attribute selection and the other module deals with the actual classification. The attribute selection involves the reduction of the number of dimensions using PCA. The selected attributes are fed to the autoencoder which learns data approximation using hidden neurons. GD technique is employed for training the resultant network using sparsity penalty which makes the hidden layer activations "zero", thus making the network represent inputs through a reduced number of activations. The authors observed that the sparse oriented attributes were able to effectively encode the data and hence led to better classification accuracy. A 3-layer ANN with softmax as an activation function was employed by the authors for training the classifier. The classifier was validated using a 10-fold CV. The ANN was compared with SVM trained using Gaussian kernel and Softmax regression on 10+ diverse data. The ANN devised by the researchers performed better than these algorithms on 11 of these datasets. Table 16.1 demonstrates certain other research works done in cancer risk estimation using ANN.

Table 16.1 Survey of ANN for cancer risk estimation.

Paper	Methods	Accuracy
[13]	Probabilistic Neural Network (PNN)	80%
[14]	Intelligent System and GA	98.6%
[15]	Perceptron Network	93%
[16]	Fuzzy classifier with boosting	90%
[17]	Multi-layer perceptron with error BP	87-93%
[18]	FG-SVM-RFE	100%
[19]	Neuro Fuzzy network	80%
[20]	Fuzzy Neural Network (FNN)	92-96%
[21]	PNN	96%
[22]	FNN	92-96%
[23]	ART1 Network	92%

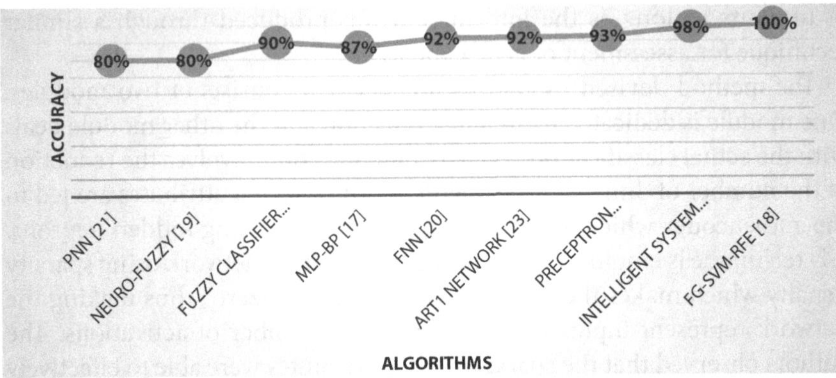

Figure 16.1 Comparison of ANN techniques for cancer risk estimation.

Figure 16.1 shows the accuracies of different forms of ANN for cancer risk estimation.

From Table 16.1 and Figure 16.1 following points can be observed:

- Most of the "ANN" demonstrated remarkable outcomes to categorize malignancy cells precisely. Particularly "MLP" revealed more than 97% "accuracy" and "PNN" exhibited 96% accuracy,
- "Perceptron" with 93% and "ART1" revealed 92% accuracy as well. After eliminating the "missing values" from "datasets" the trial findings improved.
- Findings of ANN can be improved by appropriate settings of its parameters. Though "ANN" methods deliver a great categorization ratio, the "training time" taken by them remains high.

16.2.1 ANN Technique for Early LC Detection

In [24], the authors performed experiments for detecting lung cancer. It is supposed to be a common form of cancer that occurs in the majority of individuals. Compared to other kinds of cancer, the death rate associated with lung cancer is maximum. Hence detection during its initial stages is very critical. This work employed ANNs for analysing data of 190+ patients that were collected from Poland. Around 40 features like age, gender, history of patients, outcomes of medical tests, etc., are utilized as input features while the target value to be predicted is the binary class of the patient suffering from lung cancer or not. All these input features were gathered based on background experience. The dataset was divided into three categories of 90+ records as training data, 40+ records as test data, and validation data each, respectively.

In [24], the ANN technique was devised successfully to obtain higher accuracy for predicting whether the patient has lung cancer or not from a healthy population. In other words, ANN can be effective for choosing a small population of people who are facing maximum risk of lung cancer. The work comprises factors of demography like age, residence, education, etc., and medical measurement data like creatinine, haemoglobin, etc.; outcomes achieved from spirometry tests, data collected from interview (like whether lung cancer is hereditary in family, loss of weight, etc.) and medical examination like X-Ray, etc., performed by doctors of primary care and specialists. The patients for examination were chosen between the ages 20 and 80 years approximately. The middle age of the patients under examination was around 60 years with 70% being men and 90% being smokers of cigarettes. 100+ patients were reported with lung cancer.

The Statistica software was utilized for building the ANN. The 40+ data attributes were assigned a particular number denoting the attribute of interest. Prior to starting the computations, the entire data was subject to scaling between the range of 0 to 1 since the large, valued attributes would dominate the smaller ones. The target value was encoded as 0 (indicating that the person did not have lung cancer) or 1 (denoting that person has lung cancer). ANN with 1input, 1 hidden, and 1 output layer was utilized since this architecture has been utilized prominently in several works [25–27]. Primarily, the ANNs were devised to obtain the optimal network with configured parameters. Subsequently, the ANN was iterated with 10 models and the best model was selected. The hidden neuron number had to be chosen carefully along with the transfer function as their impact on the result was more. The trials were done by altering the hidden neuron number from 3 to 15. The ANN was tested with three diverse transfer functions, namely linear, tangent and logistic. The weights were tuned during the process of training by utilizing diverse BP techniques like GD, BFGS, and SCG [28, 29]. The MSE was employed as the metric for assessing the error.

The different ANN models used with corresponding test accuracies are shown below:

- "48-9-2" → "97.91"
- "48-5-2" → "97.91"
- "48-10-2" → "70.83"
- "48-7-2" → "83.3"
- "48-15-2" → "93.75"

- "48-6-2" → "81.25"
- "48-15-2" → "97.91"
- "48-4-2" → "93.75"
- "48-14-2" → "91.66"
- "48-15-2" → "97.91"

The best ANN model was selected based on the test accuracy. The primary network architecture is displayed in Figure 16.2.

The accurate patient classification was achieved by the ANN architecture displayed in Figure 16.2. There were 48 input neurons corresponding to the 48 input attributes that described the 190+ patients, 9 hidden layer neurons that were selected experimentally, and 2 layers of output signifying whether the patient had lung cancer or not. The supervised BFGS technique was applied. The function of activation was utilized to scale the layer signal, and the hidden layer comprised 9 neurons with linear activation function. The output layer was created by 2 neurons using the tan function shown in eq. 16.1.

$$\frac{e^{a}-e^{-a}}{e^{a}+e^{-a}} \tag{16.1}$$

Where 'a' denotes the activation neuron ranging from -infinity to +infinity. Table 16.2 indicates the participant characteristics used in the experimentation [30].

The superior outcomes concerning "training, validation and test data", were achieved with "ANNs" utilizing "linear" and "tangent" functions in the "hidden" and "output layers", correspondingly. "BFGS" and "conjugate

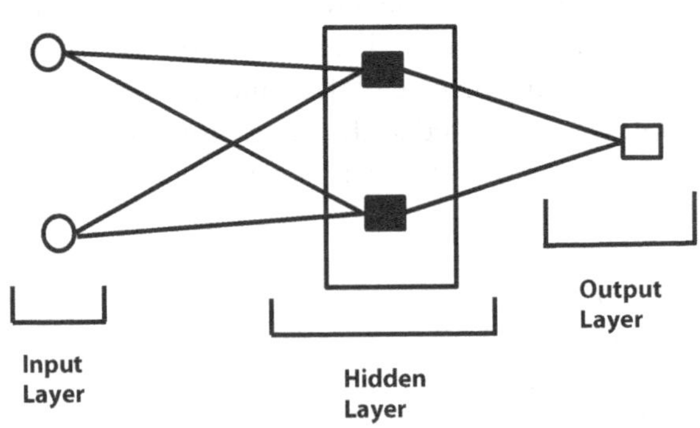

Figure 16.2 Architecture of the applied Artificial Neural Network.

Table 16.2 Characteristics of the participants.

Attributes	Subjects	Percentage of subjects with LC
Total	190 plus	100 plus
Gender - Male - Female	 50 plus 130 plus	 20 plus 70 plus
Age - Median - Range Age	 60 plus 20 to 90	 60 plus 40 to 90
Residence site - Capital of the state. - Urban - Town - Destitute	 70 plus 30 plus 50 plus 1	 30 plus 20 plus 30 plus 1
Education - Elementary School - Vocational - High School - University	 60 plus 35 plus 25 plus 5 plus	 30 plus 20 plus 9 plus 1 plus
Smoking Cigarette - Yes - No	 170 plus 15 plus	 95 plus 3 plus

gradient" procedures were employed for determining the optimal "weight values". It produced the superior outcomes concerning performance and lowest number of "training" iterations. Subsequently the iterations were halted to avoid "over-fitting" challenges.

The proposed ANN model is efficient in predicting lung cancer risk. Hence, our model can improve the current diagnosis and prognosis methods. The main advantage of ANN is commercially available software which can be used in any desktop computer or standard laptop directly in the doctor's office. Although ANN will never replace expert opinion of the doctor, they can help in screening and can be used by experts to double-check their diagnosis. Diagnostic tests such as computed tomography (CT) or chest radiographs scans are not effective for early detection, therefore probably ANN can be useful in the initial diagnostic testing.

16.2.2 ANNs in Image Processing for Early Diagnosis of BC

Early detection of BC helps in its successful treatment. Hence, suitable screening techniques for detecting the initial signs of BC are significant. Segmentation of images has been widely employed for the identification and categorization of BC. Training of several techniques of image segmentation centred on attributes of the histogram, identification of edges, growth of regions, or categorization of pixels has been done.

a. Mammography

It represents an efficient technique utilized in various medical centres and hospitals for the prior identification of BC. It has been found that around 30% of the death rate has been reduced due to this method [31]. The key aim of this technique is to identify the tumour growth and eliminate it before it grows into large proportions [31–33]. Because of the absence of explicit behaviour, BC is hard to detect during its early stages. For example, there are subtle growths of tissues during the early stage of BC which are hard to be differentiated from the normal breast tissues [31]. It has been observed that only around 3% of the needed material is discovered during the mammogram process in which a segment of the questionable area is enclosed with normal tissues and vessels. Due to this, radiologists face difficulty in detecting BC. Hence, CAD is devised to eliminate this drawback by aiding radiologists in comprehending mammograms in an effective fashion [32]. ANN techniques are widely utilized in CAD for understanding and analysing the results of mammography along with the making of decisions with respect to biopsy. Two methods of ANN exist for interpreting mammography. In the first method, ANN is applied directly to the interest region (ROI) which is represented by image information. In the second method, pre-processed features are analysed to comprehend the scenario from the signals of the image [33]. Figure 16.3 demonstrates the structure of ANN used by the authors. The data comprising multiple attributes are fed to the ANN composed of numerous hidden layers [33, 34].

The process of calcium gathering in the regions of the tissues of the breast which are soft is called microcalcification. These regions are miniature in size and number. They are normally detected in a group of lines/circles along with the additional activity of cells located in the region of the breast [35]. Several scientists have devised CAD systems using ANN to identify microcalcification. In [36], the authors have designed the structure of images utilizing the second-order statistics of gray-level. Their categorization was centred on perceptron implementation based on ANN with three layers. The network employed the BP algorithm during the training

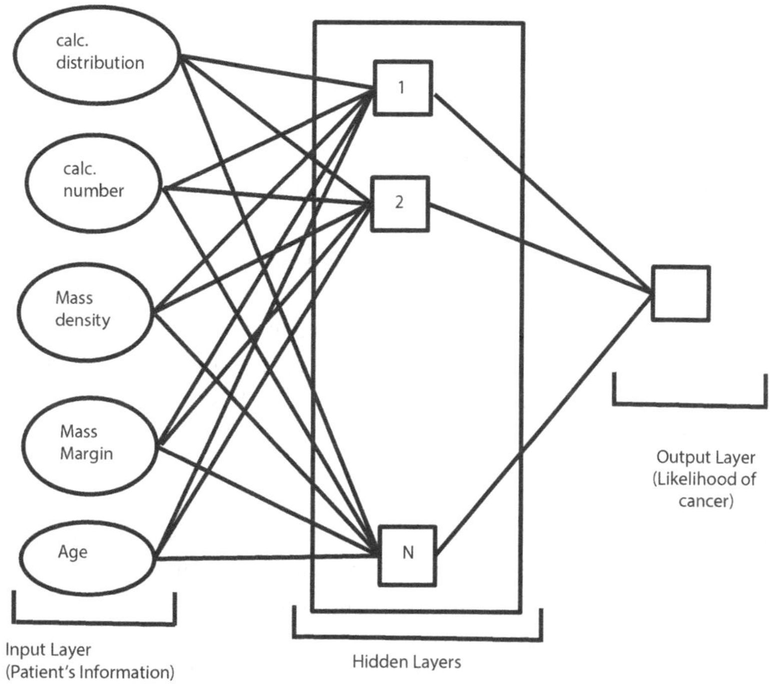

Figure 16.3 Standard ANN design for categorizing tumours in breasts.

process that has been utilized successfully for several applications associated with the recognition of patterns and classification [30, 36]. The feature of entropy has major power of discrimination in terms of classification [36]. Research related to the potential of correlation between the second-order attributes of histogram and cancer was explored. Numerous ANN models were devised for examining the attributes extracted from segments that exhibited prominent outcomes for the classification of mammographic images into benign and malignant classes on the extracted features [37]. Segmentation of images is performed on images that are raw in nature for identifying bright, local, and small spots. In [38], the authors employed ANN after the process of segmentation to discriminate the segmented entities as microcalcifications or non-microcalcifications. The ANN's accuracy was estimated on test images for assessing the TP and FP rates. Cascade correlation (CC) was employed by the ANN for the classification of patterns. The authors used an ANN which was able to organize by itself along with the supervised method. The authors demonstrated that the CC ANN technique provided better results for the identification of microcalcification [38]. BC diagnosis can be implemented using methods related to the registration of images in which ANN has to be found to perform effectively. In [39], GLCM attributes were mined and fed for ANN training for detection of the BC. Subsequently, the mined attributes were matched with additional ANN architectures for the classification of images into benign and malignant. Results above 80% were achieved for the FF-BP-ANN while 60+ % accuracy was obtained for CF-BP-ANN after the optimization of neurons and layers.

Overfitting of data is seen to occur when ANN is used in CAD mammography. Hence, BBN was utilized to compare ANN for the detection of positive regions of mass centred on computed attributes in CAD. GA optimization was done using GA to assess the accuracy and reliability of ANN and BBN. The outcome demonstrated no major difference between ANN usage and BBN for detection of CAD if the network is suitably optimized.

b. Ultrasound

Images of ultrasound can be employed as inputs to ANN for identifying BC. The ANN capacity to determine and identify malignant and benign regions is investigated here. In [38], the authors utilized ANN in BC detection using images of ultrasound. The authors performed training for malignant and benign instances separately. First, the pictures comprising lesions of benign were fed, and then the pictures comprising the malignant lesion were fed. The topology related to "spider web" was introduced by the authors which generated two vectors employed in the process of classification. The first vector signifies the effects of localization in a neighbourhood that is defined. The second vector denotes the features which are global. The strategy provides higher benefits as this topology is susceptible to smaller areas and therefore offers improved outcomes in small regions by placing additional weights on the specified regions. Additional parameters related to shape and texture may be considered for enhancing the performance of the method. The regions of benign and malign vary slightly in their shape and texture.

In [39], the scientists executed the ANN using both shape and texture parameters. The matrix of the probability of transition and length of run on the texture feature has been utilized for quantification of image consistency. The parameter associated with shape demonstrates the abnormality existing in the lesion border. Moreover, a three-layer ANN comprising neurons in the input, hidden and output layers were employed. They also implemented the process of training by dropping one out prior to learning and utilizing the omitted data for testing. From the outcome, the implementation of texture obtained a very reasonable result on lesions of liquid and solid. The task of classification is significant as it is widely explored to classify breasts as cancerous and non-cancerous classes. Breasts which are having high density are at a higher risk of getting cancer. In [26], the authors described the significance of texture images in the categorization of tissues related to density. CNN was utilized to substitute the BP technique where the pictures were directly fed to ANN. For estimating the texture thickness, the statistical difference in terms of grey-level and attributes are utilized, whereas attributes related to spatial information will exhibit the distribution of elements. The strength of this technique is that no picture of the tumour is given to the network. Likewise, the division is not performed prior; but threshold value is utilized. The disadvantage is that the cost of computation is high which makes the method unsuitable for operation in real time.

ANN classifier technique has been utilized with autocorrelation attributes to categorize the BC as malignant. In [27], ANN has 25 inputs with multiple layers. It comprises 24-D vector of features from the image of ultrasound. It is also composed of a threshold that is defined before the input layer to categorize cancer. The system demonstrated a relatively higher precision for categorizing cancers. It could also assist naïve operators in identifying complex images of ultrasound. One of the benefits of Ann is that it can undergo additional optimization by feeding it with large ultrasound images since ANN can be trained well for big data.

The Self-organizing map (SOM) is a kind of ANN that learns in an unsupervised fashion. It is one of the popularly utilized classifiers in previous years. It expresses lesser dimensional information effectively when compared to the complex data with high dimensions as done in [28]. With regard to SOM, the training is comparatively better as the expected outcome is not required. SOM chooses the nearest input vector. The categorization of malignant and benign lesions was performed automatically. The information related to the analysis of texture constitutes the training data. Yet, the accuracy of this classifier is marginally lesser than the ANN of the work [27]. The SOM techniques have a forecasting accuracy of around 90+% that assists in avoiding benign surgeries. In [29], the authors introduced the Hybrid Neural Network (HNN) system to extract texture attributes from four ultrasound images. The HNN consisting of two phases was employed to integrate the texture data from the ultrasound imagery. The work [32] proved that using the four images together produced a better outcome in comparison to their independent usage.

In [35], bootstrap strategy with ANN was utilized for BC classification. The bootstrap was employed for randomly doing sampling after which the instances were computed again. This method is effective when a huge amount of data is unavailable as less data is enough for training.

By using the bootstrap technique, less data is balanced by adding a component of image analysis. In [32, 33], the ANN was tested on a testbed with five structural attributes. Using these attributes, the characteristic difference was highlighted between malignant and benign as follows: (i) with respect to Spiculation benign was larger while malignant was smaller; (ii) with regard to the pattern of branch and benign is fewer while malignant is larger; (iii) nodule brightness was large for benign and small for malignant; (iv) shape of ellipsoid was small in case of benign and large in case of malignant; and (v) lobulation number in terms that it is fewer concerning benign and large with respect to malignant. Recent research has shown the usage of ensemble/hybrid techniques to enhance the traditional ANN technique to identify the tumours of BC. In [34], the hybrid of k-means and BPNN demonstrated effective performance for the diagnosis of BC.

c. Thermal Imaging (TI)

The identification of BC at an early stage using TI has been researched for a very long time [36]. TI is an effective technique since it can warn the women 10 years before the BC happens [30]. Certain researchers have employed ANNs for manipulating and categorizing information from IR, by considering them as input to ANN [37]. In [38], LCNN delivered improved diagnosis for medical experts on IR images that are multispectral in nature. Transformation of wavelets is also effective using ANN in case of attributes with multiple dimensions for the IR image, particularly when the breast temperature is impacted by several factors of pathology including the state of mind [37]. Discrimination of right and left breast in an asymmetric fashion was done to generate attributes based on statistics like average temperature and standard deviation that were employed as ANN parameters [36]. In [30], Complementary Learning Fuzzy NN (CLFNN) which employed fuzzy learning was devised for analysing the thermographic images. CLFNN took into consideration several aspects like the history of family and the difference of temperature of the statistical attributes among contralateral breasts. This technique is being widely employed in numerous countries presently.

ANN helps in the early identification of BC through TI. ANN facilitates the identification of relevant features from thermal images. This assists in the early detection of BC. ANN performs better than other classifiers such as SVM, and Bayesian Classifiers for the identification of BC via TL. The extraction of relevant features from the thermal images forms the basis of classification.

d. Pancreatic Cancer (PC) prediction through ANN

Early identification of PC is daunting due to the occurrence of cancer symptoms only during advanced stages. Furthermore, dependable tools of screening to recognize high-risk patients are not present. To resolve this issue, ANN was devised, tested, and trained to utilize the data of health of around 8 lakh residents captured from the NIHS and PLCO datasets along with 800+ patients detected with PC. Forecasting of PC risks was done on a person-by-person basis by using 18 attributes into ANN [40]. PC forms the fourth-leading cause related to cancer in both women and men in the USA even though its rate of incidence is lower [41]. Around 50,000+ PC cases and 43,000+ associated deaths were recorded in 2017 alone in the USA. Around 3% of the new cancer cases were PC cases. Also, 7% of overall deaths were associated with PC. It is observed that PC may become the second most common reason for cancer-related death by 2030 [42, 43]. As discussed previously, the rate of mortality for PC is higher since its symptoms appear during advanced stages for 80% of the patients [41–43].

Besides PC, several types of cancer-related to the lung [44], prostate [45], endometrial [46], and colorectal [47] have been investigated using ANN, DT, NB, LDA, and LR. Results have demonstrated the superior performance of ANN when matched with other methods with respect to sensitivity, AUC, and specificity. Hence, ANN was utilized with two hidden layers each comprising 12 neurons apart from the input and output layers. The input and output attributes (in the range of 0 to 1) were divided into 70% as training data and 30% as test data while maintaining the proportion of malignancy instances to non-malignancy instances was kept constant. From the training data, CV of 10-fold was employed to assess the model performance that was learnt on diverse datasets. Once the superior model is selected, training is performed on the training data and then assessed on the test data. Logistic activation function and the SSE cost function were utilized. Standard BP algorithm with GD was employed for training purposes. To make the ANN converge faster, momentum was utilized. The model was subjected to batch training using the entire dataset in place of online training [48]. Five thousand iterations were run for training. The ANN output was a fractional value in the range of 0 to 1. Values closer to 1 indicated that the patient had a higher likelihood of getting PC. This value could be converted into the status of cancer in terms of Yes or No using the threshold value below which cancer status would be "NO", else it would be "YES". An array of values for threshold were employed for assessing the specificity and sensitivity after training. The chosen value of threshold from the training data was utilized to calculate the specificity and sensitivity for the test data.

The average of the performance on the validation data was utilized as the performance metric for evaluating the trained models. Particularly, AUC was employed as the performance measure since it was a good metric for categorizing the data into different groups of risk [49]. After the superior prototype was chosen, a performance assessment on both training and test data was done to assess the capability of the score of risk to discriminate among the PC and Non-PC persons. Apart from AUC, the forecasted likelihoods from the model and the predicted results were done to assess the model training.

The ANN technique was assessed in a clinic for exhibiting its potential. The technique was applied to classify the population into three classes of risk, *viz.*, low, medium, and high. The limits of the low-, medium- and high-risk categories were chosen carefully from the training data so that not more than 1% of the population without PC and with PC would be classified into risk categories of high and low, respectively. Nevertheless, the boundary between medium and high risk might be chosen to classify additional persons with PC in the high-risk class with regard to potential risks in diagnosing patients with no cancer. After the selection of these boundaries, the classification scheme was applied on testing data to exhibit the model's application in the clinic. According to this method, screening of individuals with high risk would be performed without delay while the individuals falling in low- and medium- risk categories would undergo a smaller number of frequent screenings.

16.3 Datasets Used in Cancer Risk Estimation

This segment demonstrates popular cancer datasets which are used for experimentation purposes by various researchers around the world.

16.3.1 Datasets Related to Breast Cancer

The dataset [50] comprises more than 23 Lakhs mammograms of women gathered from the Association of Breast Cancer (BC). BC is primarily utilized for exploring the changes in associations, factors of risk, or other statistical challenges like the impact of data imputation for omitted values or alternate estimation models. The cancer outcome represented the target attribute which formed vast factors of risks for cross-categorization. No previous BC identification or imaging of breasts was performed on the women nine months earlier to the screening mammogram, i.e., the mammogram was done on the women for the past five years but not during

```
Class: no-recurrence-events, recurrence-events
age: 10-19, 20-29, 30-39, 40-49, 50-59, 60-69, 70-79, 80-89, 90-99.
menopause: lt40, ge40, premeno.
tumor-size: 0-4, 5-9, 10-14, 15-19, 20-24, 25-29, 30-34, 35-39, 40-44,
            45-49, 50-54, 55-59.
inv-nodes: 0-2, 3-5, 6-8, 9-11, 12-14, 15-17, 18-20, 21-23, 24-26,
           27-29, 30-32, 33-35, 36-39.
node-caps: yes, no.
deg-malig: 1, 2, 3.
breast: left, right.
breast-quad: left-up, left-low, right-up, right-low, central.
irradiat: yes, no.
```

Figure 16.4 Format of the BC dataset used in ML and data analytics.

the previous nine months. It is associated with the registry of cancer and pathology information. In 2012, the second adaptation of this dataset was made available. The main difference between the first and second versions was that the second version constrained only one observation per woman. The remaining characteristics of the data were the same.

Another type of BC dataset comprised 250+ records with one target outcome (class) [51], 200+ records belonging to one category and 80+ records of another category. The dataset consists of nine features where some are linear while some are nominal in nature. The input attributes are age, menopause, size of tumour, inv-nodes, node-caps, the scale of malignancy, size of the breast, and quadrant of the breast. The target attribute is the class attribute consisting of two classes which are events that recur and events that do not recur. The format of this dataset is shown in Figure 16.4.

16.3.2 Dataset Related to Lung Cancer

The dataset [52] comprises 30+ records of data type integer. There are more than 50 features that consist of values that are missing. Feature-1 represents the class label. All input features are nominal in nature consisting of integer values ranging between 0 and 3. The job is to classify each record into one of three classes. There are 9 records belonging to class-1, 13 records of class-2, and 10 records of class-3.

16.3.3 BC Coimbra Data Set and BC Wisconsin (Diagnostic) Data

The BC Coimbra Data Set [53] comprises 110+ instances of integer type. There are nine numeric input attributes. The task is to classify the data into two classes, *viz.*, 1 and 2. Value of 1 denotes healthy control class while value-2 indicates Patients class, indicating the absence or presence of cancer, respectively. The information was gathered during the analysis of blood during routine check-ups. The researchers can experiment on the dataset for cancer detection. Accurate results can then be employed as biomarkers of breast cancer. The BC Wisconsin (Diagnostic) Data [54] consists of 500+ records with 30+ real-valued features. The attribute measurements were done using a breast mass image. They explain the cell nuclei traits existing in the image. The 32 features comprise 30 real-valued attributes along with ID and diagnosis. The researchers can utilize this dataset for binary classification purposes in which each record can be classified into two classes (class B indicates benign, and Class M denotes malignant). Around 350+ instances are allocated to Class 'B' while 200+ records are assigned to class 'M'.

The type of cancer dataset to be used in ML relies on the type of application. Hence, the model selection cannot be generalized.

16.3.4 Comparison of ANN Techniques with Other Methods for Cancer Risk Estimation

This section compares ANN with other machine learning and computational intelligence models for estimating cancer risk.

The researchers [55] performed a comparison between LR and ANN for forecasting the risk of BC. The AUC of LR was around 0.963 and that of ANN was around 0.965. Both LR and ANN exhibited better performance with a P value less than 0.001 when compared to radiologists. However, no substantial difference was observed between the AUCs of LR and ANN. A review of LR and ANN models was performed where the authors demonstrated their application with respect to forecasting the risk of BC. They demonstrated the utility of statistical and ML models with the assistance of medical experts for a better understanding of the risk factors associated with cancer and precise diagnosis. The LR and ANN models exhibited high discrimination with respect to accuracy and comparable performance with ANN outputting a higher AUC. Both techniques produced a higher AUC for all levels of threshold when compared to the radiologists, which implies that the models had a greater power of discrimination than the radiologists.

In [56], the authors tested the ANN technique using a 10-fold CV. The area under curve (AUC), sensitivity, specificity, and accuracy were utilized for comparing the models. The AUC and ROC were utilized to assess how the model differentiated between patients suffering from the risk of having thyroid cancer and those who were not suffering from this risk. The models were able to effectively identify the risk of individuals possessing thyroid cancer. The authors constructed a diagnostic model using the ANN to forecast the malignancy in patients of multinodular goiter possessing unspecified cytology. The ANN model achieved values of around 70% for sensitivity, 94% for specificity, 99% for positive predictive value, and 85% for negative predictive value while the related number for the LR model was 66%, 78%, 70%, and 44%, respectively.

Prediction of BC cancer using different ML methods has been reviewed in [57]. The main goal of this work was to bring out the comparison between data mining and ML methods for BC prediction. The key aim of the work was to bring out the best performing ML model for BC forecasting. Several works related to BC prediction using LR, LDA, NB, k-NN, SVM, DT, RF, AdaBoost algorithms were reviewed by the authors. The authors observed that most of the works utilized a mixture of linear and non-linear algorithms or the blend of ensemble and non-linear methods. The review paper was categorized into different segments that provided a comparative analysis of the techniques with respect to the rate of accuracy. The authors finally highlighted the most suitable ML model for BC prediction among the compared ones. Table 16.3 shows the comparison of ANN with other ML techniques for BC prediction [57].

Figures 16.5 to 16.11 show the comparison of ML methods with ANN on different datasets.

From Table 16.3 and Figures 16.5–16.11, it can be observed that there are three vital approaches for the forecast of BC: ML, "Ensemble", and "Deep Learning (DL)" methods. Every method is appropriate for numerous scenarios and datasets. It can be inferred that SVM was found to be the most appropriate ML technique for the prediction of BC. Various scientists have demonstrated the superiority of "SVM" on the "WDBC" dataset. Certain scientists have demonstrated experimental outcomes of ML techniques on two diverse tools like "Weka" and "Spark". Again, SVM demonstrated superior performance with 98.03% on "Weka" and 99.68% on "Spark" in comparison to other ML techniques. There exist diverse ML and DL techniques that are being recently explored for predicting malignancy. The precisions of these procedures, however, differ for several datasets. Consequently, there is a need to explore certain sophisticated models which could enhance the prediction accuracy. Generalization of ML techniques on any type of dataset is also required. The key hurdle faced by most ML techniques is the accessibility of medical information. Every technique needs a huge volume of "training data" for performing experiments. In this

Table 16.3 Comparison of different ML methods with ANN for BC prediction.

Reference	Algorithms	Data	Number of attributes	Data type	Validation technique	Accuracy
[58]	C4.5 ANN SVM	Iranian Center ICBC	22	Numeric, Mixed Type	k-fold	93.6% 94.7% 95.7%
[59]	MLP k-NN CART NB SVM	UCI WDBC	32	Images, Binary values	10-fold	99.12% 95.61% 93.85% 94.73% 98.24%
[60]	NB ANN CART	SEER	151,886 records	Numeric, Mixed	10-fold	84.5% 86.5% 86.7%
[61]	J48 NB MLP LR SVM k-NN	WBC	9	Numeric, Mixed	10-fold	95.59% 96.79% 94.78% 96.79% 97.59% 95.19%
[62]	ANN SVM	Wisconsin Hospital	11	Numeric, Discrete	SMO	95.4% 96.9%

(Continued)

Table 16.3 Comparison of different ML methods with ANN for BC prediction. (*Continued*)

Reference	Algorithms	Data	Number of attributes	Data type	Validation technique	Accuracy
[63]	NB RF LR MLP k-NN	UCI-USA	10	Numeric, Discrete	MCC, ROC Area	71.6% 69.5% 68.8% 64.6% 72.37%
[64]	CART RBF NB J48	WDBC	32	Images, Discrete	10-Fold	92.8% 93.6% 92.6% 92.9%

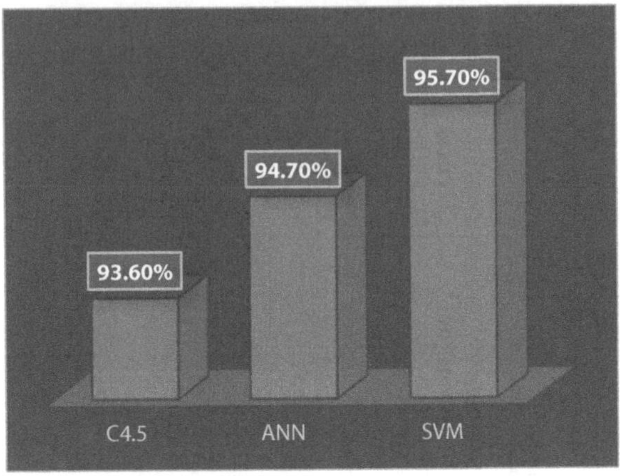

Figure 16.5 Comparison of algorithms on Iranian Center ICBC dataset.

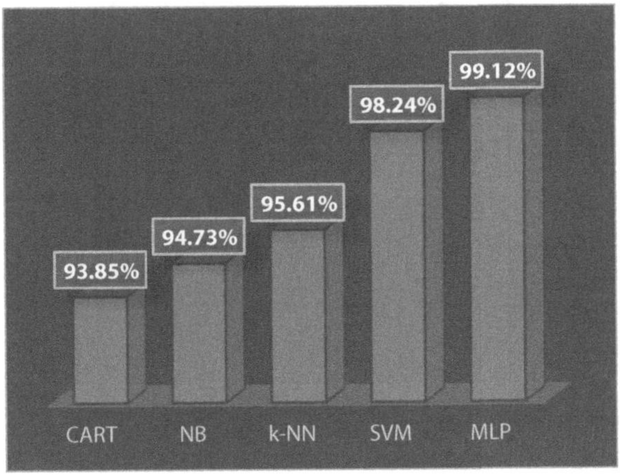

Figure 16.6 Comparison of algorithms on UCI WDBC data.

regard, numerous researchers are gathering medical images for the successful detection of malignancy. To resolve this issue of small datasets, several scientists are exploring the approaches of information augmentation such as "cropping, filtering, rotating, cleaning", etc., which could assist in obtaining large-sized datasets.

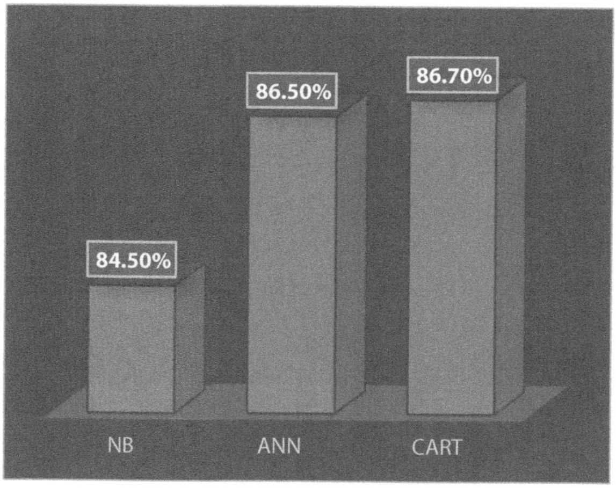

Figure 16.7 Comparison of algorithms on SEER data.

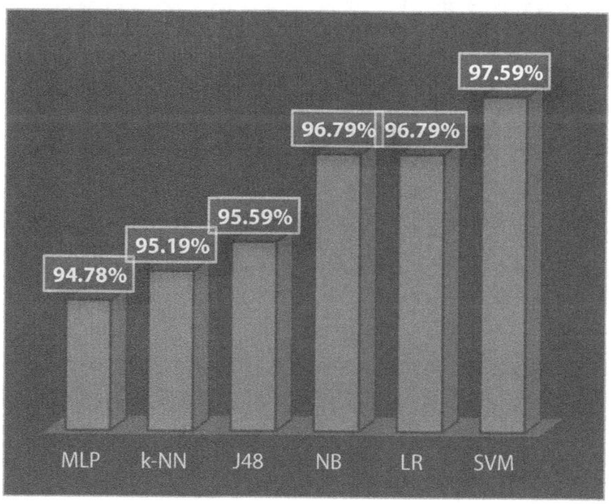

Figure 16.8 Comparison of algorithms on WBC data.

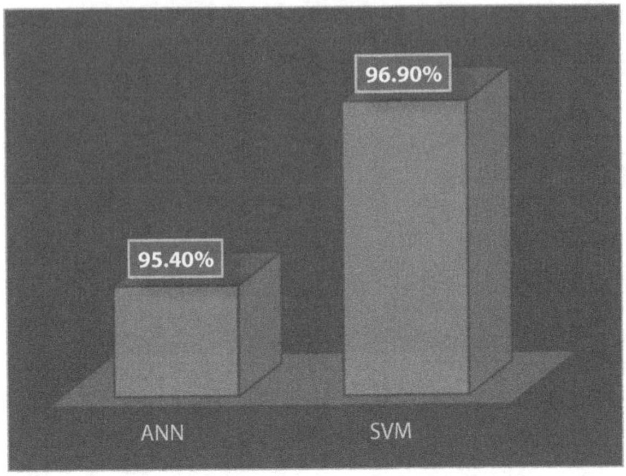

Figure 16.9 Comparison of algorithms on Wisconsin data.

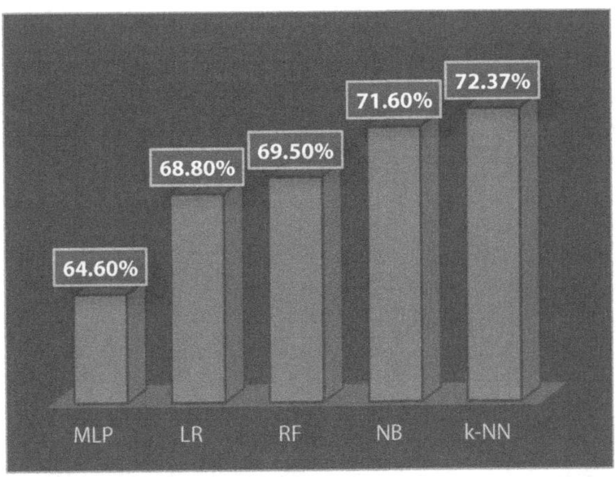

Figure 16.10 Comparison of algorithms on UCI-USA data.

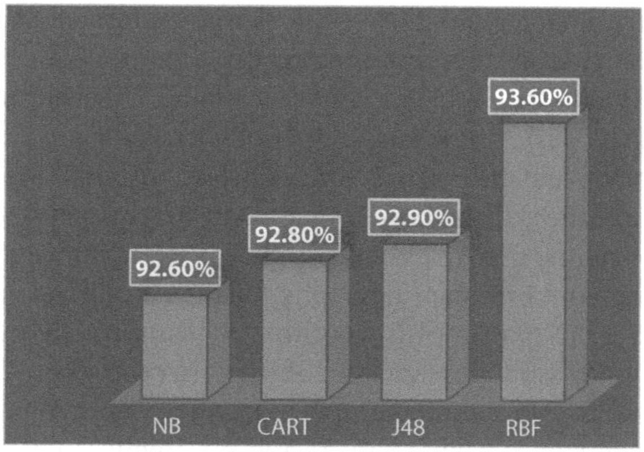

Figure 16.11 Comparison of algorithms on WDBC data.

16.4 Discussion

This segment discusses the importance of ANN in the estimation of different types of cancer like lung cancer (LC), BC.

With respect to diagnosis of LC, it was observed that the most vital parameter for the accurate classification of patients into those having LC and those not having LC was occupational exposure. For instance, occupational exposure to industrial elements for longer periods was found to be the most common risk factor for LC. Exposure to particular carcinogens such as arsenic, exhaust from diesel, asbestos, were found to be associated with LC [65, 66]. Furthermore, LC was found to attain its peak after 25 to 30 years of exposure to such toxic substances [67]. Red Blood Cells (RBC), weakness in muscles, cigarette smoking, the score associated with Zubrod were also observed to be significant attributes for the analysis of LC. The coefficients with higher values of sensitivity collected from RBCs helped to assess illnesses linked with progressive neoplasma. Patients suffering from an advanced stage of cancer were found to have a lower level of erythrocyte which was associated with secondary anaemia. The Zubrod score/ECOG scale is a coefficient with a higher value of sensitivity from the medical angle. It permits to ascertain the overall patient condition and their quality of life when suffering from cancer. It is also utilized in other critical and chronic illnesses. It ranges from 0 degrees indicating normal performance to 5 signifying deaths. These degrees specify the level of physical efficiency of the patient.

For the past few years, numerous CAD methods have been devised for examining BC to aid radiologists in analysing the overall image and to emphasize irregular regions that require more investigation. The system can aid radiologists in detecting tumour that cannot be identified by the human eye. Due to the rapid growth of technologies, several scientists are focussing on the design of smart methodologies that can be employed in mammography to enhance the accuracy of classification. These ML methods make use of the skills of humans more effectively than statistical techniques. Based on the outcomes of research, ANN has shown to be a useful classification method in mammography for the categorization of masses. The three-layer ANN using BP algorithm is being widely used in ANN mammography. Several implemented ANNs are centred on the notion of improving the TP rate and reducing the FP and FN rates for achieving the best possible outcomes. The specificity and sensitivity can be enhanced by utilizing wavelet based ANNs like PSO wavelet ANN.

Concerning the applications related to ultrasound, CAD techniques are being widely explored for ascertaining BC malignancy. This is mainly because the images from ultrasound possess non-radiation features, they are less expensive, and provide fast outcomes with enhanced accuracy. Recently, an enhanced form of BC diagnosis utilizing images from ultrasound has been designed, which employs 3D ultrasound imaging techniques for providing in-depth information on the region of the breast relative to 2-D imaging. The 2-D features are joined to form 3-D imaging. Additionally, the susceptibility of the ultrasound images is handled well by techniques based on ANN. Promising outcomes are obtained by most of the research works that have employed ANN. Techniques that hybridize two or more ANNs are being lately devised for BC diagnosis. A two-phase ANN which is hierarchical in nature has been observed to generate more consistent outcomes than utilizing the analysis of images independently. Improvement in ANN accuracy has been seen by providing it with many inputs for the successful detection and categorization of BC. Nevertheless, the hidden neuron number does not significantly influence the accuracy of the system. The ANN performance is dependent on the application type and the numerous attributes that are considered. It has been observed that most ANN methods demonstrated improved results with respect to several performance metrics. One more benefit of utilizing ANN in detecting BC is that ANN training can be done to enhance precision. Also, ANN can be hybridized with other methods of signal processing like wavelets to generate improved outcomes.

Diverse methods for the identification and categorization of BC exist depending on the input parameters. For example, the case of imaging related to IR, BC detection relies heavily on the statistical attributes of the tissue thermal behaviour like mean, SD, etc., along with the disproportionate separation of the contralateral breasts. Hence, it has been observed that the classification techniques for images focussed on ANN are beneficial in thermography. The "MRI Scan" is a dependable method for the localization of tumours along with a prior diagnosis of cancer since it is usually advised for the identification of soft tissues. Several techniques for the segmentation of images and 3D extraction are being employed in applications related to MI. Lately, several types of classification methods have been devised with several precise specifications for imaging of breasts based on MRI. The diagnosis incidence related to the cellular microfollicular ranges from 6% to 26% when compared to other aspirates. These comprise lesions that are follicular/irregular classes that vary from 10% to 40% in terms of the rate of cancer. Most medical experts advise operation for these patients since even a 10% risk of cancer is thought to be very high. On the contrary, around 80% of the patients may undergo needless thyroidectomy. Several techniques have been recommended to enhance the detection accuracy of thyroid lesions in patients. These comprise measurements of cytometric deoxyribonucleic acid, studies associated with morphometric, and the application of markers on the aspirates of the cell. But none of these techniques are relevant in medical practice. In this aspect, ANNs have been found to act as a better alternative to traditional methods based on statistics for forecasting complex biological processes.

In short, ANNs constitute a category of mathematical models which are nonlinear in nature. They comprise complicated neuron structures which are interconnected to each other through layers. These entities which are computational in nature, combine an input series using the additive operation and generate an outcome, such as the presence/absence of cancer. ANNs can be employed to perform modelling in a nonlinear fashion thus providing a novel option to LR which forms the most utilized technique for devising forecast models for predicting binary classes in the medical domain. Several benefits are offered by ANNs which comprise less training in statistics, capability to identify complex relationships among dependent and independent features which are nonlinear in nature, capability to recognize all probable associations among the predictor features, and the multiple algorithms for training are available. Surgeries that are not required for patients suffering from multinodular goiter may be avoided by utilizing ANN. ANNs acts as an effective tool for identifying cancer and its related tumours. However, more research is needed for devising optimal ANN models that may be successfully applied for practical medical applications [56].

Many medical datasets are not freely available which makes the design and development of ML and DL models for cancer prediction a very difficult task. Every technique needs a large dataset for training purposes and measurements associated with computation. To overcome this major challenge, several researchers have attempted to obtain image data of cancer patients that contain confidential information about the patients. Several of these datasets have been made open source in the form of raw images so that they can be accessed by a wide range of researchers. Another method of handling the issue of limited data is to augment the information using many techniques that comprise certain vital aspects like rotating, cropping, cleaning, filtering, etc. These schemes are supposed to assist in obtaining more available information about the patients.

16.5 Future Scope

Presently, ML techniques are yet in the stage of testing and experimentation for cancer detection. As more and more datasets of better quality are becoming available, researchers are devising accurate and precise models. In the upcoming years, cancer biopsy may be done by first performing clinical tests either at home or a clinic. The information may be given as input to the ML system. The detailed report may be received through the mail by the patient/pathologist offering an accurate/precise prediction about the potential of the patient getting cancer. Even though the job of a pathologist is not replaced currently, in the future ML may replace his/her job or assist them. Nevertheless, a lot of work needs to be done in the development of sophisticated ML algorithms in the medical domain where accurate prediction is critical. As more and more medical data become available accurate and sophisticated ML models may be built in the near future. Numerous challenges like data with multiple dimensions along with data heterogeneity are faced in the development of sophisticated ML algorithms for cancer risk estimation. As mentioned previously, researchers need to work on the issue of the limited dataset by employing techniques related to the augmentation of data. Imbalanced datasets need to be resolved as they can lead to bias in the prediction results. Another significant concern that needs to be resolved is the number of imbalanced BC images against patches of affected regions for accurate diagnosis and forecasting of BC.

16.6 Conclusion

Cancer forms a serious disease that is found to cause many deaths worldwide. Because of cancer's widespread destruction of lives around the world, early detection of cancer and estimation of risks associated with cancer is critical. In this regard, several ML models are being utilized for cancer risk estimation. Among the ML models, ANN has been found to be very beneficial for working on cancer detection and risk estimation. This chapter presented an extensive review of how ANN is being employed for cancer risk estimation. The focus of this chapter was to explore the various avenues in which ANN can be used for cancer risk estimation. The key aim of this chapter was to summarize the prior research work done for predicting BC, LC, PC, etc., using ANN. This chapter offers all essential knowledge to the novices who would like to examine the ANN algorithm to achieve the base of DL. This chapter demonstrates that ANN is still predominant among other ML techniques for cancer risk estimation. When compared to classic statistical techniques, the ML methods have exhibited considerable improvement in forecasting accuracy in most cases of a cancer diagnosis. Even though most research works have been performed exhaustively and validated extensively, the quality and quantity of medical data need to be improved along with the experimental strategy. Thus, suitable enhancements in the experimental strategies with better data quality and quantity would improve the general quality of several ML classifiers.

References

1. Gupta, A., Kaushik, D., and Verma, A., Machine Learning model for Breast Cancer Prediction. *4th International Conference on I-SMAC (Palladam, India)*, 472-477. doi: 10.1109/I-SMAC49090.2020.9243323, 2020.
2. Shandilya, S., Chandankhede, C., Survey on recent cancer classification systems for cancer diagnosis. In *International Conference on Wireless Communications, Signal Processing and Networking (WiSPNET)*, 2590-2594. doi: 10.1109/WiSPNET.2017.8300231, 2017.
3. Park, Y.S., Lek S., Multilayer Perceptron for Ecological Modeling. Developments in Environmental Modelling. *Ecological Model Types*, 28, 123-140. doi:10.1016/B978-0-444-63623-2.00007-4, 2016.
4. Khan, Farrukh., Khan, M. A., Abbas, S., Athar A., Siddiqui, Y., Khan, A.H., Saeed, M.A., Hussain, M., Cloud-Based Breast Cancer Prediction Empowered with Soft Computing Approaches. *Journal of Healthcare Engineering*, 16, https://doi.org/10.1155/2020/8017496, 2020.
5. Shikha, A., Jitendra, A., Neural Network Techniques for Cancer Prediction: A Survey. *Procedia Computer Science*. 60, 769-774. 10.1016/j.procs.2015.08.234, 2015.

6. Oustimov A, Vu V., Artificial neural networks in the cancer genomics frontier. *Transl Cancer Res.* 3(3), 191-201, doi: 10.3978/j.issn.2218-676X.2014.05.01, 2014.
7. Khan J, Wei J.S, Ringnér M, Saal, L.H., Ladanyi, M., Westermann, F., Berthold, F., Schwab, M., Antonescu, C.R., Peterson, C., Meltzer, P.S. Classification and diagnostic prediction of cancers using gene expression profiling and artificial neural networks. *Nature Medicine*, 28, 673-679. https://doi.org/10.1038/89044, 2001.
8. Pal NR, Aguan K, Sharma A, *et al.*, Discovering biomarkers from gene expression data for predicting cancer subgroups using neural networks and relational fuzzy clustering. *BMC Bioinformatics*, 8(5), https://doi.org/10.1186/1471-2105-8-5, 2007.
9. Chang KH, Miller N, Kheirelseid EA, *et al.*, MicroRNA signature analysis in colorectal cancer: identification of expression profiles in stage II tumors associated with aggressive disease. *Int J Colorectal Dis.*, 26(14),15-22. doi: 10.1007/s00384-011-1279-4, 2011.
10. Petalidis LP, Oulas A, Backlund M, *et al.*, Improved grading, and survival prediction of human astrocytic brain tumors by artificial neural network analysis of gene expression microarray data. *Mol Cancer Ther*, 7:1013-1024, doi: 10.1158/1535-7163.MCT-07-0177, 2008.
11. Fakoor R, Ladhak F, Nazi A, *et al.*, Using deep learning to enhance cancer diagnosis and classification. In: *Proceedings of the 30th International Conference on Machine Learning, JMLR*, 28, doi: 10.1109/ICSCAN.2018.8541142, 2013.
12. Raina R, Battle A, Lee H, *et al.*, Self-taught learning: transfer learning from unlabeled data. In: *Proceedings of the 24th International Conference on Machine Learning.* ACM, 759-66, https://doi.org/10.1145/1273496.1273592, 2007.
13. Rui Xu, Xindi Cai, Donald C, Wunsch II., Gene Expression Data for DLBCL Cancer Survival Prediction with A Combination of Machine Learning Technologies. In *Proceedings of the IEEE International Conference on Medicine and Biology*, 894-897, DOI: 10.1109/IEMBS.2005.1616559, 2005.
14. Bevilacqua, V., Mastronardi, G., Menolascina, F., Pannarale P., & Pedone., A Novel Multi-Objective Genetic Algorithm Approach to Artificial Neural Network Topology Optimisation; The Breast Cancer Classification Problem. In *Proceedings of International Joint Conference on Neural Networks (IJCNN '06).* 1958 – 1965. DOI: 10.1109/IJCNN.2006.246940, 2006.
15. L. Ziaei, A. R. Mehri, and M. Salehi., Application of Artificial Neural Networks in Cancer Classification and Diagnosis Prediction of a Subtype of Lymphoma Based on Gene Expression Profile. *Journal of Research in Medical Sciences* 11(1), 13-17, 2006.
16. Hiro Takahashi, Yasuyuki Murase, Takeshi Kobayashi, Hiroyuki Honda., New cancer diagnosis modelling using boosting and projective adaptive resonance theory with improved reliable index. *Biochemical Engineering Journal*, 33, 100–109. https://doi.org/10.1016/j.bej.2006.08.004, 2007.

17. Sung-Bae Cho, Hong-Hee Won., Cancer classification using ensemble of neural networks with multiple significant gene subsets. Springer Science and Business Media. *Applied Intelligence*, 26, 243–250, https://doi.org/10.1007/s10489-006-0020-4, 2007.
18. Yuchun Tang, Yan-Qing Zhang, Zhen Huang, Xiaohua Hu, Yichuan Zhao., Recursive Fuzzy Granulation for Gene Subsets Extraction and Cancer Classification. *IEEE Transactions on Information Technology in Biomedicine*, 12(6), 723-730. DOI: 10.1109/TITB.2008.920787, 2008.
19. Joshi D.M, Rana N.K., Misra V.M., Classification of Brain Cancer using Artificial Neural Network. In: *Proceedings of International Conference on Electronic Computer Technology (ICECT)*,112 - 116, DOI: 10.1109/ICECTECH.2010.5479975, 2010.
20. Rajeswari P., Reena G.S., Human Liver Cancer Classification using Microarray Gene Expression Data. *International Journal of Computer Applications*. 34(6), 25-37, 2011.
21. Barnali, S., Mishra D., A Novel Feature Selection Algorithm using Particle Swarm Optimization for Cancer Microarray Data. Proceedings of International Conference on Modelling Optimization and Computing (ICMOC-2012), *Procedia Engineering*. 38, 27-31, doi: 10.1016/j.proeng.2012.06.005, 2012.
22. Swathi S., Babu G.A, Kumar. Sendhil R, Sreenu, Naik B., Performance of ART1 Network in the Detection of Breast Cancer. In: *Proceedings of International conference on Computer design and Engineering (ICCDE)*, 49(19), 100-105. DOI:10.7763/IPCSIT, 2012.
23. Dev Jayshree, Dash Sanjitkumar, Dash Sweta, Swain Madhusmita., A Classification Technique for Microarray Gene Expression Data using PSO-FLANN. In: *Proceedings of International Journal on Computer science and Engineering*, 4(9), 1534-1535, 2012.
24. Goryński, K., Safian, I., Grądzki, W. *et al.*, Artificial neural networks approach to early lung cancer detection. *Central European Journal of Medicine*, 9, 632–64, https://doi.org/10.2478/s11536-013-0327-6, 2014.
25. Marszałł M.P., Krysiński J., Sroka W. D., Nyczak Z., Stefanowicz M., Waśniewski T., Romaszko J., Buciński A., ANN as a prognostic tool after treatment of non-seminoma testicular cancer. *Central Eur. J. Med.*, 7, 672-679, https://doi.org/10.2478/s11536-012-0027-7, 2012.
26. Luenberger D.G., Ye Y., Linear and nonlinear programming. *International Series in Operations Research & Management Science*. 116 (Third ed.), New York, Springer, pp. xiv+546, 2008.
27. Knyazev, A.V., Lashuk I., Conjugate Gradient Methods with Variable Preconditioning. *SIAM J. Matrix Anal.* A 29(4), 1267-1280. doi:10.1137/060675290, 2008.
28. M. M. Mehdy, P. Y. Ng, E. F. Shair, N. I. Md Saleh, C. Gomes., Artificial Neural Networks in Image Processing for Early Detection of Breast Cancer.

Computational and Mathematical Methods in Medicine. vol. 2017, Article ID 2610628, 15 pages. https://doi.org/10.1155/2017/2610628, 2017.

29. L. Hadjiiski, B. Sahiner, M. A. Helvie *et al.*, Breast masses: computer-aided diagnosis with serial mammograms. *Radiology*. 240(2), 343-356, https://doi.org/10.1148/radiol.2401042099, 2006.
30. B. Widrow, R. G. Winter, R. A. Baxter., Layered neural nets for pattern recognition. *IEEE Transactions on Acoustics, Speech, and Signal Processing*. 36(7), 1109–1118, doi: 10.1109/29.1638, 1988.
31. J. Dheeba, N. Albert Singh, S. Tamil Selvi., Computer-aided detection of breast cancer on mammograms: a swarm intelligence optimized wavelet neural network approach. *Journal of Biomedical Informatics*. 49, 45-52, https://doi.org/10.1016/j.jbi.2014.01.010, 2014.
32. W. Xu, H. Li, P. Xu., A new ANN-based detection algorithm of the masses in digital mammograms. In *Proceedings of the IEEE International Conference on Integration Technology (ICIT '07)*. 26–30, DOI: 10.1109/ICITECHNOLOGY.2007.4290471, 2007.
33. A. Alayliogh, F. Aghdasi., An artificial neural network for detecting microcalcifications in wavelet-enhanced digitised mammograms. In *Proceedings of the South African Symposium on Communications and Signal Processing (COMSIG '98)*. 127–132, 1998.
34. T. Ayer, Q. Chen, E. S. Burnside., Artificial neural networks in mammography interpretation and diagnostic decision making. *Computational and Mathematical Methods in Medicine*. vol. 2013, Article ID 832509, 10 pages. https://doi.org/10.1155/2013/832509m, 2013.
35. T. Balakumaran, I. L. A. Vennila, C. G. Shankar., Detection of microcalcification in mammograms using wavelet transform and fuzzy shell clustering. *International Journal of Computer Science and Information Technology*. 7(1), 121-125, 2010.
36. P. Dhawan, Y. S. Chitre, M. Moskowitz, G. Eric., Classification of mammographic microcalcification and structural features using an artificial neural network., In: *Proceedings of the Annual International Conference of the IEEE Engineering in Medicine and Biology Society*, 13, 1105-1106, 1991.
37. Y. Chitre, A. P. Dhawan, M. Moskowitz., Artificial neural network-based classification of mammographic microcalcifications using image structure and cluster features. *Digital Mammography*. 31-40, DOI: 10.1109/IEMBS.1994.411887, 1994.
38. S. W. Kevin, C. C. Doss, L. P. Clarke, R. A. Clark, M. Lee., A neural network approach to microcalcification detection. In *Proceedings of the Conference Record of the IEEE Nuclear Science Symposium and Medical Imaging Conference (USA)*, 2, 1273-1275, doi: 10.1109/NSSMIC.1992.301506, 1992.

37\. S. Saini, R. Vijay. Back propagation artificial neural network. In: *Proceedings of the 5th International Conference on Communication Systems and Network Technologies*. 1177–1180, 2015.

38. Adam Buciński, Tomasz Bączek, Jerzy Krysiński, Renata Szoszkiewicz, Jerzy Załuski., Clinical data analysis using artificial neural networks (ANN) and principal component analysis (PCA) of patients with breast cancer after mastectomy. *Reports of Practical Oncology & Radiotherapy.* 12(1), 9-17, https://doi.org/10.1016/S1507-1367(10)60036-3, 2007.
39. Marszałł MP, Sroka W, Adamowski M, Słupski P, Jarzemski P, Siódmiak J, Odrowąż-Sypniewska G., Engrailed-2 protein as a potential urinary prostate cancer biomarker: a comparison study before and after digital rectal examination. *Eur J Cancer Prev.* 24(1), 51-56, doi: 10.1097/CEJ.0000000000000046, 2015.
40. Muhammad Wazir, Hart Gregory R., Nartowt Bradley, Farrell James J., Johung Kimberly, Liang Ying, Deng., Pancreatic Cancer Prediction Through an Artificial Neural Network. *Frontiers in Artificial Intelligence.* 2, pages 2, Doi:10.3389/frai.2019.00002., 2019.
41. Pannala, R., Basu, A., Petersen, G.M., Chari, S. T., New-onset diabetes: a potential clue to the early diagnosis of pancreatic cancer. *Lancet Oncol.* 10, 88-95, doi: 10.1016/S1470-2045(08)70337-1, 2009.
42. Klein, A. P., Lindström, S., Mendelsohn, J. B., Steplowski, E., Arslan, A. A., Bueno-De-Mesquita, H. B., *et al.*, An absolute risk model to identify individuals at elevated risk for pancreatic cancer in the general population. *PLoS ONE.* 8: e72311, doi: 10.1371/journal.pone.0072311, 2013.
43. Boursi, B., Finkelman, B., Giantonio, B. J., Haynes, K., Rustgi, A. K., Rhim, A. D., *et al.*, A clinical prediction model to assess risk for pancreatic cancer among patients with new-onset diabetes. *Gastroenterology.* 152, 840–850 doi: 10.1053/j.gastro.2016.11.046, 2017.
44. Hart, G. R., Roffman, D. A., Decker, R., and Deng, J., A multiparameterized artificial neural network for lung cancer risk prediction. *PLoS ONE.* 13: e0205264., doi: 10.1371/journal.pone.0205264., 2018.
45. Roffman, D., Hart, G., Girardi, M., Ko, C. J., Deng, J., Predicting nonmelanoma skin cancer via a multi-parameterized artificial neural network. *Sci. Rep.* 8, 1701, doi: 10.1038/s41598-018-19907-9, 2018.
46. Hart, G.R. Nartowt, B. & Muhammad, Wazir & Liang, Ying and Huang, Gloria and Deng, J., Endometrial Cancer Risk Prediction and Stratification Using Personal Health Data. *International Journal of Radiation Oncology, Biology, Physics.* 105: E134-E135. 10.1016/j.ijrobp.2019.06.2170., 2019.
47. Nartowt Bradley J., Hart Gregory R., Muhammad Wazir, Liang Ying, Stark Gigi F., Deng Jun., Robust Machine Learning for Colorectal Cancer Risk Prediction and Stratification. *Frontiers in Big Data.* 3, pages 6., DOI:10.3389/fdata.2020.00006, 2020.
48. Roffman, D. A., Hart, G. R., Leapman, M. S., Yu, J. B., Guo, F. L., Ali, I., *et al.*, Development and validation of a multiparameterized artificial neural network for prostate cancer risk prediction and stratification. *JCO Clin. Cancer Inform.* 2:1-10, doi: 10.1200/CCI.17.00119, 2018.

49. Metz, C. E., Basic principles of ROC analysis. *Semin. Nucl. Med.* 88: 283-298, DOI: 10.1016/s0001-2998(78)80014-2, 1978.
50. BCSC, Risk Estimation Dataset [online] Available at: https://www.bcsc-research.org/data/rfdataset [Accessed 2 11.2020], 2006.
51. UCI Machine Learning Repository, Breast Cancer Data Set [online], Available at: https://archive.ics.uci.edu/ml/datasets/breast+cancer [Accessed 7 11.2020], 1998.
52. UCI Machine Learning Repository, Lung Cancer Data Set [online] Available at: http://archive.ics.uci.edu/ml/datasets/Lung+Cancer [Accessed 9 11.2020], 1992.
53. UCI Machine Learning Repository, Breast Cancer Coimbra Data Set [online] Available at: https://archive.ics.uci.edu/ml/datasets/Breast+Cancer+Coimbra [Accessed 14.11.2020], 2018.
54. UCI Machine Learning Repository, Breast Cancer Wisconsin (Diagnostic) Data Set. [online] Available at: https://archive.ics.uci.edu/ml/datasets/Breast+Cancer+Wisconsin+(Diagnostic)https://archive.ics.uci.edu/ml/datasets/Breast+Cancer+Coimbrahttp://archive.ics.uci.edu/ml/datasets/Lung+Cancer [Accessed 12 11.2020], 1995.
55. Ayer, Turgay & Chhatwal, Jagpreet & Alagoz, Oguzhan & Kahn, Jr, Charles & Woods, Ryan & Burnside, Elizabeth. Comparison of Logistic Regression and Artificial Neural Network Models in Breast Cancer Risk Estimation. *Radiographics: a review publication of the Radiological Society of North America, Inc.* 30: 13-22. doi:10.1148/rg.301095057, 2009.
56. Saylam, Barış & Keskek, Mehmet & Ocak, Sönmez & Akten, Ali & Tez, Mesut. Artificial neural network analysis for evaluating cancer risk in multinodular goiter. *Journal of Research in Medical Sciences: The Official Journal of Isfahan University of Medical Sciences*, 18(7), 554-557, 2013.
57. N. Fatima, L. Liu, S. Hong, H. Ahmed., Prediction of Breast Cancer, Comparative Review of Machine Learning Techniques, and Their Analysis. *IEEE Access.* 8, 150360-150376. doi: 10.1109/ACCESS.2020.3016715, 2020.
58. L. G. Ahmad, A. T. Eshlaghy, A. Poorebrahimi, M. Ebrahimi, A. R. Razavi., Using three machine learning techniques for predicting breast cancer recurrence. *J. Health Med. Inform.* 4 (124):3. DOI: 10.4172/2157-7420.1000124, 2013.
59. Bataineh, A comparative analysis of nonlinear machine learning algorithms for breast cancer detection. *Int. J. Mach. Learn. Comput* 9(3), 248-254., doi: 10.18178/ijmlc.2019.9.3.794, 2019.
60. Bellaachia, E. Guven. Predicting breast cancer survivability using data mining techniques (2006). In *Proc. SIAM Int. Conf. Data Mining.* 58. 100-110.
61. R. Kumar, G. Ramachandra, K. Nagamani., An efficient prediction of breast cancer data using data mining techniques., *Int. J. Innov. Eng. Technol.* 2(4), 139, 2013.

62. A. Bayrak, P. Kirci, T. Ensari., Comparison of machine learning methods for breast cancer diagnosis. In: *Proc. Sci. Meeting Elect. -Electron. Biomed. Eng. Comput. Sci. (EBBT)*, 1-3. DOI: 10.1109/EBBT.2019.8741990, 2019.
63. S. Bharati, M. A. Rahman, P. Podder., Breast cancer prediction applying different classification algorithm with comparative analysis using WEKA. In: *Proc. 4th Int. Conf. Electr. Eng. Inf. Commun. Technol. (iCEEiCT)*. 581-584. DOI: 10.1109/CEEICT.2018.8628084, 2018.
64. S. Aruna, S. Rajagopalan., A novel SVM based CSSFFS feature selection algorithm for detecting breast cancer. *Int. J. Comput. Appl*, 31(8), 1-7. doi: 10.5120/3844-5346, 2011.
65. Chen T.M., Kuschner W.G., Non-tobacco related lung carcinogens: Lung cancer principle and practice. In: Harvey P. *et al.*, editors. 3rd ed, Lippincot Williams and Wilkins., 61-73, 2005.
66. Bij S., & Hendrik Koffijberg H., Lenters V., Lung cancer risk at low cumulative asbestos exposure: meta-regression of the exposure–response relationship. *Cancer Cause. Control*, 24: 1-12. doi: 10.1007/s10552-012-0107-7, 2013.
67. Alberg A.J., Samet J.M., Epidemiology of lung cancer. *Chest*. 123, 21S-49S., 2003.

17

Applications and Advancements in Data Science and Analytics

T. Mamatha[1]*, A. Balaram[2], B. Rama Subba Reddy[3], C. Shoba Bindu[4] and M. Niranjanamurthy[5]

[1]Dept. of CSE, Sreenidhi Institute of Science and Technology, Hyderabad, TS, India
[2]Dept. of CSE, CMR Institute of Technology, Hyderabad, TS, India
[3]Anantha Lakshmi Institute of Technology and Sciences, Anantapur, Andhra Pradesh, India
[4]Dept. of CSE, JNTUA University, Anantapuramu, AP, India
[5]Dept. of AI&ML, BMS Institute of Technology and Management, Bangalore, India

Abstract

Data science is an interdisciplinary field that utilizes logical techniques, cycles, calculations and frameworks to separate information and experiences from uproarious, organized and unstructured information, and apply information from information across a wide scope of use spaces. This chapter focused on data science and analytics role in software testing, data analytics tools helped in extracting enormous amounts of data and analyzing that data with the help of so many techniques and keeping that data at a centralized place which testing people can utilize those data at any time for applying on product to test their validity. And how selenium, a software testing tool helpful in data science for performing image scraping. Selenium is one of the popular open source tool available in market for testing application and this selenium enhanced to provide its capability in data analytics in extracting data from web sources. And this chapter also lists out the applications of data science in various sectors like health, banking, manufacturing, agriculture, gaming, recommended system, E-commerce etc. This chapter also elaborates data science current existence in the market with the help of challenges and advancements and lastly the chapter will end with tools available in data science and analytics.

**Corresponding author*: mamathat@sreenidhi.edu.in

Kukatlapalli Pradeep Kumar, Aynur Unal, Vinay Jha Pillai, Hari Murthy, and M. Niranjanamurthy (eds.)
Data Engineering and Data Science: Concepts and Applications, (409–440) © 2023 Scrivener Publishing LLC

***Keywords*:** Software testing, data science, software testing tools, machine learning, data science tools

17.1 Data Science and Analytics in Software Testing

The current business scenario is changing at a rapid pace by collaborating advent automated technologies such as Data analytics, machine learning, data science etc. To streamline their work process and bring good results as outcome they incorporating testing as one of the most integral part to gain more quality in the product. Software testing is one of the phase in software development which developing rapidly. The importance of testing increased due to customer interest towards quality goods [6], it's due to growing competition in the industry to produce excessive quality and reliable code which works for longer time, this in turn gives confidence to deliver in the market in a very small span of time to maintain a competitive environment in the industry. The people involved to make this happens in the whole testing process are tester to perform black box Testing, Developer for performing white box testing, SQA (Software quality analytics) people to check the quality levels of the software through reviews and Customer to perform acceptance testing. Data analytics [1, 2] has crucial role in testing process in gathering a huge amount of data from various origins and examine it in a systematic way by using tools like SAS, WEKA, etc. are highly developed by companies to carry out the software testing process in a smooth or bug free manner. Industries are widely applying this technology to derive required interfaces and make informed business decisions.

Data analytics execute and process the gathered data by carrying out some sequence of steps by applying various algorithms during process. In Software development data in generated frequently that is analyzed by the analytics tool [3] to make the testing process easy and efficient. Thus reduce the error rate in the process and can give the product to the customer within a given time span. Giving correct data as input which in turn produce quality product to the market. This happens when it enables industry to understand the customer sentiments and current market value in developing the applications as per customer requirements and needs. This we can call as customer centric and concentrating on factors such as functional, performance, logical etc. Any technologies may be of Data science, big data [5] and AI need to test for functional, security and performance levels than only we can assure for quality of the product. In that way in market we have so many tools are available. We have classification of tools in this regard first one is Test management tools which helps to store testing process information like testing artifacts, bug tracking formats, test case management, test plan management etc. In order to store information we need take the help of tools that available in the market example Test director, Test link, Jira, qTest, practiTest, xQual, TestRail. Second comes the Functional Tools, used to test the functionality of the application by giving input and observing the output. The functional tools include UFT (Unified Functional Testing), QTP, ranorex, selenium, Testcomplete etc. Third category is performance Testing tool, used to test performance levels of the product like speed, scalability, security, response time, reliability etc. The performance tools include LoadRunner, JMeter, LoadNinja, WebLoad, LoadUI Pro etc. All the mentioned category wise tools enabled testers to deliver the software product as per customer expectations by analyzing their feedback and developing testing solutions accordingly. Any software produced by any area should be tested for quality checks that happened only with software testing tools, without software testing we can't assure the quality of the product and we can't give the product with confidence to the customer.

Software testing needs large amounts of test data that enables the data analytics to examine various bugs during the testing with the help of developers by applying multiple cases and actions to fix the bugs to increase product quality. Data analytics also helping testing in identifying the weak spot and actual effect of the defects to carry out the further testing process accurately. This has majorly contributed in optimizing workflows of the development process and prioritizing testing activities.

Software testing people get humorous support from data analytics [4] to perform their testing process with data they kept at centralized place such as scripts, log files, etc. which in turn gives potential solutions to detect early defects and take deterrent actions over bugs. Which has brought the scope of issues to peak level. This boosted in improving quality of the process and can deliver secure and reliable product to the end user.

17.2 Applications of Data Science and Analytics

Data Science and analytics which is dominating in all the way and helping the industry to carry out the data. So data is needed in every sector to carry out their work therefore data science and analytics has greater importance in the current market place. Below are some of the applications which are shaping our industry with help of data science and analytics.

1. Health Sector

- Analyzing Medical Image
- Identifying and predicting disease
- Personalized healthcare recommendations
- Drug Discovery
- Bio Informatics
- Virtual Assistant

By considering the medical image analysis, data science has been able to bloom the technology in such a way that X-Rays, MRI's and CT-Scans can be analyzed automatically but earlier, the search process for the clues in the medical images is done by the doctors and medical examiners manually. However, this manual process has been changed with strong sphere of influence in analyzing medical images with so much of research in implementing new technologies [2], now we are in a position where data scientists have produced robust image recognition tools that can automatically find flaws in the image and generate the reports.

Drug discovery playing vital role in almost all the sectors especially in sports. The competitors are using drugs to enhance their immunity system. To detect this kind of fraud competitors drug discovery been introduced. This field making use of data science and analytics to draw up or prepare new medicines [2]. Data Science simplifying this complex process of discovering drugs and will provide early insight of success rate. With the help of machine learning, we can compare and analyze various amalgamations of drugs and can be able to discover a new drug. We can also find the effects of the newly discovered drug on different gene structures to predict the outcome of it.

To analyze textual data, data science providing us one more beautiful technology called as Natural Language Processing simply we call as NLP [2], is a technology of data science mainly focused on the analysis of textual information with the help of virtual assistant or bots that can respond for user questionnaire. These intelligent bots are magnified and extended to use in healthcare domain to answer questions of patients and provide them with proper diagnostic guidelines.

2. **Web Surfing**

    - The search Engine gives search results within a fraction seconds by using Data science technology.

    We have many web surfing engines like google, yahoo, bing, etc. All these utilizing the current data science technology by applying various algorithms to produce the accurate outcome for our googled query in very less span of time like in seconds or milli seconds. Google consumes and exceeds twenty petabytes of data on daily basis.

3. **Manufacturing industries**

    - Automating Manufacturing Units
    - Predicting and monitoring systems
    - Abnormality Detection

    Data researcher has gained a major role in production side and being used in large scale in manufacturing area for upgrading this area with intention to reduce cost and encourage earning more profits with the collaboration of other areas like the **IOT**, Data science etc. Data science has helped industries to detect the issues and to manage whole software to predict the continual flow of data. In addition, data science can help manufacturing industries to reduce their hours in production and can monitor their energy costs. Data science also help industries by conducting analysis with help of data science tool and produced customer reviews for company to take suitable precautions and enhance the standard of their systems. Data science also showing interest to automate industrial work with new innovative techniques. Industries developing autonomous systems with the help of existing and real-time data. And also it removed unessential jobs and launched powerful machines that can use ML technologies.

4. **Recommendation Systems**

    - Recommending products
    - Analyzing Reviews

- To create a recommendation system. Such as, "suggested friends" on Facebook and Instagram or "suggested videos" on Netflix or YouTube or any other video streaming platforms, everything is done with the help of Data Science.

Data Science has become the backbone of recommendation systems, in filtering the data as per the customer reviews and making industries choice in inventing customer recommended product to gain more market value. With the help of recommended system, E-commerce industry are making profits by providing understanding to the customers by showing the reviews of the particular product, these will be considered from consumer old purchases on one particular item style (on the basis of ranking out of 5 or 10). These hybrid recommendation systems are useful for the companies to provide better service to the end-users that ie customers.

Companies also utilizing sentiment analysis by using NLP (natural language processing)to study and understand the assessment given by the client. And also used to study texts and online surveys. Machine learning plays vital role in detecting fraud merchants.

5. Voice and Image Identification

- Voice or speech identification software now in the market are "Siri", "Google assistant", "Cortana" and "Alexa" works on the basis of Data science techniques.
- The famous face book identifies your friend when you upload a photo with them, this happens with data science only.

Various industries need the application of facial and speech recognition to carry out necessary aspects of the organization. This process works with a huge amount of data that is combined with advanced algorithms to identify various patterns and execute face and speech recognition process in an efficient manner. The industries are shifting towards the use of highly advanced data science applications that works with face recognition algorithms.

Data science refers to a combination of technology, algorithms and data inference that is used to the source of information and convert the data into valuable resources for creating IT and business strategies. Data science applications are generally used for mining structured and unstructured data by using various scientific methods and algorithms, to identify different patterns and enhance the company's efficiency. Data science applications works by using statistical algorithms and a range of innovative methodologies, such as data mining, machine learning, cluster analysis, etc.

These applications are majorly contributing in the speech recognition process. It works by developing methodologies and technologies that enables the industries to convert spoken language into text by computer systems. Data science applications are used for converting the spoken words or phrases into a machine-readable format. This process is executed by the applications by gaining insights from the data that is gathered and allows the systems to gather hidden information from the data. These insights are used to create various patterns and algorithms and execute those patterns efficiently to convert the data into a readable format. These applications can be applied to large datasets to execute the process efficiently, by identifying voice interfaces and structuring document production.

Data science also enhances image recognition process by creating separate categories for each person in a database and analyzing the data by using definite algorithms. These applications help in optimizing the image detection process. Data science works by using high bandwidth data services and provides optimum security in carrying out the whole process.

Large volume of digital data is being created with the growing use of social media and this data is being used by the companies to deliver better services to its clients. Data science applications are being applied to this data to identify and detect various attributes of an image by producing symbolic information and converting the information into machine readable or detectable format. The data science applications are particularly used for detecting multiple faces, logos, etc. and provide precise attributes and characters by reading that image.

6. Gaming world

- Data science technology is used by EA Sports, Sony, Nintendo and helps in improving the user's gaming experience. Using Machine Learning techniques many games are developed and it aids in updating automatically when users move to higher levels.

  Gaming became one of the important entertainment industries of the globe because of the most exciting and learning based applications of data science can be seen in gaming. Gaming industry utilizing data analytics to implement and scrutinize, design and implement hypotheses, and structure experiments to test various hypothesis. make predictions and empower machine learning algorithms, identify patterns and trends to guide service maps and improve gaming models The main theme of the gaming development is to keep the player engaged and interested in playing by capturing performance levels of the player, tune the game as per player specification, awarding medals and points to encourage them. EA Sports, Zynga, Sony, Nintendo, Activision-Blizzard have made gaming experience to another level with the help of data science. Gaming industry utilizing machine learning algorithms to improvise themselves depending on the player moves from one level to another level.

7. Online Price Comparison

- PriceRunner, Shopzilla, Junglee are few examples which rely on the Data science mechanism. Here, data is retrieved from the appropriate websites using APIs.

  Web sites which are populated with so many of pages, each page can handle lots of data, here the data may be images or text or videos etc, which can be inherited by using "APIs" and "RSS Feeds". If you are familiar with these sites like "PriceGrabber", "Price Runner", "Junglee", "Shopzilla", "DealTime", than you know the comfort of relating the price of a product from multiple vendors at one place. "PriceGrabber", "Price Runner", "Junglee", "Shopzilla", "DealTime" are few price comparison websites, now available in the market. These website can be found in domains such as technology, automobiles, hospitality, apparels etc.

8. **Banking**

- Fraud Detection
- Debit and Credit risk management
- Customer Lifetime value

Banking is one of the major application in data science which can handle their resources effectively and keep up with the growing competition. In addition to this, smarter decisions are taken by banks through fraud detection. Banks can collaborate with companies to find the frauds that involve a credit card, insurance, and accounting. Data science also helps banking sector in analyzing customer cycles, expenditure patterns and suggest customers several offers or benefits that suit them accordingly. Data science makes banking work very flexible in calculating customer lifetime value and manages the number of customers they have.

Further, banking sectors have the capability to take risk in making models with data science to assess their performance. In real time environment banks make use of machine learning techniques like predictive analytic to improve their analytics strategy.

9. **Transport**

- Car Monitoring system
- Self Driving Cars
- Predicting safety Precautions

In transportation sector data science is playing an important role for the safety of drivers with help of safety driving environments. Introduced self-driving cars which became most trending topic and replaced humans with data science reinforcement learning able to create intelligent vehicles and other flavors like cloud computing and IoT to identify traffic lights, stop signs, etc. Data Scientists enhancing working levels of the vehicle which adds value to the drivers. Through analysis of fuel consumption patterns, driver behavior and active vehicle monitoring, data science has created a strong foothold in the transport industry. Industries can create better logistical routes using a variety of variables such as "consumer profile", "location", "economic indicators", and "logistics".

For example transportation companies for instance uber is using data science for price optimization to provide better user experience. Predictive tools, which use elements like a "weather pattern", "availability of transport", "customers", etc. to accurately calculate the price.

10. Finance

- Strategic Decision Making
- Risk Analysis
- Customer Clustering

Data scientists have taken crucial role in automating various strategic financial tasks. Similar to banks which is mostly used for risk management, risk analysis, real-time analytics and consumer analytics. Every larger companies has risk during their business process, and it should be mandatory to analyze risk before taking decision so automated risk analytics to be included to take decisions for the industry. This increase the security and trustworthiness of the company. The first step in managing the risk is finding, supervising and prioritizing the risks with help of machine learning algorithms. These algorithms improve cost and design stability through training on the huge available data like economic transactions and customer data, which helpful in creating model and optimize cost.

Financial institutions predict customer lifetime value and their stock market moves by using predictive analytics which is the part of machine learning. Continuous analysis on the goods can helpful to financial companies to take proper decisions on customer interests and can create a personalized relationship with them, depending on the choices of consumer preferences on choosing good and modification of preferences. Data Science also plays a key role in algorithmic trading.

In the earlier days, data were prepared and analyzed in batches means one by one, the disadvantage is that data became old by the time it was processed and analyzed so it require real time data to satisfy current circumstance.

This boosted to use real-time analytics of customers which increases the personalization. We can analyze customer reviews through multiple analysis techniques like sentiment analysis and machine learning algorithms which boost them to increase them business depending on customer interest shown on product. Also, the advanced machine learning techniques like **natural language processing** and data mining have contributed to the transformation of information for smarter governance that helps to increase the profitability of businesses.

11. **Airline Route Planning**

- Anticipating the delay of flights
- Deciding which class to buy
- Decision making depending on situation to reach destination, we can halt or take break between the travel.
- Implementing end-user loyalty programs.

Excluding some airline services, other airline industries across the world faced huge problems to handle their operations and to increase occupancy ratio due to high raise in air fuel prices and offering discounts which made the situation worse. But this situation is not persisting for longer period because airline industries incorporated the data science to estimate and take strategic decisions in areas which they faced loses. Airlines industries extensively using data science to decide which class to buy, they can forecast flight hindrance etc. Top airline industries like Alaska airlines and Southwest airlines making use of data science and bringing changes to make more profits.

17.3 Selenium Testing Tool in Data Science

17.3.1 Basic Techniques for Testing Voice-Based Applications

This approach makes use of Speech Recognition Engine (SRE) for automatic control of the voice-first devices just like any human being would. Voice-based apps are proving the interaction between humans and machines can be frictionless. This is indeed a new method for the brands and businesses for engaging better with the consumers by rendering an easy, familiar, and fast voice experience.

For ensuring a superior quality voice-based app, you require having an extensive testing strategy, which covers all the phases of the lifecycle of voice requests such as VUI, interactions in between voice platforms and voice apps, and back-end services. We face different challenges when testing voice apps than when we test GUI apps. For instance, GUI apps limit the number of possible interactions a user might do. Voice, on the other hand, allows a much richer and complex set of spoken interactions, increasing the difficulty of testing. Additionally, the backend behind voice apps includes several components not owned by developers. These AI-powered elements are constantly learning and evolving by gathering insights from the myriad of interactions they receive. This is why they get constant updates and improvements, which requires us to keep up on our side by doing continuous verification to be sure nothing has broken, and that our app continues to deliver great voice experiences to our users.

Example of some of the Voice Based Application for Testing:

- Alexa, siri and virtual assistant Testing.
- Google Home apps testing .
- Voice-enabled web apps and websites.

Voice App Testing Layers

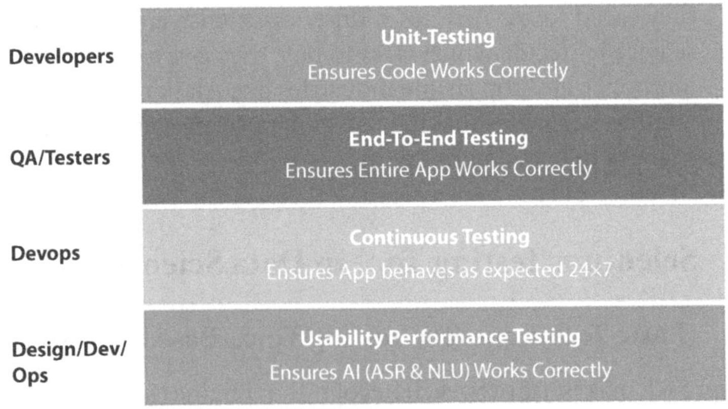

Figure 17.1 Voice app testing layers.

Unit Testing

This type of test is targeted to voice app developers. They need to do unit testing to ensure the code is working correctly in isolation. As you perform unit testing while coding, you need it to be fast, so it doesn't interrupt your coding pace. Unit testing is focused on making sure your code and logic are correct so there is no need to hit the cloud (where most voice apps backend lives), or call real external services. It is important that the unit testing tool you choose supports mocks and preferably runs locally. The above Figure 17.1 represents the Voice App testing layers.

End-to-End Testing

This type of test is targeted to QA teams. The tests should be comprehensive and if possible easy to write, read and maintain. As closely as possible, we want to imitate what real users do in their everyday interactions with voice applications.

Continuous Testing (Monitoring)

This type of test is targeted to Ops people and is to ensure that a service, once deployed, works flawlessly. This kind of testing verifies your voice app on a regular interval. It should be easy to set up and the results should be delivered immediately when the voice app has stopped behaving as expected.

Usability Performance Testing

Usability Performance Testing makes sure that the AI components of the app are working correctly. It is targeted to Product Managers and developers. The main goal is to identify issues with the speech recognition and NLU behavior of the assistant and your app. It consists of comprehensive testing of the interaction model, creating a baseline set of results. These results, in turn, are the basis for making improvements to the interaction model and the code of the skill - once revisions are completed, additional tests is run to ensure everything is working as expected.

17.3.2 Image Web Scraping Using Selenium

Example:

Web scraping is a technique to grasps an huge worth of data about electric good, mechanical goods, people, places, equipments etc. Crawling or Spidering are the synonyms for web scraping which extract data from the web sources automatically that is from a website. This crawling is an easy and relatively it takes very little span of time to extract more volumes of data, while doing this process it can keep more stress on the server where the source is hosted. So these sources are from websites such as Electronic commerce portal, social media channels to analyze end-users purchasing patterns and customer's sentiments etc..

Software testing tool [7] i.e Selenium is used to perform web scraping. Selenium is an web based testing tool which is open-source tool. Initially used for testing web applications developed in IT industry but it extends it capabilities to be used for crawling. To perform this use chrome but can work almost same in any browser, Selenium is a bundle of flavors like selenium IDE, Selenium RC, Web Drivers and Grid etc..

Initial Setup:

1. Install selenium package using pip or conda
2. Install web driver
3. Run web driver for confirmation

Step1: Import libraries (Requirements)

```
import requests
import os
import selenium
import io from selenium import webdriver
import time from PIL
import image from webdriver_manager.chrome
import ChromeDriverManager from selenium. Common.exceptions
import  ElementClickInterceptedException
```

Step 2: Web Driver Installation

```
Dri=webdirver.Chrome(ChromeDriverManager().install())
```

Step 3: Choose url chrome or any other then Copy and paste search URL at dri.get("")Ex:labtop

```
S_url=http://www.google.com/
Dri.get(s_url.format(q='labtop'))
```

Step 4: The search open new browser for the link specified

Step 5: Select the last page

```
dri.execute_script("window.scrollTo(0,document.body.scrollHeight);")
Time.sleep(4)
```

Step 6: Perform scraping on web page to localize the image

```
imgR=dri.find_element_by_xpath("//img[contains@class,'Q4LuWd')]")
totalR=len(imgR)
```

Step 7: Extract all the image links in the page to a "list", which stores those links. To perform this, select inspect element by right click on the browser.
Step 8: Than select any class attributes Ex: class, id, etc., this attribute should be common for all images.

Step 9: Individual image link will be extracted.

```
img_u=set()
for i in range(0,len(imgR)):
  img=imgR[i]
   try:
    img.click()
    time.sleep(2)
   act_imgs=dri.frd_elements_by_css_selector(img.n3VNCb')
   for act_imag in act_imags:
    if act_img.get_atribute('src') and 'http' in act_img.get_attribute('src'):
            img_u.add(act_img.get_attribute('src))
                              exceptElementClickInterceptedExceptionor
ElementNotInteractableExceptionas erro:
  print(err)
```

Above code describe the following task:

- Iterating through each thumbnail and then clicking it.
- Let browser sleep for 2 sec.
- Finding unique HTML tag that corresponds to image and locate it on page.
- Get multiple results for a each picture. Extract just the image that you want.
- We go through each execution cycle to crop "src" and check whether this src have http.

Step 10: In one particular directory save and download each image

```
os.chdir(C:/Qurantine/Blog/WebScrapping/Dataset1')
for i, urlin enumerate(img_u):
file_n=f"{i:150}.jpg"
try:
 img_com=requests.get(url).content
except Exception as e:
   print (f"ERROR-  COULD NOT DOWNLOAD {url}-{e}")
try:
  img_f=io.BytesIO(img_con)
  image=Image.open(img_f).convert('RGB')
  file_p=os.path.join(baDir, file_n)
  open(file_path, 'wb') as f:
   image.save(f,"JPEG",quality=90
   Print(f"Saved –{url}-At:{file_p}")
  Except Exception as e:
    Print(f"Error-Could Not Save {url} –{e}")
```

Step 11: Successfully extracted the images for your work.

17.4 Challenges and Advancements in Data Science

Challenges:

- Need of Automated machine learning for transforming data science with better data management.
- Heterogeneous information & data is required for accurate analysis
- Not adequate pool available in Data science
- Management of the financial support for a data science team
- Access data permissions
- Data Science results not effectively used by business decision makers
- Unable to reach Security and Privacy issues
- As technology enhancing facing lack of domain expert
- Inadequate support of data science team for small organizations
- Choosing the Right Algorithm
- Improper Data Processing

Advances:

In the upcoming years this technology is evolving continuously, there will be massive growth in the AI and ML field [1]. There is already a considerable amount of data to be managed, and with new technological advancements, we can utilize big data in many ways. For that, we have to stay up to date with the latest trends in data science.

- **Usage of Intelligent Applications**
  Now artificial intelligence is growing gradually in all the sectors including medical, army, banking etc. this gradual improvement can see more intelligent and advanced systems of AI in all domains. Artificial Intelligence has become the accepted stream for small scale to larger scale industries, and it will bloom in the next few years because it promoting firms to step forward to improve their business procedures and handle the customers and their data in efficient manner.

  The future scope is automated machine learning this helps data science to metamorphose for better data handling. For this we need to have expertise knowledge on deep learning to execute. Also in 2020 we found advanced applications in Machine learning, Data analytics, Deep learning , data science etc. with use of intelligent technology.
- **Internet of Things Improvements:**
  The new trend in the market is usage of smart devices like Google assistant, Cortana by Microsoft and alexa, allowing the end user to automate their regular routines and grabbing consumer interest and attention through these smart devices. Thus it will give boosting toward business encouragement to invest in information technology, especially in smart devices. The investment in IoT will target one trillion dollars by the end of 2020 according to the IDC report. Before 2020, we used devices that allow controlling our home appliances like TV, oven, AC, lights, doors etc. Some examples, IoT played remarkable work for agriculture side which helping the farmer to detects soil types, soil moisture etc. In telecommunications, Vodafone gained much value in market by incorporating IoT technology to connect and assists people with learning disabilities with initiation of a a

project called "Connected Living" which improve individual life to learn more.

- **Evolution of Big data analytics**
Big data analytics [5] is one which plays major role in data science to gain business competitive edge over data. Many technologies like python, which analyze large data to use in enterprises to enhance their business in market and to attract consumers to increase their sales rate. Most of the companies trying to forecast the business issues and identifying the reasons behind certain uncertainties with the help of predictive analytics. It helps companies in identifying what's going to happen in the future. For instance, we analyze the interest of the customer by their browsing or purchase history to predict intelligent strategies to target new purchasers with the help of predictive analysis.
- **Need of Data science Security Professionals**
The rate of professional ratio in data science security is demanding for more to process data to customer securely. Even though in the market already we have security analytic experts who gain good knowledge in data science. As both Artificial Intelligence and Machine learning entirely depends on data and to handle this data securely and efficiently data scientists must have expertise in this field. For that data security scientists must be well versed with the latest technologies of data science or big data analysis. For example, python is amongst the most used languages in data science and data analysis, so having a clear understanding of python concepts can help you tackle the problems related to data science security.
- **Advancements in Machine Learning**
The consistent advancements in Machine Learning are leading to many technological developments [3] in different sectors. The voluminous and emerging datasets, affordable data storage alternative and ease in computational processing are some of the major factors which are making machine learning more reliable. Machine learning techniques are being used in different sectors like agriculture, gaming, e-commerce, health and many more. The ease of identifying patterns and accurate predictions has added to its increased importance.

In the recent times, Machine learning has been playing a pivotal role in the modern-age agriculture. The use of different machine learning algorithms, both supervised and unsupervised, like K-nearest neighbor, Support Vector Machines, Random Forest as well as advanced image processing techniques have helped to identify better patterns and improved prediction in this sector. Crop Management, Plant Disease Prediction, Agriculture Robot, Soil Management, Livestock Management are some of the major applications of machine learning in the field of agriculture. These algorithms have helped in better analysis and effective decision making. For instance, machine learning acts as an important tool for crop yield prediction. It helps to identify information regarding what crop to grow and predict the yield based on different factors like soil type, environment, moisture, humidity and soil pH value. The usage of advanced pre-processing techniques and Deep Neural Networks (DNN) has ameliorated the accuracy of the prediction thus leading to better decisions. It uses the Support Vector Machine, a supervised machine learning algorithm, and TwoInnerProduct layer, a deep learning method, to estimate the crop yield. This usage of these algorithms helped to achieve high accuracy levels.

Machine Learning also gained importance in the E-Commerce sector. In E-Commerce, it is acting as a catalyst to take the customer experience to a new-level. It is helping to gain increased revenue by recognizing customer behavior patterns. For example, many e-commerce websites like Amazon, Netflix use the Collaborative Filtering Model for implementing recommendation systems. It is an unsupervised machine learning technique to similar users or items based on simple rating and use. This helps in effective decision making in E-Commerce industry and customer relationship management. The usage of machine learning algorithms in virtual assistants has also gained immense importance in the developing world. The voice assistants like Siri, performs work based on voice based instructions. These instructions are recorded, transmitted to the server through cloud and decoded using complex Machine Learning algorithms to act accordingly.

In the Business and Banking sector machine learning has its application in fraud detection, Chatbots and many more. For illustration, in Credit Card Companies, Machine Learning is used to identify any fraudulent activities and unauthorized accessing of cards to prevent financial loses. It also helps in identifying fake id's and fake accounts that intend to steal money in the middle of transactions. The usage of Bayesian Network for the data processing is required, without experimentation. This supervised learning algorithm helps in effective modeling, in turn used for pattern identification. It also gives good accurate results. Stock market prediction is also another example for the application of Machine Learning in the Business Sector. These adaptable understanding algorithms help to identify patterns and insights which are undiscoverable by human power. It uses Logical Regression, K-Nearest Neighbor and advanced techniques like LSTM to predict the future stock price for the company or other financial instrument traded on an exchange. It gave results with accuracy rate as high as 85%. Furthermore, the machine learning techniques have great potential to improve decision making and faster prediction in the business sector.

Machine learning methods have also gained importance in the automobile industry. To exemplify, consider the self-driving cars. The self-driving car allows smooth movement of vehicles on the road without human intervention. It uses unsupervised machine learning techniques to detect people and objects while driving. Another example for the machine learning application in this sector is traffic- flow prediction. The Support Vector Machine, a complex Machine Learning method, is used to predict the number of vehicles flowing through a given time. The introduction of deep learning techniques helped to achieve more accurate results.

The usage of machine learning techniques in the health sector is growing at a faster rate. It is majorly used for disease diagnosis. The diagnosis of life threatening diseases at the preliminary stages is more important. Advancements in machine learning help to develop better methods for prediction in the early stages. Early diagnosis of the disease helps to reduce severity and reduce expenses. The usage of machine learning techniques helps in better effective decision. Several unsupervised and supervised learning methods like "K-Nearest Neighbor", "Support Vector Machines", "Random Forest", "K-means", "MLPClassifier" and other image processing techniques are being used to detect various diseases. These techniques are mainly used for the prediction of cancer, heart disease, diabetes and personalized medicine etc. For instance, Support Vector Machine, a supervised machine leaning algorithm, gives great accuracy for the prediction of heart disease. This algorithm helps in early diagnosis of heart disease. SVM gives reliable output and accuracy and is the best method for disease and peat prediction. Another example is role of machine learning to detect breast cancer. Breast Cancer has become a major concern amongst women in modern times. Several women die every year due to late diagnosis or wrong diagnosis. The introduction of machine learning techniques helped in predicting the type of tumor thus helping in early diagnosis. The use of Random Forest algorithm helped in early prediction and has given accuracy levels as high as 97%. Thus machine learning techniques have immense importance in the health care sector. It helps to provide health care professionals to use data and analytics to provide better services, improve processes, and automate tasks. The essence of machine learning in healthcare is due to its ability to process huge datasets beyond the scope of human capability, and then reliably convert analysis of that data into clinical insights that aid physicians in planning and providing care, leading to better outcomes, lower costs of care, and increased patient satisfaction.

17.5 Data Science and Analytics Tools

The above Figure 17.2 shows the Data Science and related Analytical Tools. Most associations utilize more than one investigation device, incorporating accounting sheets with measurable capabilities, factual programming bundles, modern information mining instruments, and prescient demonstrating apparatuses

1. SAS

- Full form of SAS is Statistical analysis system.
- SAS data is available in "HTML", "PDF", "Excel", "RTF" and other depending on our output requirement.

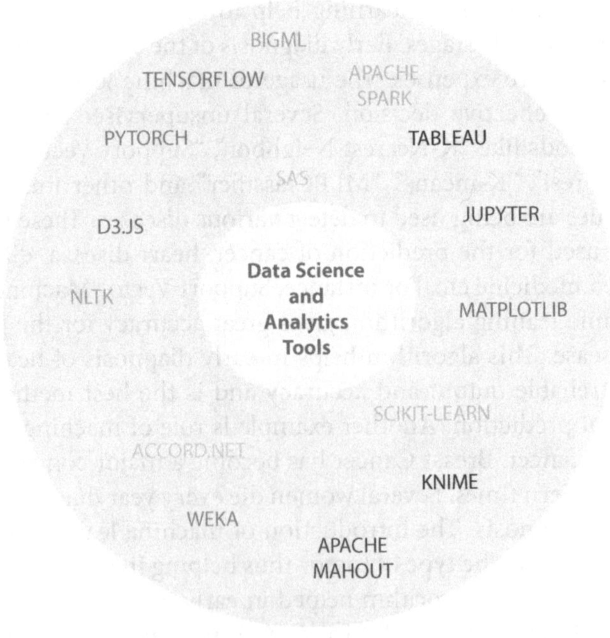

Figure 17.2 Data science and analytical tools.

- SAS makes the work flexible for non-technical people in the form of graphical and user interface representation.

Advantages

- It is Easy to learn.
- SAS syntax is very easy to learn.
- It can learn easily without any programming skill so anyone can learn it.
- The coding is in simple form.
- It is like giving instructions to the machine what to do.
- It is Easy to Debug.

Disadvantages

- Cost of SAS is very expensive.
- It is not Open source.
- Difficult in Text Mining.
- It is Difficult than R.

2. Apache Spark

- It is an open-source tool.
- It is based on distributed processing system which is used for big data workloads.
- It is fast and used for processing large scale of data.

Advantages

- High Speed
- Ease to Use
- It usesAdvanced Analytics
- It is Dynamic in Nature
- Apache Spark is powerful
- It is Open-source

Disadvantages

- It cannot do automatic optimization process
- It as File Management System
- It as Small Files Issue

3. BIG ML

- Big ML is a machine learning platform design to do developers to build enterprise-level applications are used in real-time predictions.
- It has a user-friendly and visually engaging interface.
- As a result, they can be productive and efficient regard.

Advantages

- It as Integration & Automation
- It as Scalability
- It is a One-Line Automation
- It is used in Real-time Predictions

4. D3.js

- Simply called it as D3, D3 abbreviated as " data driven documents".
- Is a Javascript library. Can be used with any JS framework of a choice like "Angular.js", "React.js" or "Ember.js".
- Used for producing data visualization in web browsers.

Advantages

- Is an open-source and it can use source code and you can add own.
- Works with web standards like HTML, CSS and SVG
- D3 works directly with web standards

Disadvantages

- D3 doesn't always older browsers.
- D3 has some data-source limitations.
- D3 cannot easily conceal original data.

5. Mat-lab

- Matlab stands for "Matrix Laboratory"
- Has more than four million users
- It is a proprietary language for technical computing.

- Familiar mathematical notation is used to express problems and solutions.
- uses: Math and computation.

Advantages

- Implement and test is easily.
- Computational codes easily.
- Debug easily.
- Symbolic computation is easily
- Platform independent
- Has pre-defined functions

Disadvantages

- It is costly
- Slow due to interpreted language
- MATLAB Online cannot interact with other h/w's.
- "xlsread" and "xlswrite" operate in basic level.

6. Ggplot2

- Is a data visualization package for R
- Used to generate graphics.
- gg abbreviated as "grammar of graphics".
- ggplot need three arguments i) data, ii)aes iii) geom to create plot.

Advantages

- Automatically deals with spacing, text, titles but also allows you to annotate by "adding"

Disadvantages

- "magic" evaluation model
- probably even harder to customize than.

7. Tableau

- It is flexible to understand than python.

- It can use or handle millions data in easiest way.
- Available with more option to visualize the data to enhance the user experience.

Advantages

- Data visualization
- Quickly Create Interactive visualizations
- Ease of Implementation
- Use of other scripting languages in Tableau

Disadvantages

- Scheduling or notification of reports:
- It as Static and single value parameters
- No Custom Visual Imports
- Custom formatting is done

8. Tensor Flows

- Tensor flow offers a JS library that helps in machine learning development. Its APIs will help you to create and train the models.
- Helps to build and train your models.
- It can be used as a model converter and can run existing models.
- It is free software.
- We can easily train and deploy models in cloud.
- It plays vital role in numerical calculations and in neural network.
- With help of API's like keras, we can build machine learning models effectively.

Advantages

- Utilize as script tags or as by installing through NPM

Disadvantages

- Complex to understand

9. JUPYTER

- The Jupyter Notebook is the ideal method when you wish to share with others the thought process behind your insight.
- Is a widely used machine learning note book .
- Processing speed is very high and efficient platform.
- Supports three languages i) Julia ii) R iii)Python.

Disadvantages

- Jupyter had code auto-completion feature like PyCharm.

10. NLTK

- Full form of NLTK is Natural Language Toolkit
- 32 universities in the US and 25 countries using NLTK in their courses
- Supports classification, tokenization, stemming, tagging, parsing, and semantic reasoning functionalities

Advantages

- It can hold up many languages compared to others.

Disadvantages

- Difficult to understand and process
- It is slow

11. Scikit-learn

- It provides library and used for machine learning development in python.
- It helpin data mining and data analysis in generating models and algorithms for Classification, Regression, Clustering, Dimensional reduction, Model selection, and Pre-processing.

Advantages

- Easily understandable documentation is available.

- Easily we can edit parameters of any algorithm while calling objects.

12. KNIME

- Full form of KNIME is Konstant information miner
- Is a tool for data analytics, reporting and integration platform.
- Using the data pipelining concept, it combines different components for machine learning and data mining.
- It can integrate the code of programming languages like C, C++, R, Python, Java, JavaScript etc.
- It can be used for business intelligence, financial data analysis and CRM.

Advantages

- It can work as a SAS alternative.
- Easy to install.
- Understand easily.

Disadvantages

- Unable or difficult to develop complex models.
- Visualizing and exporting capabilities are limited.

This are other machine learning tools

- IBM WATSON
- Amazon machine learning
- open NN

13. PyTorch

- Is a Lua based processing structure, scripting language and machine learning library.
- It helps in developing neural networks.
- Optimization algorithms are extensively used for building neural networks.
- Can be utilized on cloud computing area.

- It provides distributed training, various tools, and libraries.

Advantages

- It helps in creating computational graphs.
- Ease of use because of the hybrid front-end.

14. Apache Mahout

- Helps mathematicians, statisticians, and data scientists for executing their algorithms.
- It provides algorithms for Pre-processors, Regression, Clustering, Recommenders, and Distributed Linear Algebra.
- For general mathematical processing java libraries are included.
- Distributed linear algebra methods are inherited.

Advantages

- Works efficiently for complex datasets.
- Very simple
- Can be enhancible

Disadvantages

- Less documentation available.
- Few algorithms are undefined like Accord.Net

15. Accord.Net

- For image and audio processing Accord.Net provides machine learning libraries.
- It provides algorithms for:
    - Numerical linear algebra
    - Numerical optimization
    - Statistics
    - Artificial Neural networks.
    - Image, audio, & signal processing.
    - It also provides support for graph plotting & visualization libraries.

Advantages

- All Library files are available through source code and by executable installer & NuGet package manager.

Disadvantages

- It works only for .Net languages.

16. Weka

- It is an open source tool
- WEKA stands for "Waikato Environment for Knowledge Analysis"
- Weka tool helps in data mining for solving real time mining problems.
- Data preparation
- Uses classification and regression algorithms to apply on data
- Clustering technique is used to extract similar dataset
- Weka is Visualization tool to inspect the data
- Uses association rules for mining.

Advantages

- Algorithms are easy to understand.
- It is good and suits to student as well.

Disadvantages

- For better understanding, detailed documentation and online support not available.
- It can suits for small datasets.

17.6 Conclusion

Chapter focused on Data science and analytics advancements and applications. Applications cover most of all the domains where data science and analytics concepts can apply. High level AI is a field of software engineering that ganders at how to further develop processing power by permitting projects to advance as they run, without extra programming. The fate of ML obviously demonstrates the expanded use of AI across different industry verticals. Generative AI is an AI branch that spotlights on producing content like composing text, creating pictures, text to picture age and making music. As per Gartner, Generative AI is an essential AI innovation pattern for current year. Progresses in this innovation have considered late forward leaps that advance quicker and more productive business knowledge, abilities to utilize going from facial acknowledgment to normal language handling. The objective of genuine man-made brainpower (a PC or program that thinks and conveys like a person) has not yet been accomplished. The role of testing on data science and sample web Scraping to show how it applied in data science. Chapter also discussed data science and analytics tools available in the current market.

References

1. Murugan Anandarajan, Chelsey Hill, Thomas Nolan - "Practical Text Analytics" Springer Nature Switzerland AG 2019 Number of Pages XXVIII-285 doi.org/10.1007/978-3-319-95663-3 2019
2. N. S. Godbole and J. Lamb, "The triple challenge for the healthcare industry: Sustainability, privacy, and cloud-centric regulatory compliance," 2013 10th International Conference and Expo on Emerging Technologies for a Smarter World (CEWIT), 2013, pp. 1-6, doi: 10.1109/CEWIT.2013.6713768.
3. C. L. Philip, Q. Chen and C. Y. Zhang (2014), inData-intensive applications, challenges, techniques and technologies: A survey on big data, Information Sciences, 275, pp.314-347.
4. M. K.Kakhani, S. Kakhani and S. R.Biradar (2015), Research issues in big data analytics, International Journal of Application or Innovation in Engineering & Management, 2(8), pp.228-232.
5. J. Manyika, B. Brown, J. Bughin, R. Dobbs, C. Roxburgh, A. Byers (2011), from Big data: The Next Frontier for Innovation Competition and Productivity pp:1-20 May 2011
6. Ansari, M. B. Shagufta, A. Sadaf Fatima and S. Tehreem, "Constructing Test cases using Natural Language Processing," 2017 Third International Conference on Advances in Electrical, Electronics, Information, Communication and Bio-Informatics (AEEICB), 2017, pp. 95-99, doi: 10.1109/AEEICB.2017.7972390.
7. P. Ramya, V. Sindhura and P. V. Sagar, "Testing using selenium web driver," 2017 Second International Conference on Electrical, Computer and Communication Technologies (ICECCT), 2017, pp. 1-7, doi: 10.1109/ICECCT.2017.8117878.

About the Editors

Kukatlapalli Pradeep Kumar, PhD, is an associate professor and the Program Coordinator for Data Science at Christ University, Bangalore, India. He has 13 years of research and academic experience. He has published in many journals, and presented numerous conferences papers.

Aynur Unal, PhD, educated at Stanford University (class of '73), has taught at Stanford University for almost 40 years and established the Acoustics Institute. Her work on "New Transform Domains for the Onset of Failures" received a prestigious research award.

Hari Murthy, PhD, is a faculty member in the Department of Electronics and Communication Engineering, CHRIST University, Bengaluru, India. He finished his PhD from the University of Canterbury, New Zealand where his thesis was on novel anticorrosion materials. He has authored book chapters and published papers in international journals and conferences and has served as part of the program committees for several international conferences.

Vinay Jha Pillai, PhD, is an associate professor in the Department of Electronics and Communication Engineering at CHRIST University, Bangalore, India. He has 12 years of academic experience and holds two patents. He has also completed two funded projects as principal investigator.

M. Niranjanamurthy, PhD, is an assistant professor in the Department of Computer Applications, M S Ramaiah Institute of Technology, Bangalore, Karnataka. He earned his PhD in computer science at JJTU, Rajasthan, India. He has over 11 years of teaching experience and two years of industry experience as a software engineer. He has published several books, and he is working on numerous books for Scrivener Publishing. He has published over 60 papers for scholarly journals and conferences, and he is working as a reviewer in 22 scientific journals. He also has numerous awards to his credit.

Index

Accuracy (A) 352, 366,
Activation function 97, 113, 114, 342, 377, 380, 387
Akaike Information Criterion 210
Amazon Redshift 51, 52, 53, 54, 55
ANN Modeling 89, 98
ANN technique 120, 373, 378, 379, 382, 383, 386, 388, 390, 398
ARIMA Model 10, 19, 303, 309, 313, 314
Artificial Neural Networks 8, 94, 96, 113, 201, 227, 437
Associated Risks 1, 8, 21
Autoregressive model 98,

Bagging 139, 140, 146,
Bagging Meta Estimator 137, 138, 147, 149, 150, 152, 168, 169
Bayesian regularization 87, 114
Big Data 28, 49, 50, 201, 283, 284, 290, 291, 295
Biological Neuron Model 95, 117
Boosting 137, 138, 146, 147
Breast Cancer 388, 390, 442

Canopy K-Means 303, 304, 305
Chest Radiograph 351, 352, 381
Classification 388, 390, 397, 398, 399, 410, 435, 438,
Classification Technique 25, 30, 34, 35, 38, 141, 181, 292, 352, 399
Cloud Data Warehouse 49, 50, 51, 53, 54, 55, 57, 59, 60, 61, 63, 64
Cluster Analysis 71, 72, 73, 74, 78, 81, 82
Clustering Methodology 38
Clustering Techniques 28, 72, 73, 82, 179, 369,
COPD 279, 280, 281, 285, 287, 289, 290, 295, 298, 352, 354

Data Analysis 2, 8, 71, 435
Data collection 2, 74, 89, 93, 260, 290, 326
Data exploration 93
Data Masking 8, 9, 21, 64,
Data Mining 23, 24, 25, 26, 27, 28, 29, 30, 35, 38, 52, 71, 72, 201, 267, 307, 438
Data Models 8
Data processing 74, 92, 93, 428
Data Science life cycle 89, 92
Data Warehouse 49, 50, 51, 65, 72
Decision Tree 23, 26, 27, 28, 34, 35, 36, 38, 139, 147, 149, 258, 294
Dynamical System 85, 86, 87, 89, 115

Elapsed Time (ET) 352
Emotional Analysis 227, 230, 233, 236, 248, 249
Evaluation Metrics 20, 352, 364, 365

Face Alignment 232
Feature extraction 93, 173, 174, 176, 179, 180, 182, 201, 348, 355
Feedforward ANN 96, 97

F-N system 99, 101, 102, 103, 117, 124
Forecasting 10, 201, 307, 308, 309, 312, 387
Fuzzy Set Theory 255, 256, 275

Glioblastoma 195, 196, 197, 199
Google BigQuery 58, 59, 60, 62, 65

IICS 51, 64, 66
Image Preprocessing 185, 321
Image Segmentation 319, 322, 352, 382

K-Means 304, 305, 307, 335, 345, 355, 386, 429
K-means clustering 303, 305, 345, 369,
KNN Clustering 81

LESH Vector 324, 325
Linear Regression 10, 279, 288, 293, 294, 295, 297, 335, 339, 341, 343, 355, 358, 359
Linearization 88, 90, 99, 111
Logistic Regression 27, 140, 164, 198, 289, 335, 342, 359, 369
Lung Cancer 281, 317, 318, 319, 324, 330, 351, 352, 358, 361, 369, 374, 378, 379, 389

Machine Learning 9, 20, 23, 30, 39, 55, 58, 87, 89, 92, 94, 112, 137, 138, 139, 146, 175, 179, 181, 182, 227, 279, 295, 297, 326, 335, 336, 338, 348, 351, 354, 359, 374, 410, 415, 418, 427, 428
Matthews Correlation Coefficient (MCC) 352, 367
Mean squared error 36, 38, 116, 200, 295, 341, 343, 352
Mean Squared Error (MSE) 116, 352, 366
Metrics in Evaluation 20
Model order reduction 86, 106
MOR of F-N System 102
Moving Average 305, 307, 310, 312

Naïve Bayes 23, 28, 139, 143, 344, 362
NARX neural network 98
Network Topologies 96
Nonlinear Autonomous 89
Novel Framework 23, 26, 29, 38
Numerical techniques 88

Optimization 57, 87, 92, 115, 242, 418

Performance evaluation 93, 94, 365
Performance gain 82
POD-ANN 87, 107, 108, 115, 116, 118, 120, 121, 123, 125, 126, 128, 129
POD-ANN method 88, 108, 120
Pose Estimation 225, 226, 231, 232, 235, 238, 243, 245, 247, 249
Precision (P) 352, 366
Prediction in Data Science 9
Product Quality 3, 4, 8, 40, 411
Proper orthogonal decomposition (POD) 86

Quality Assurance 1, 2, 3, 8, 20, 21

Random Forest 368, 369, 428, 429
Recall (R) 352, 366
Recommendation Engine 303
Recurrent ANN 96, 97, 98
Reinforcement Learning 338, 347, 348, 357, 417
Resource consumption 74
RFM 303, 304, 306, 307
Ring Oscillator 85, 88,89, 99, 104, 105, 106, 107, 117, 120, 121, 122, 123, 125
Ring Oscillator model 85, 104, 106, 107, 117, 120

Semi Supervised Learning 30, 338, 347, 348
Snowflake 52, 55, 56, 57, 58, 65
Social Media 23, 24, 25, 27, 28, 32, 34, 35, 38, 39, 45, 231, 282, 415
Spiking Model 99, 116, 117
Supervised Learning 115, 116, 293, 335, 338, 339, 345, 347, 348, 429
Support Vector Machine (SVM) 28, 363

Telegraph equation 99, 108, 110, 112, 125, 128
Test Automation 7, 8
Testing Metrics 4
Transaction database 71, 77, 82

Unsupervised Learning 293, 335, 338, 345, 347, 348

VLSI Interconnect 89, 99, 108, 109, 111, 112, 117, 125

Also of Interest

By the same editors

MATHEMATICS AND COMPUTER SCIENCE VOLUME 2, Edited by Sharmistha Ghosh, M. Niranjanamurthy, Krishanu Deyasi, Biswadip Basu Mallik, and Santanu Das, ISBN: 9781119896326. This second volume in a new multi-volume set builds on the basic concepts and fundamentals laid out in the previous volume, presenting the reader with more advanced and cutting-edge topics being developed in this exciting field.

MATHEMATICS AND COMPUTER SCIENCE VOLUME 1: Concepts and Applications
Edited by Sharmistha Ghosh, M. Niranjanamurthy, Krishanu Deyasi, Biswadip Basu Mallik, and Santanu Das, ISBN: 9781119879671. This first volume in a new multi-volume set gives readers the basic concepts and applications for diverse ideas and innovations in the field of computing together with its growing interactions with mathematics.

DATA WRANGLING: Concepts, Applications, and Tools, Edited by M. Niranjanamurthy, Kavita Sheoran, Geetika Dhand, and Prabhjot Kaurk, ISBN: 9781119879688. Written and edited by some of the world's top experts in the field, this exciting new volume provides state-of-the-art research and latest technological breakthroughs in next-data wrangling, its theoretical concepts, practical applications, and tools for solving everyday problems.

ADVANCES IN DATA SCIENCE AND ANALYTICS, Edited by M. Niranjanamurthy, Hemant Kumar Gianey, and Amir H. Gandomi, ISBN: 9781119791881. Presenting the concepts and advances of data science and analytics, this volume, written and edited by a global team of experts, also goes into the practical applications that can be utilized across multiple disciplines and industries, for both the engineer and the student, focusing on machining learning, big data, business intelligence, and analytics.

WIRELESS COMMUNICATION SECURITY: Mobile and Network Security Protocols, Edited by Manju Khari, Manisha Bharti, and M. Niranjanamurthy, ISBN: 9781119777144. Presenting the concepts and advances of wireless communication security, this volume, written and edited by a global team of experts, also goes into the practical applications for the engineer, student, and other industry professionals.

MEDICAL IMAGING, Edited by H. S. Sanjay, and M. Niranjanamurthy ISBN: 9781119785392.Written and edited by a team of experts in the field, this is the most comprehensive and up-to-date study of and reference for the practical applications of medical imaging for engineers, scientists, students, and medical professionals.

SECURITY ISSUES AND PRIVACY CONCERNS IN INDUSTRY 4.0 APPLICATIONS, Edited by Shibin David, R. S. Anand, V. Jeyakrishnan, and M. Niranjanamurthy, ISBN: 9781119775621. Written and edited by a team of international experts, this is the most comprehensive and up-to-date coverage of the security and privacy issues surrounding Industry 4.0 applications, a must-have for any library.

Check out these other related titles from Scrivener Publishing

CONVERGENCE OF DEEP LEARNING IN CYBER-IOT SYSTEMS AND SECURITY, Edited by Rajdeep Chakraborty, Anupam Ghosh, Jyotsna Kumar Mandal and S. Balamurugan, ISBN: 9781119857211. In-depth analysis of Deep Learning-based cyber-IoT systems and security which will be the industry leader for the next ten years.

MACHINE INTELLIGENCE, BIG DATA ANALYTICS, AND IOT IN IMAGE PROCESSING: Practical Applications, Edited by Ashok Kumar, Megha Bhushan, José A. Galindo, Lalit Garg and Yu-Chen Hu, ISBN: 9781119865049. Discusses both theoretical and practical aspects of how to harness advanced technologies to develop practical applications such as drone-based surveillance, smart transportation, healthcare, farming solutions, and robotics used in automation.

MACHINE LEARNING TECHNIQUES AND ANALYTICS FOR CLOUD SECURITY, Edited by Rajdeep Chakraborty, Anupam Ghosh and Jyotsna Kumar Mandal, ISBN: 9781119762256. This book covers new methods, surveys, case studies, and policy with almost all machine learning techniques and analytics for cloud security solutions.

Printed in the USA/Agawam, MA
September 18, 2023

851750.035